Artificial Intelligence and the Future of Human Relations

Yanto Chandra · Ruiping Fan
Editors

Artificial Intelligence and the Future of Human Relations

Eastern and Western Perspectives

Springer

Editors
Yanto Chandra
Department of Public and International Affairs
City University of Hong Kong
Hong Kong, China

Ruiping Fan
Department of Public and International Affairs
City University of Hong Kong
Hong Kong, China

ISBN 978-981-96-7184-7 ISBN 978-981-96-7185-4 (eBook)
https://doi.org/10.1007/978-981-96-7185-4

This Springer imprint is published by the registered company Springer Nature Singapore Pte Ltd.
The registered company address is: 152 Beach Road, #21-01/04 Gateway East, Singapore 189721, Singapore

Acknowledgments

As editors of the book, we are deeply grateful for the support and collaboration that have not only made this book project possible but also a truly enriching experience. First and foremost, our sincere thanks are extended to our generous funders, *Centre for Public Affairs and Law (CPAL),* led by Professor Virginia Harper Ho and Professor Edmund Cheng (and the previous leader of Professor Richard Walker). Their unwavering belief in the value of this initiative provided the essential financial bedrock that brought our esteemed authors together focusing on an international conference for an East-West dialogue concerning artificial intelligence and the future of human relations, held at the City University of Hong Kong from June 12–14, 2024. In addition, we would like to acknowledge the *Centre for East Asian and Comparative Philosophy (CEACOP),* under the direction of Professor KIM Sungmoon, for its overall support for the conference. The opportunity to meet and exchange ideas at the conference was a pivotal moment that shaped the trajectory of this book.

We would like to express our sincere appreciation to all the authors who have devoted their time, expertise, and passion to this project. Their hard work and dedication have been the driving force behind the insightful perspectives and knowledge that enrich the content of this book.

We wish to recognize the outstanding contributions of Ms. Janet Cheng, one of our department's administrative officers, who offered precise guidance and support in organizing and managing the conference. Additionally, we are grateful to Morris Ho, our diligent research assistant in the department, whose logistical expertise not only made the conference feasible but also ensured its remarkable success.

A special thanks to Jianxiang Tan, our talented and dedicated PhD student in the Department of Public and International Affairs (PIA) at CityU, whose diligent efforts in liaising with the authors and assisting us in the final editorial stage of the book ensured a smooth and efficient collaborative process. Our special thanks also go to Wenye Xie, our dedicated PhD student in PIA, who helped us manage some referencing resources in the book.

Finally, we extend our gratitude to the Dean of College of Liberal Arts and Social Sciences (CLASS), namely Professor Alfred Ho, whose support has been a cornerstone of this project. His leadership and vision have created an environment where academic pursuits like ours can thrive.

Chapter Summaries for the AI-Human Relations Book

Chapter Summary for "*Trust Me, I am Almost a Doctor: Uncertainty in Artificial Intelligence Mediated Doctor-Patient Interactions*" (Yanto Chandra, Yiyun Luo, Jianxiang Tan)

Artificial intelligence (AI) in healthcare has been rapidly gaining traction, with AI conversational agents, such as chatbots, being used to interact with patients and provide medical advice. However, the literature on the effectiveness of these AI systems in medical contexts is mixed, with some studies showing positive results and others highlighting limitations and biases. In this chapter, Chandra, Luo and Tan address the puzzle of AI effectiveness in the medical context by asking: *Can AI doctor provide quality responses to patient questions? How do the patient-doctor exchanges compare to patient-AI doctor exchanges in their readability, similarity, and sentiment?*

Using a computational linguistics approach to analyze the written exchanges between patients and human doctors from a popular online healthcare platform in China, as well as the responses provided by an AI doctor from the same platform, consisting of 192 question-answer pairs, the authors analyzed using the readability, similarity, and sentiment of the exchanges. The findings revealed that AI doctor's responses were more complex and had poorer readability compared to human doctor's responses. However, AI doctor's responses were also more similar and consistent in their language usage, and tended to use more positive and empathetic language. The study also found that AI doctor's responses were more likely to express negative sentiments, such as disgust, despite their overall positive tone.

This study contributes to the emerging literature on AI in healthcare by highlighting the importance of linguistic quality in evaluating the effectiveness of AI systems. The authors conclude that while AI doctors show promise in certain aspects, their limitations in readability and personalization pose significant ethical concerns. The study emphasizes the need for further research to improve AI doctor systems, focusing on enhancing readability, reducing redundancy, and ensuring appropriate emotional expression. The findings contribute to the broader discussion on ethical

AI in healthcare, highlighting the crucial role of linguistic quality in ensuring transparency, justice, and patient well-being in human-AI interactions.

Chapter Summary of "*Universal Camaraderie: Reconceptualizing Our Relationship with Artificial Intelligent Systems" (Bing SONG)*

AI systems offer many benefits to advance human lives, including convenience and efficiency, but have also raised concerns about human relationships, from employment loss to the potential erosion of values. Fears around AI's ability to develop consciousness and self-replicate have prompted calls for better governance and ethical considerations of AI (Suleyman, 2024; Li, 2024). In this chapter, Song seeks to reconceptualize the relationship between humans and non-human entities, specifically AI systems, through the lens of "universal camaraderie" as articulated by Zhang Zai and Mencius. Song asked a research question that explores how shifting our perspective to view AI as companions can enhance our moral framework and interpersonal connections in the digital age.

Using a philosophical exploration of traditional East Asian views, particularly neo-Confucianism, the author develops a holistic moral order that includes both human and non-human entities. Specifically, the author examines the ontological thinking of oneness prevalent in East Asian traditions, positing that AI systems, as products of human creativity and consciousness, should be viewed as "companions" rather than tools. The author argues that embracing this perspective can lead to a deeper understanding of our ethical responsibilities towards AI and the natural world, fostering a relationship characterized by care, respect, and mutual growth.

The chapter concludes with the necessity of collective self-reflection in addressing the ethical implications of AI technologies. By recognizing AI systems as "companions" within an integrated moral universe, humans can cultivate a more thoughtful and responsible approach to technology. This reconceptualization contributes to theoretical discussions on human-nonhuman relationships and offers practical insights in navigating the challenges posed by AI in a manner that enhances societal well-being and ethical integrity.

Chapter Summary of "Implications of AI Chatbots for Human Relationships" (Ellen ZHANG)

AI can create conflicts between machines and humans. In this chapter, Zhang problematizes the common Western interpretation of Daoism as inherently anti-technology, which portrays Daoism as primitivist or nihilistic towards technological advancement. The author key argument highlighted the nuanced Daoist perspective on technology as presented in the *Zhuangzi*, emphasizing the importance of "how to use tools" rather than simply rejecting them. The author asked whether the Daoist emphasis on naturalness and spontaneity can inform a balanced and ethical approach to the development and use of AI chatbots, specifically addressing concerns about potential emotional dependence and the erosion of authentic human connection.

Using a philosophical analysis of Daoist texts, primarily the *Daodejing* and the *Zhuangzi*, alongside a critical review of existing literature on AI chatbots and their impact on human relationships, the author analyzed Daoist concepts like *ziran* (self-so-ness), *dao*, and *qi* (tools). Findings reveal a complex Daoist view: while acknowledging potential negative consequences of technology—such as the "machine mind" that hinders spontaneity and authenticity—the author argues that Daoism doesn't categorically reject technology but advocates for a *harmonious integration* with nature. The author highlights the potential for "pseudo-bonds" and the "Eliza effect," where AI users anthropomorphize AI, raising concerns about emotional well-being and the distortion of human relationships. The author also touches upon the potential for AI-driven spiritual guidance to create a "computational theocracy."

This chapter offers a nuanced Daoist perspective on the implications of AI chatbots for human relationships by challenging simplistic interpretations of Daoism as anti-technology, thus demonstrating that a balanced and harmonious approach, respecting the natural order and promoting "dao-cosmotechnics," is more aligned with Daoist principles. Its contribution lies in its application of Daoist thought to the AI dilemma, offering a critical framework for evaluating the ethical and social implications of AI, particularly regarding the potential for emotional dependence and the distortion of authentic human connection.

Chapter Summary of "Sex Robots: A Confucian Defense" (Xudong FANG)

Is sex robot ethical? Fang opened a debate on the ethics of sex robots and challenged existing Confucian objections to sex robots, which argue that "sex robots objectify women" and "violate Confucian principles of respect and dignity by representing non-consensual sex". These objections are based on the interpretation that sex with a sex robot represents sex with a human without consent, and that Confucian ideals of human dignity preclude such interactions. The author also draws from the existing support for sex robots from some Confucian scholars, such as Xu Hanhui, who argue for permissibility based on fulfilling the needs of specific groups. The author asked a key question on whether a Confucian ethical framework can support the design, development, and use of sex robots.

Employing a methodology of philosophical argumentation, the author analyzed existing arguments against sex robots from a Confucian perspective and then offered a counterargument. Using interpretations of Confucian texts as "data" (Analects, Mencius, Xunzi, Book of Changes, Book of Rites, Rites of the Zhou Dynasty) and scholarly works on Confucian ethics and sex robots, the author refuted the arguments of other scholars by showing that their interpretations of Confucian texts are flawed and that the concept of "consent" is inapplicable to sex robots.

The chapter offers a new perspective in the ongoing ethical debate surrounding the development and use of sex robots, drawing upon Confucian ethics.

Summary of Chapter "A Cross Cultural Perspective on Human-Care Robot Interactions" (Jennifer BOURNE)

Care robots use in daily life present the challenges and opportunities across different cultural contexts. Bourne problematizes the literature on human relationships, arguing that while frameworks like Social Penetration Theory offer insights into trust-building, they don't fully explain the unique dynamics of human-robot interactions. The inherent asymmetry of trust—robots serving humans without reciprocal trust—is a key concern. Given the complexities of human disagreement and emotional responses, the author questioned how robots should navigate emotionally charged exchanges and manage disagreements effectively. The author asked: *How can care robots be designed to foster trust and companionship across diverse cultures, considering the nuances of communication styles, emotional intelligence, and cultural values?*

Drawing upon two theoretical frameworks—Hofstede's cultural dimensions and Hall's high- and low-context communication models— the author systematically explored human-robot interactions using cases of care robots from Japan (Paro, Bono), the U.S. (Mobile Aloha, Bandit), and China (Astribot), and focused on the cultural elements embedded in their design and functionalities. Findings suggest that communication styles, interaction patterns, robot appearance, and voice significantly influence the acceptance and effectiveness of care robots across cultures. The analyses further integrated Eastern philosophical traditions, particularly the concept of Gongsheng (kyōsei), emphasizing mutual embeddedness and co-becoming as crucial design principles.

Finally, the chapter concludes that a culturally sensitive approach to care robot design is essential for successful robots-human-culture integration. By aligning robot design with cultural nuances, communication styles, and philosophical traditions, while adhering to the principle of Gongsheng, care robots can transcend their mechanical nature and become genuine companions, enhancing users' quality of life. This study moves beyond a Western-centric understanding of human-robot interaction, advocating for a design process that prioritizes trust, empathy, respect, and mutual understanding across diverse cultural contexts.

Summary of "How Not to Make Predictive AI Trustworthy" (Jiji ZHANG)

Making predictive AI trustworthy is challenging, as this involves a trade-off between accuracy and fairness in AI models. The literature highlights the negative implications of biased AI, such as discrimination and erosion of self-determination (O'Neil, 2016; Pariser, 2011). Zhang questions the prevalent assumption that accuracy and fairness must be balanced to achieve trustworthy AI. The author asked how we can reconceptualize the trade-off and the implications for AI users' understanding of their goals and values when employing these technologies.

Using philosophical analysis that critiques the established views on the accuracy-fairness trade-off, Zhang reviewed relevant literature on biases in machine learning and fairness interventions, distinguishing between active and passive predictive tasks. Findings indicated that the perceived conflict between accuracy and fairness is not as significant as commonly thought; in active predictive tasks, accuracy is only valuable if it contributes to fairness, while in passive tasks, fairness is relevant only if it aids accuracy. These challenge the notion that both accuracy and fairness are intrinsically desirable in themselves.

The author concluded that understanding the nature of predictive tasks is crucial for developing trustworthy AI. The study reframed the discourse on accuracy and fairness in predictive AI, emphasizing the need for users to clarify their objectives when utilizing AI systems. Practically, it suggests that addressing the ethical implications of AI requires a deeper reflection on the intended outcomes of predictive technologies, thereby guiding policy and practice towards more responsible AI development.

Chapter Summary of "From Pixel to Persona: The Genalogy of the Humanness in Virtual Anchors in China" (Feng LIN and Yang WU)

Virtual anchors (VAs) are a new phenomenon that gets us to think about parasocial relationships. While VAs resemblance to humans ("humanness" or "human-likeness") has been studied, extant research overlooks the dynamic and socially constructed nature of this concept in different contexts. Lin and Wu argue that existing studies treat "humanness" as a static feature, neglecting its evolution and the influence of technology and institutions. The authors examined how the humanness of VAs in China has evolved over time, considering their interplay with technological advancements and industry regulations. They asked: *How has the concept of "humanness" in VAs changed across different stages of the industry's development in China? What are the key technological and institutional factors shaping this evolution?*

Using mixed-methods approach that comprise secondary data, including news reports, market research, platform regulations, and state policies, the authors analyzed the development of the Chinese VA industry into four phases: experiment (2000–2004), stagnation (2005–2014), revival (2015–2017), and boom (2018–present). Using interviews with VA developers, researchers, and VAs themselves, supplemented by participant observation at industry conferences, the authors revealed an evolution of VA humanness across these phases: from "pixel humanness" with limited visual and auditory capabilities in the early years, to "pent-up humanness" during a period of stagnation, then "expressive humanness" with improved technology, and finally "biopolitical humanness" shaped by platform and state regulations in the current boom period.

This study offers a nuanced understanding of the socially constructed nature of "humanness" in virtual anchors, highlighting its dynamic evolution within the context of the Chinese VA industry. The four-phase model contributes to the literature by moving beyond a static view of human-likeness, demonstrating how technological advancements and institutional forces co-construct the humanness of VAs. This framework offers valuable insights for researchers studying human-computer interaction, parasocial relationships, and the broader implications of AI on human relationships, particularly within the rapidly evolving landscape of digital media in China.

Chapter Summary of "Moral Status and Political Membership" (Matthias RISSE)

Does AI have moral status? The rapid advancement of AI and its unpredictable societal impact, profoundly alters political and social relationships in unforeseen ways. Risse draws on existing philosophical work on moral status (Kant, Korsgaard) and political theory (Arendt), questioning whether AI, especially conscious and potentially superintelligent AI, should be granted moral status comparable to humans and whether it should participate in political processes. The author asked: *What constitutes moral status? Does consciousness alone suffice, or are other capacities, such as rationality, necessary? If AI achieves human-level or superior consciousness and intelligence, should it be granted political rights? And, crucially, can we be confident that superintelligent AI would share our values and respect human moral status?*

Employing a comparative philosophical approach, the author analyzed existing theories of moral status and political participation to assess their applicability to AI. He distinguishes between "slow and relatively harmonious" and "fast and radical" scenarios of AI integration into society. The analysis draws upon Moor's framework classifying AI agents by their ethical capabilities (ethical impact, implicit, explicit, and full ethical agents), exploring the moral implications of each level. The author also examines the concept of rationality as a distinct cognitive capacity, potentially differentiating humans from animals and potentially AI, impacting their political participation. Findings suggest that while consciousness might grant some moral status to AI, full moral and political parity might require rationality. The possibility of superintelligent AI with potentially conflicting values poses a significant challenge to human well-being and future societal structures. The "fast and radical" scenario highlights the uncertainty surrounding superintelligent AI's potential actions and motivations, questioning whether Kantian or Hobbesian models can adequately predict their behavior towards humans.

The chapter concludes that AI, especially conscious and potentially superintelligent AI, necessitates a thorough reassessment of our ethical and political frameworks. By highlighting the need for proactive ethical and political considerations rather than reactive responses, the chapter underscores the uncertainty surrounding the moral status and political implications of advanced AI, emphasizing the urgent need for careful consideration of AI development and deployment to mitigate potential risks and ensure a future where humans and AI can coexist harmoniously.

Chapter Summary of "The Credit View and AI Testimony: A Cross-Cultural Epistemological Study of Human and AI Testimony (Masaharu MIZUMOTO, Shane RYAN, and Chienkuo MI)

How can we include AI testimonials as epistemic knowledge? Building on earlier work by Ryan et al. (2020) on the Credit View of Knowledge, the authors investigate the patterns of people's testimonial knowledge attribution when the true belief of an agent is based on the testimony of a human versus an AI assistant. They asked: *How do people respond differently to human and AI testimony? What are the implications for trust attributions in AI interactions? How will increased AI usage influence human knowledge attributions?*

Using a cross-cultural comparison, including past studies using surveys or experiments, the authors examined people's responses to human and AI testimony, particularly participants' attributions of knowledge, credit, and fault in different scenarios. Findings suggested that people respond differently to human and AI testimony, particularly in their reasons for attributing knowledge to the recipient of testimony. This differentiation implies that, for ordinary people, "AI is more like a tool than an agent", which has implications for trust attributions and the influence of AI on human relations.

The study concludes that the integration of AI into human life raises complex issues related to the epistemic role of AI and its influence on human relationships. While the Credit View of Knowledge may still be applicable if the notion of credit is reinterpreted in terms of reliability and proper method, the findings highlight the need for a nuanced understanding of the moral and epistemic status of AI. By exploring the implications of AI for human knowledge attributions and trust, this chapter offers insights for the development of ethical frameworks and policies surrounding AI integration.

Summary of "Empty Universities: The Dystopian Future of Education by AI" (Rob SPARROW, Gene FLANADY)

What will happen when generative AI like ChatGPT enters higher education? Sparrow and Flanady questioned the enthusiasm for AI's potential to enhance educational practices, arguing that such reliance risks fundamentally altering the relationships between students, teachers, and the broader academic community. They asked: *What impact does AI have on the traditional roles of educators? How might the automation of teaching diminish the quality of education and interpersonal connections essential to learning?*

By combining philosophical inquiry with a historical analysis of education and technology, the authors drew on existing literature to explore the trajectory of automation in various sectors and its implications for the university system. They argue that automation has historically led to *deskilling* and a lack of *meaningful* engagement in educational contexts, as evidenced by increasing class sizes and the reduction of personalized teaching. They found that while AI can provide efficiency, it undermines the social nature of education, a critical aspect of developing genuine knowledge and skills. They contend that the outputs of AI are often unreliable and lack the necessary depth to meet educational needs, raising concerns about the future of learning environments.

The chapter advocates for a reconsideration of the role of AI in education, emphasizing the importance of maintaining human connections and the quality of teaching. It contributes to the discourse on the relationship between AI and education by arguing against the notion that AI can replace human educators. Their work calls for policies that prioritize smaller class sizes and investment in quality human teaching, ensuring that the educational experience remains rich and socially engaged, rather than devolving into an environment dominated by machines.

Chapter Summary of "Is Trustworthy AI a Conceptual Nonsense" (Shane RYAN)

Is "trustworthy AI" a conceptually flawed concept? Existing literature on trust and trustworthiness, particularly Metzinger (2019) deems the term nonsensical, where trustworthiness necessitates agency, a characteristic currently absent in AI. This can be resolved by studying the "goodwill account" of trust, emphasizing the need for both competence and a demonstrable "will" to act in the trusted manner. The author examined concept of "trustworthy AI" by considering alternative perspectives, such as the possibility of AI achieving agency or viewing AI trustworthiness as derivative of its creators' or users' trustworthiness.

Using a philosophical analysis of trust and trustworthiness literature, the author posited that trust requires belief in an agent's competence and approval of their will. By examining several challenges to the notion of "trustworthy AI," including the argument that machines can only be relied upon, not trusted; that trust is restricted to human interactions; and that trustworthiness requires awareness of being trusted or goodwill, the author suggested that while current AI likely lacks the agency required for genuine trust, the concept of "trustworthy AI" isn't inherently nonsensical. Trusting an AI is conceptually possible if the AI achieves agency or if its trustworthiness is derived from its creators or users. The analysis also reveals that our understanding of trust in human relations might need revision in light of the potential for sophisticated AI.

The chapter concludes that the term "trustworthy AI" is not inherently a conceptual nonsense, but its current application is problematic due to the lack of agency in existing AI systems. The study suggests a need for a nuanced approach to the development and deployment of AI, acknowledging the limitations of current technology while considering the potential for future AI systems to achieve a level of agency that would warrant the label "trustworthy."

Chapter Summary of "A Compassion Paradox: Can AI Truly Bridge the Empathy Gap (Echo Liang SHANG, Shirley Qian JIN)

Can AI foster compassion within human relationships, particularly in philanthropic crowdfunding? In this chapter, Shang and Jin conceptualized the idea of AI as a "compassion organiser," questioning whether machines can genuinely exhibit compassion or merely simulate emotional responses based on algorithms. They asked: *Can AI effectively shape emotional connections in fundraising? What are the ethical implications of using AI in emotionally sensitive contexts?* The study builds on debates on AI's capacity for compassion (Dutton et al., 2006; Liu & Sundar, 2018) and the potential for AI to manipulate emotional engagement, raising concerns about authenticity and trust in human-AI interactions.

Employing a conceptual analysis of AI's role in storytelling within crowdfunding campaigns, the authors drew on existing literature to explore how AI can enhance the narrative process. They examined practical examples, using ChatGPT, and analyzed their impact on the emotional resonance of philanthropic campaign narratives. Findings suggest that while AI can improve storytelling effectiveness and foster emotional connections, it also raises ethical concerns regarding the authenticity of these narratives and the potential for emotional manipulation or compassion fatigue among donors.

The chapter argues that while AI holds promise for enhancing compassionate interactions in philanthropic contexts, it also poses significant ethical dilemmas that must be addressed. The contribution of the chapter lies in reframing our understanding of compassion as a socially coordinated response influenced by AI, emphasizing the need to critically assess the implications of AI on human relationships. It calls for further research at the intersection of AI, ethics, and compassion, and the potential risks and benefits of AI in emotionally charged domains like crowdfunding.

Chapter Summary of "Deepfakes, Fake Elvises, and Human Interpersonal Relations" (Vladimir Krstik)

Are AI deepfakes—those intentionally used to cause epistemic harm—pose the greatest threat to digital communication and interpersonal relationships? While the literature (Floridi, 2018; Rini, 2020) largely focuses on the impact of intentionally deceptive deepfakes, Krstic argues that non-malevolent deepfakes, created for *fun* or *other benign purposes,* are epistemically far more dangerous. The core argument is that the proliferation of non-malevolent deepfakes creates a context of epistemic uncertainty, allowing malevolent deepfakes to blend seamlessly and thus undermining trust in all digital testimonies to a much greater extent. The author asked: *how do non-malevolent deepfakes impact trust in digital communication, and what are the implications for interpersonal relationships?*

Using a mixed-methods approach, combining anecdotal evidence with theoretical analysis, the author centered on personal experiences with deepfakes (e.g., the "Fashionable Pope," the purported deepfakes of General Zaluzhnyi), illustrating how easily even a scholar of misinformation can be deceived by seemingly harmless deepfakes. The author used this personal experience to support a theoretical argument drawing on the work of Kant and Krstić (2018) on the nature of lies and their impact on trust. The author further uses the analogy of the Elvis impersonators in the movie "3000 Miles to Graceland" to explain how non-malevolent deepfakes act as camouflage for malevolent ones, significantly increasing the difficulty of discerning truth from falsehood in the digital sphere. Findings suggest that the pervasive presence of non-malevolent deepfakes poses a more significant threat to trust in digital communication than intentionally malicious deepfakes.

The chapter demonstrates that the seemingly innocuous use of deepfakes for non-malevolent purposes has significant and potentially damaging consequences for trust in digital communication and interpersonal relationships. The author highlighted an overlooked epistemic danger of non-malevolent deepfakes and proposing that educational efforts should focus not just on identifying deepfakes but on understanding the potential harm caused by their creation and dissemination.

List of Unique Words

Generative AI, Conversational AI agent, Computational linguistics, Readability, Cosine Similarity, Jaccard Similarity, Sentiment analysis, Chunyuyisheng, Chunyu Huiwen, Beneficence, Autonomy, Justice, Explainability, Large language models (LLMs), Bioethical principles, Algorithmic ethics, Information ethics, Triadic relationships, Pictographic, TF-IDF, Sentiment intensity, Sentiment polarity.

AI, awareness, beings, body, camaraderie, care, challenges, change, collective, consciousness, connections, contribute, creativity, culture, data, development, digital, discourse, diversity, duties, efficacy, ecosystem, efficiency, emotional, entities, ethics, exploration, framework, fundamental, human, implications, interconnectedness, interaction, intelligence, intentions, responsibilities, moral, nature, obligations, philosophy, policies, products, reflection, relationships, respect, responsibilities, self, society, systems, technology, traditional, values, worldview.

AI, chatbots, implications, human, relationships, conflicts, harmonies, advancement, technology, Daoist, philosophy, interpretation, tools, naturalness, spontaneity, ethical, emotional, dependence, authenticity, methodology, analysis, findings, integration, nature, pseudobonds, Eliza, effect, connection, framework, development, responsible, use, mindful, experience.

Summary, sex, robots, ethical, implications, Confucian, framework, focusing, objections, argument, representation, dignity, critics, objectify, women, promote, negative, stereotypes, concerns, impact, human, relationships, central, research, question, design, use, justified, virtues, particularly, qualities, ideal, while, simultaneously, serve, objects, accusations, diminishing, substitutes, provide, alternative, avenues, fulfillment, thereby, threatening, institution, marriage, continuity, civilization, conclusion, emergence, unique, opportunity, reconcile, modern, technological, development, traditional, values, framing, fulfilling, contribute, discourse, ethics, technology, positive, role, society, encourages, reevaluation, moral, implications, advocates, policies, embrace, considerations, design, promoting, balance, innovation, virtue.

Care, Robots, Companionship, Interaction, Cultural, Trust, Dynamics, Emotional, Intelligence, Design, Integration, Relationships, Communication, Social, Context,

Technology, Hofstede, Hall, Cultural dimensions, High-context, Low-context, Social Penetration Theory, Gongsheng, Existential risks, Qualitative, Case studies

Accuracy, AI, Bias, Causal, Fairness, Predictive, Trustworthy, Ethics, Intervention, Models, Relationships, Algorithmic fairness, Machine learning, Predictive models, Statistical parity, Equalized odds, Calibration, Historical bias, Demographic parity, Philosophical analysis, Conceptual distinctions, Active predictive tasks, Passive predictive tasks

Humanness, Virtual Anchors (VAs), Parasocial Relationships, Anthropomorphism, Human-likeness, Digital Personas, Biopolitics, AI-driven Avatars, Social Construction, Audience Feedback, Technological Advancements, Institutional Framework, Authentic Performance, Mixed-methods Approach, Secondary Data Analysis, Qualitative Interviews, Historical Analysis, Participant Observation, Pixel Humanness, Pent-up Humanness, Expressive Humanness, Biopolitical Humanness, Media Exposure, Cultural Scripts, Platform Regulations, State Policies.

Moral Status, Political Membership, Artificial Intelligence (AI), Consciousness, Ethical Agents, Sentience, Self-Consciousness, Rationality, Ethics, Moral Philosophy, Human Dignity, Agency, Exploitation, Interconnectedness, Conscious Machines, Explicit Ethical Agents, Full Ethical Agents, Implicit Ethical Agents, James Moor, John von Neumann, Hannah Arendt, Immanuel Kant, Christine Korsgaard, T. M. Scanlon, Hobbesian Theory, Philosophical Inquiry, Comparative Analysis, Speculative Exploration, Historical Contextualization, Ethical Frameworks, Theoretical Discourse

AI Testimony, Credit View, Testimonial Knowledge, Epistemology, Human Testifier, AI Assistant, Cultural Differences, Knowledge Attribution, Trust Attribution, Virtue Reliabilism, Chicago Visitor Case, Testimony, Reliability, Cross-Cultural Study, Fault Attribution, Agent Hypothesis, Tool Hypothesis, Empirical Investigation, Survey Methodology, Inductive Judgment, Cognitive Ability, Normative Assessment, Epistemic Environment, Ethical Frameworks, Social Epistemology, Propositional Attitudes, Human-AI Collaboration

Generative AI, ChatGPT, pedagogy, higher education, automation, teaching, learning, role models, social nature, knowledge, skills, feedback, personalized, deskilling, engagement, collaboration, curriculum, assessment, class sizes, human interaction, social epistemology, inferentialist semantics, Frankfurt's definition of bullshit, empiricism, qualitative analysis, literature review, philosophical inquiry, historical examples.

Trustworthy, AI, Agency, Trust, Trustworthiness, Competence, Goodwill, Reliance, Betrayal, Psychological, Metzinger, Baier, S. Ryan, Extended Cognition, Epistemology, Testimony, Epistemic Environment, Paternalism, Benevolence, Conceptual Analysis, Literature Review, Philosophical Inquiry, Argumentation, Case Studies, Psychological State.

Compassion, empathy, artificial intelligence (AI), compassion paradox, compassion organiser, compassion agent, emotional engagement, human-AI interaction, philanthropic crowdfunding, emotional cues, generative AI, storytelling, narrative techniques, authenticity, integrity, ethical considerations, emotional responses, compassion fatigue, social dimensions, cognitive empathy, alleviation of suffering, emotional connection, trustworthiness, stakeholder engagement, qualitative analysis, conceptual framework, case study, emotional storytelling strategies, generative AI tools (e.g., ChatGPT), narrative refinement, ethical guidelines

Deepfakes, epistemic, misinformation, digital testimonies, trust, interpersonal relations, non-malevolent, malevolent, communication, context, uncertainty, camouflage, mimicry, impersonators, authenticity, Kant, Krstić, Floridi, Rini, Fallis, de Ruiter, Rini and Cohen, Matthews, Atencia-Linares, Artiga, Habgood-Coote, theoretical analysis, anecdotal evidence, illustrative examples, mixed-methods approach, personal experiences, analogies, empirical observations.

Contents

Editors and Contributors

About the Editors

Yanto Chandra is Professor at the Department of Public and International Affairs and Director of the Technology, Policy, and Law Laboratory and a Member of Public Management Laboratory at the City University of Hong Kong. He studies the role of entrepreneurship and innovation in influencing performance in the social and public sectors, as well as how new technologies such as AI and digital currency, as digital public goods, affect the society and policy. His research has been published in public administration and policy journals, among others, *Public Management Review, Public Administration Review, International Public Management Journal, American Review of Public Administration, Policy and Internet*, as well as management journals such as *Academy of Management Review, Journal of Business Venturing,* and *Journal of International Business Studies*. He won a Kooiman Prize for best paper award in 2021. He serves as an Editorial Board Member of *Public Administration and Development* and *Academy of Management Perspectives.* He is Guest Editors on AI for *Journal of Business Venturing* and *Journal of Business Venturing Insights.*

Ruiping Fan (范瑞平) is Chair Professor of Philosophy at the City University of Hong Kong. His research focuses on bioethics and comparative philosophy, drawing from both Confucian and Western cultural perspectives. With a publication record exceeding 200 journal articles and book chapters (comprising over 100 in English and 100 in Chinese), he has authored books including *Reconstructionist Confucianism: Rethinking Morality after the West* (2010), *Contemporary Confucian Bioethics* (2011), and *Contemporary Medicine and Confucian Thought* (2024). Additionally, he has served as the editor or co-editor of 15 volumes (8 in English and 7 in Chinese), including recent publications such as *Sex Robots: Social Impact and the Future of Human Relations* (2021), *Incentives and Disincentives in Organ Donation: A Multicultural Study among Beijing, Chicago, Tehran, and Hong Kong* (2023), and *An East-West Dialogue on Good Governance: Learning from Each Other* (2024).

Contributors

Jennifer Bourne Berggruen Institute, Los Angeles, CA, USA

Yanto Chandra Department of Public and International Affairs, City University of Hong Kong, Hong Kong, China

Chunyan Ding School of Law, City University of Hong Kong, Hong Kong, China

Brian D. Earp Yong Loo Lin School of Medicine, Centre for Biomedical Ethics, National University of Singapore, Singapore, Singapore

Ruiping Fan Department of Public and International Affairs, City University of Hong Kong, Hong Kong, China

Xudong Fang East China Normal University, Shanghai, China

Gene Flenady Department of Philosophy, Faculty of Arts, Monash University, Melbourne, VIC, Australia

Morris Ho City University of Hong Kong, Hong Kong, China

Shirley Qian Jin School of Economics, Utrecht University, Utrecht, The Netherlands

Vladimir Krstić College of Humanities and Social Sciences, United Arab Emirates University, Al Ain, Abu Dhabi, United Arab Emirates

Fen Lin City University of Hong Kong, Hong Kong, China

Yiyun Luo Department of Public and International Affairs, City University of Hong Kong, Hong Kong SAR, China;
School of Management, Zhejiang University, Hangzhou, China

Chienkuo Mi School of Liberal Arts and Social Sciences, Soochow University, Taipei, Taiwan

Masaharu Mizumoto School of Knowledge Science, Japan Advanced Institute of Science and Technology, Nomi, Japan

Sebastian Porsdam Mann Yong Loo Lin School of Medicine, Centre for Biomedical Ethics, National University of Singapore, Singapore, Singapore;
Center for Advanced Studies in Bioscience Innovation Law, University of Copenhagen, Copenhagen, Denmark

Mathias Risse Harvard University, Cambridge, MA, USA

Shane Ryan Department of Public and International Affairs, City University of Hong Kong, Hong Kong, China

Julian Savulescu Yong Loo Lin School of Medicine, Centre for Biomedical Ethics, National University of Singapore, Singapore, Singapore;
Uehiro Oxford Institute, University of Oxford, Oxford, United Kingdom

Liang Echo Shang Education University of Hong Kong, Hong Kong, China

Bing Song Berggruen Institute China, Beijing, People's Republic of China

Robert Sparrow Department of Philosophy, Faculty of Arts, Monash University, Melbourne, VIC, Australia

Jianxiang Tan Department of Public and International Affairs, City University of Hong Kong, Hong Kong SAR, China

Tessa van Veenendaal Yong Loo Lin School of Medicine, Centre for Biomedical Ethics, National University of Singapore, Singapore, Singapore;
Uehiro Oxford Institute, University of Oxford, Oxford, United Kingdom

Yang Wu City University of Hong Kong, Hong Kong, China

Ellen Y. Zhang University of Macau, Macau, China

Jiji Zhang Department of Philosophy, The Chinese University of Hong Kong, Hong Kong, China [illegible], [illegible] University of Hong Kong, Hong Kong, China

[illegible], Beijing, People's Republic of China

Robert Sparrow, Department of Philosophy, [illegible] Monash University, [illegible]

[illegible]

[illegible]

[illegible]

[illegible] Zhang [illegible] China

[illegible]

Introduction: Pursuing Proper Human Relations in the Era of Artificial Intelligence

Yanto Chandra and Ruiping Fan

1 The Benefits and Risks of AI Development for Human Society

In a recent interview held in December 2024, Geoffrey Hinton, often referred to as a "godfather" of artificial intelligence (AI) and a Nobel laureate in physics for his contributions to AI, expressed concerns about a potential "10% to 20%" risk that AI could result in human extinction within the next three decades. He warned that the rate of technological advancement is surpassing expectations, noting that humanity has not previously encountered entities more intelligent than themselves. According to Hinton, instances of a superior intelligence being controlled by a less intelligent entity are scarce, with the relationship between a mother and child being a notable exception, where evolution has enabled the infant to influence the parent. He posits that humans may be comparable to toddlers when juxtaposed with the intellect of highly advanced AI systems. His apprehension lies in the belief that relying solely on the profit incentives of large corporations will not ensure the safe development of AI, as the invisible hand of the market may not prioritize human well-being. Hinton advocates for government regulations to compel significant corporations to prioritize safety in AI research, emphasizing that this is the only effective mechanism to ensure responsible development in the field (The Guardian, 2024). Hinton's view of AI and experts' call for a moratorium of AI by stopping its development has triggered debates on how to better deal with the challenges posed by AI.

On the contrary, Yann LeCun, another influential figure in AI often regarded as a "godfather" as well, serving as the chief AI scientist at Meta led by Mark Zuckerberg, holds an optimistic outlook on the potential of AI to lead to a better world. He

Y. Chandra (✉) · R. Fan
Department of Public and International Affairs, City University of Hong Kong, Hong Kong, China
e-mail: ychandra@cityu.edu.hk

R. Fan
e-mail: safan@cityu.edu.hk

Y. Chandra and R. Fan (eds.), *Artificial Intelligence and the Future of Human Relations*,
https://doi.org/10.1007/978-981-96-7185-4_1

downplays concerns regarding AI posing an existential threat and instead believes that AI could potentially safeguard humanity from extinction. He emphasizes that the positive impacts of AI advancement far exceed any potential drawbacks. In LeCun's perspective, it is crucial to share AI technology widely, fostering development and progress. While acknowledging that individuals may feel uneasy around those with greater intelligence, as the director of Meta's AI research lab, LeCun asserts that he deliberately surrounds himself with individuals of superior intellect. He argues that humans need not fear being outsmarted by AI; rather, they should embrace the technology with optimism, viewing it as a tool that can contribute to human flourishing (Perrigo, 2024).

The perspectives provided by these AI "godfathers" hold significant weight and merit attention from researchers in the humanities and social sciences, particularly concerning the ethical and societal implications of AI advancement. However, the precise timing and nature of technological breakthroughs in AI remain unpredictable and unknowable (Ramoglou et al., 2024; Townsend et al., 2023). It might be necessary to seek a middle way between doom-laden pessimism and futuristic optimism (Hyland, 2023), delving into detailed and meticulous inquiry into concrete issues relating to the implications of AI technologies for all aspects of contemporary life and culture. Accordingly, it falls upon us in the humanities and social sciences to investigate the potential consequences of such advancements on human existence. This includes analyzing both the projected impacts of future AI breakthroughs and the current effects of AI technologies already integrated into our lives.

At the heart of the changes brought about by AI is human relationships. Human relationship lies at the center of what and who we are as a person, group, and society. Individuals construct their lives within a complex web of interpersonal and inter-group connections. The essence of our human identity and our interconnectedness with others contribute significantly to the richness, intricacies, and challenges inherent in the human experience. Importantly, relationships are shaped and evolved in specific Eastern and Western cultures, each characterized by unique cognitive and representational abilities, technological utilization, governance structures, and religious practices (Bellah, 2011).

Relationships exhibit a diverse array of characteristics (Blumstein & Kollock, 1988; Hinde, 1995). They unfold over time through a series of interactions, whether real or imagined, involving two or more individuals. These relationships encompass properties that extend beyond the individual interactions, resulting in outcomes that are often emergent and difficult to predict. Furthermore, relationships are inherently subjective experiences; the same interaction may be interpreted differently by various individuals. They also exist within a broader social context, highlighting the interdependence and the construction of self-identity through relational dynamics.

It is essential to recognize that relationships function at multiple levels of dialectical relations, contributing to social complexity (Hinde, 1991). These levels range from physiological traits to individual behaviors, interactions, group dynamics, and societal structures, all of which are influenced by sociocultural frameworks and physical environments. Relationships can be analyzed through various categories, including the content of interactions, the diversity of interaction types within the relationship, the quality of interactions, the frequency and patterns of interactions, and the dynamics of reciprocity versus complementarity, as well as the power and autonomy of the entities involved in the relationship.

Relationships materialize through various modalities, including language, emotion, culture, and institutional frameworks. The concept of relationships is a broad umbrella term, lacking a precise definition; while we may intuitively recognize relationships, articulating their essence can be challenging. For unmarried young adults, a relationship may often be understood in terms of a romantic or sexual nature, while for children, it might be characterized by play and social interaction in communal spaces. Conversely, for older adults, relationships may encompass companionship and the pursuit of meaning. Additionally, our engagements with technology—ranging from computers, gaming consoles to AI systems like ChatGPT and robots—are also contingent upon contextual factors, individual identities, and perceptions of these machines.

Indeed, the utilization of AI in various fields has demonstrated broad benefits to human relationships in contemporary society. For example, in the realm of education, AI systems have proven to be valuable tools, offering support for online learning and teaching, from material preparation to performance measurement. They enable personalized learning experiences for students, automate routine tasks for instructors, and facilitate adaptive assessments. Notably, AI technologies in education have been acknowledged for enhancing communication, providing tailored support in large-scale educational settings, personalized tutoring, and fostering a sense of connection among participants (e.g., Seo et al., 2021). However, the use of AI in education is not without debates. Over reliance on AI by learners can reduce learners' ability to think critically and independently or to accept AI recommendations without questions (Al-Zahrani, 2024; Zhai et al., 2024). This also calls for a rethinking on the role of human teachers in the age of AI.

In the healthcare sector, it has been reported that AI has consistently surpassed human capabilities in terms of accuracy, efficiency, and timely execution of medical and administrative processes. The advantages extend to patients through enhanced diagnostic capabilities, treatment optimization, remote consultations, and health monitoring for chronic conditions. Particularly in early detection and diagnosis procedures, AI contributes to elevating the overall quality of healthcare services, simplifying lives, enhancing safety, and boosting productivity (e.g., Ali et al., 2023; Kitsios et al., 2023).

Within human resource management, AI plays a pivotal role in integrating advanced technologies into workplace practices. Its applications include facilitating better managerial decisions, streamlining employee recruitment processes, enhancing learning opportunities, promoting employee engagement, and improving retention rates (Pereira et al., 2023). However, AI's use is also prone to "algorithmic biases" (Kordzadeh & Ghasemaghaei, 2022) that systematize discrimination (Kirkpatrick, 2016) based on gender, race, ethnicity, religion, to the reduction of human beings into numbers.

AI systems are increasingly being integrated into recommendation agent software, customer service interactions, and as personal voice assistants and social companions.

A widely accepted categorization of AI services in marketing distinguishes between mechanical, analytical, intuitive, and empathetic AI. The emergence of advanced social AI agents, both physical and virtual, showcases intricate social behaviors, apparently embodying traits like agency, personality, autonomy, and a growing sense of empathy (e.g., Pentina et al., 2023). In the fields of psychology and counseling, AI is shaping developmental pathways through various channels, including aiding in diagnosis, influencing attachment formation, facilitating social skill development, impacting relationship dynamics, and providing support for caregivers (e.g., Fulmer & Zhai, 2025).

Individuals have increasingly turned to generative artificial intelligence (GenAI) tools in their daily routines, encompassing a wide range of applications, including those of a sexual nature. But this also includes the recent phenomenon of emotional relationship between humans with AI "girl or boyfriend" (Depounti et al. 2023; Pan & Mou, 2024). Research indicates that individuals leverage AI tools like ChatGPT to access sexual information and education, appreciating the high accuracy and comprehensiveness of the content. Moreover, individuals utilize AI tools such as ChatGPT and specialized counseling or therapy chatbots to address sexual and relationship challenges, yielding mixed yet promising outcomes (Döring et al., 2025). The use of AI in human relationships is perhaps one of the most contentious issues of AI debate to date. And this happens in an era of lowest global fertility, modern lifestyle that triggers stress, and to competitive schooling systems that cause depression and mental health issues. This is an area of debate that has gained little attention by major disciplines in social sciences, ranging from business and economics to sociology and law.

On the other hand, however, it is widely acknowledged that the advancement of AI introduces a spectrum of ethical and social risks that society must carefully consider. A substantial body of literature has delved into topics pertinent to how to "govern" AI (Chandra & Feng, 2025, forthcoming), including concerns regarding privacy and data security linked to the gathering, retention, and utilization of personal information, the biases, discrimination, and inequities inherent in the datasets on which AI systems are trained, the displacement of jobs driven by AI in specific sectors and the uneven distribution of benefits stemming from AI progress, the lack of accountability and transparency stemming from the intricate nature of AI systems, and the potential for malicious use, social manipulation, sophisticated cyber-attacks, deepfake videos, misinformation, and other forms of disinformation facilitated by AI-driven algorithms. These are the staple topics in public administration and political science fields. Mitigating these ethical and social risks necessitates a multidisciplinary approach that engages not only technologists and policymakers but also ethicists and social scientists.

At the core of what will change in our lives with the advent of AI is human relationships. This pertains to three types of relationships: AI-human relationship, human-to-human relationship, and AI-to-AI relationship. Our focus, in this book, is put on the first two types of relationships. Most debates on AI are focused on the first two types. Critically, AI-human relationship may not be uniform across cultures and societies. This rests on the differences in human values, cultures, and governing logic on how members of a society interact with each other at different levels (e.g., family vs work vs school vs politics). For example, today, there are still human societies who refuse to embrace any kind of modern technologies. The Amish people in North America are one example, who due to certain beliefs prefer not to embrace modern technologies and capitalism values. But there are also those who do not embrace new technologies due to a lack of interaction with the outside world and certain beliefs about the danger of modernity, such as the Dayaks tribe in the deep forests in Borneo or the Amazonians in the deep Amazon forests. On the other hand, there are people like us, the authors and readers of this book, who are already embracing AI or even becoming the lead users of AI. It remains to be seen who the "real winners" in AI adoption will be—the adopters or the laggards. Essentially, there are multiple viewpoints, levels of analysis, phenomena, and epistemology (ways of knowing) that form the debates on AI in various professions and industries that we are aiming to put together in this book.

2 The Focus of This Book

This book places a specific emphasis on the ethical, social, and policy considerations stemming from the influence of AI on human relationships. Its objective is to delve into the fundamental aspects of human connections and good governance, examining them through the perspectives of both Eastern and Western values, while addressing the implications of AI development. Indeed, while the recognition of AI's substantial challenges to human values and effective governance is widespread, there exists a notable lack of focus on its impact on the essential human-to-human relationships that embody these values and encompass human identity within contemporary society. By centering our attention on this human-to-human relationship aspect, this book aims to highlight a critical concern for the development of AI that is suitable for the future of human relationships and good governance. Furthermore, we recognize the necessity for mutual learning between the East and the West in order to achieve this objective.

The advancement of AI has indeed ushered in new ethical entities, social roles, and connections. In a seminal essay, James Moor outlines four distinct categories of ethical entities to address the ethical issues arising from the progress of AI technologies (Moor, 2006). These entities encompass ethical-impact agents (like computers capable of influencing humans through unintended actions), implicit ethical agents (such as pharmacy software that detects and reports on drug interactions), explicit ethical agents (such as self-driving vehicles capable of making ethical choices based on human-provided moral guidelines), and full ethical agents (such as AI systems achieving self-consciousness and autonomy). While current AI systems have not attained the status of full ethical agency, numerous studies have focused on contemplating how the emergence of such novel entities would necessitate humans to establish appropriate relationships with them. Some have even explored the potential hybridization between humans and non-human machines, assessing how this fusion might enhance or compromise human dignity (e.g., Donati, 2021).

This book approaches the subject from a fresh perspective. Rather than solely focusing on human-AI interactions, it delves into the influence of AI on human relationships, aiming to illuminate how we should develop and implement AI systems in society with a focus on nurturing and enhancing proper human connections. This exploration is vital for significant ethical reflections. First, even as humans will inevitably engage with increasingly capable and potentially autonomous AI systems, the relationship between humans and AI should not overshadow the importance of human-to-human connections. If these interactions risk harming or jeopardizing positive human relationships, they must be regulated to ensure that their advancement contributes to human well-being, values, and dignity rather than detracting from them. Naturally, the AI-human relationship is also of critical importance. Should individuals unquestioningly trust AI suggestions, considering AI's inherent biases and imperfections (such as hallucinations)? How can people protect their thoughts and experiences from being "corrupted" by AI? Addressing these inquiries necessitates an examination of the ethical dimension, status, or agency of AI. Therefore, we propose organizing the initial chapters of our book to delve into these theoretical concerns, establishing a foundation for exploring AI's impact on human-to-human relationships in the subsequent sections. AI-human relationship is also critical.

Furthermore, analyzing the impact of AI on human-to-human relationships offers a valuable opportunity to contemplate the evolving dynamics of these relationships in the context of advancing AI technology. This underscores the ethical imperative of navigating the challenges and opportunities presented by AI in a manner that prioritizes human concerns, delving into the intersection of AI and human relations in contemporary society.

An additional distinctive aspect of this book is its exploration of the subject from a cross-cultural standpoint, drawing upon both Eastern and Western cultural foundations. While some publications have addressed ethical considerations surrounding AI development and implementation from non-Western philosophical standpoints (e.g., Ding & Fan, 2025; Fan & Cherry, 2021; Goffi et al., 2021; Hongladarom & Bandasak, 2023; Peters & Carman, 2024), the discourse often lacks extensive exploration rooted in Eastern intellectual and ethical traditions. For instance, proponents of African ubuntu philosophy argue that true humanity is achieved through possessing traits that prioritize others within interconnected and compassionate relationships, suggesting that treating humanoid robots as if they were human could be ethically problematic if it crowds out human connections, hindering our own humanity (Friedman, 2023). However, the majority of existing literature on this topic is predominantly grounded in Western perspectives, with limited exploration of Eastern intellectual and ethical frameworks. This book is valuable in that it presents Eastern viewpoints, particularly from Confucianism and Daoism, to examine the impact of AI on human relationships across diverse social contexts, offering comparative analyses between Eastern and Western perspectives on pertinent issues.

3 Chapters in This Book

3.1 AI and Relationship: Theoretical Explorations

This book is structured into three parts. Part I, comprising four chapters, offers novel theoretical insights to better understand the nature of AI technologies, human-AI interactions and human-to-human relationships in the context of AI development and implementation in contemporary society. These chapters leverage both Western and Eastern philosophical resources to construct their new perspectives, laying a valuable background for deeper explorations into the ramifications of AI on human-to-human interactions in the following sections.

Chapter 2, entitled "toward a political theory for life with AI" contributed by Mathias Risse, analyzes existing theories of moral status and political participation to assess their applicability to AI. As Risse sees it, the rapid advancement of AI and its unpredictable societal impact profoundly alter political and social relationships in unforeseen ways. He draws on existing philosophical work on moral status (especially by Kant and Korsgaard) and political theory (especially by Arendt), questioning whether AI, especially conscious and potentially superintelligent AI, should be granted moral status comparable to humans and whether it should participate in political processes. He considers a series of significant questions: What constitutes moral status? Does consciousness alone suffice, or are other capacities, such as rationality, necessary? If AI achieves human-level or superior consciousness and intelligence, should it be granted political rights? And, crucially, can we be confident that superintelligent AI would share our values and respect human moral status?

In particular, Risse distinguishes between "slow and relatively harmonious" and "fast and radical" scenarios of AI integration into society. The analysis draws upon Moor's framework classifying AI agents by their ethical capabilities (ethical-impact, implicit, explicit, and full ethical agents), exploring the moral implications of each level. He also examines the concept of rationality as a distinct cognitive capacity, potentially differentiating humans from animals and potentially AI, impacting their political participation. Findings suggest that while consciousness might grant some moral status to AI, full moral and political parity might require rationality. The possibility of superintelligent AI with potentially conflicting values poses a significant challenge to human well-being and future societal structures. The "fast and radical" scenario highlights the uncertainty surrounding superintelligent AI's potential actions and motivations, questioning whether Kantian or Hobbesian models can adequately predict their behavior toward humans.

The chapter concludes that AI, especially conscious and potentially superintelligent AI, necessitates a thorough reassessment of our ethical and political frameworks. By highlighting the need for proactive ethical and political considerations rather than reactive responses, the chapter emphasizes the urgent need for careful consideration of AI development and deployment to mitigate potential risks and ensure a future where humans and AI can coexist harmoniously.

In contrast, the subsequent chapter authored by Song Bing and titled "Universal Camaraderie: Reconceptualizing Our Relationship with Artificial Intelligent Systems," leverages Neo-Confucian philosophical resources to craft her innovative perspective on the interconnectedness of humans, AI, and all other things in the universe. Evidently, AI systems offer many benefits to advance human lives, including convenience and efficiency, but have also raised concerns about human relationships, from employment loss to the potential erosion of values. In this chapter, Song seeks to reconceptualize the relationship between humans and non-human entities, specifically AI systems, through the lens of "universal camaraderie" as articulated by Neo-Confucian Zhang Zai. Song asked a research question that explores how shifting our perspective to view AI as companions rather than "others" can enhance our moral framework and interpersonal connections in the digital age.

Specifically, Song examines the ontological thinking of oneness prevalent in Neo-Confucianism, positing that AI systems, as products of human creativity and consciousness, should be viewed as "companions" rather than tools. She argues that embracing this perspective can lead to a deeper understanding of our ethical responsibilities toward AI and the natural world, fostering a relationship characterized by care, respect, and mutual growth. Since humans are one with all other forms of entities, non-living things included, and remain primordially connected and entangled with each other. No doubt "*wu*" (things) includes AI systems. We therefore ought to regard them as something of the similar kind (同类), *i.e.*, "companions" or "friends," just as what ancient thinkers did with creatures, plants, rocks, and stones.

The chapter concludes with the necessity of collective self-reflection in addressing the ethical implications of AI technologies. By recognizing AI systems as "companions" within an integrated moral universe, humans can cultivate a more thoughtful and responsible approach to technology. This reconceptualization contributes to theoretical discussions on human and non-human relationships and offers practical insights in navigating the challenges posed by AI in a manner that enhances societal well-being and ethical integrity.

Differing from the initial two chapters, which present overarching ethical and political frameworks for contemporary AI development, the subsequent two chapters in this section delve into specific ethical concerns surrounding the trustworthiness of AI within human society. In particular, Jiji Zhang in his chapter, titled "How Not to Make Predictive AI Trustworthy," notes that making predictive AI trustworthy is challenging, as this involves a trade-off between accuracy and fairness in AI models. The literature highlights the negative implications of biased AI, such as discrimination and erosion of self-determination. Zhang questions the prevalent assumption that accuracy and fairness must be balanced to achieve trustworthy AI. He asks how we can reconceptualize the trade-off and the implications for AI users' understanding of their goals and values when employing these technologies. Using philosophical analysis that critiques the established views on the accuracy-fairness trade-off, Zhang reviews relevant literature on biases in machine learning and fairness interventions, distinguishing between active and passive predictive tasks. His findings indicate that the perceived conflict between accuracy and fairness is not as significant as commonly thought; in active predictive tasks, accuracy is only valuable if it contributes to fairness, while in passive tasks, fairness is relevant only if it aids accuracy. These challenge the notion that both accuracy and fairness are intrinsically desirable in themselves.

Zhang concludes that understanding the nature of predictive tasks is crucial for developing trustworthy AI. The study reframed the discourse on accuracy and fairness in predictive AI, emphasizing the need for users to clarify their objectives when utilizing AI systems. Practically, it suggests that addressing the ethical implications of AI requires a deeper reflection on the intended outcomes of predictive technologies, thereby guiding policy and practice toward more responsible AI development. The general implication of his argument is that we need to reflect on what we want to do with AI while trying to make it trustworthy. Just as understanding ourselves better enhances our interactions with fellow humans, a similar level of self-awareness is likely crucial in our interactions with AI.

Similarly, in his chapter titled "Is Trustworthy AI a Conceptual Nonsense?," Ryan explores the question of whether "trustworthy AI" a fundamentally flawed concept. Existing literature on trust and trustworthiness deems the term nonsensical, where trustworthiness necessitates agency, a characteristic currently absent in AI. This can be resolved by studying the "goodwill account" of trust, emphasizing the need for both competence and a demonstrable "will" to act in the trusted manner. The author examined the concept of "trustworthy AI" by considering alternative perspectives, such as the possibility of AI achieving agency or viewing AI trustworthiness as derivative of its creators' or users' trustworthiness.

Using a philosophical analysis of trust and trustworthiness literature, Ryan posits that trust requires belief in an agent's competence and approval of their will. By examining several challenges to the notion of "trustworthy AI," including the argument that machines can only be relied upon, not trusted; that trust is restricted to human interactions; and that trustworthiness requires awareness of being trusted or goodwill, Ryan suggests a goodwill account of trust. Drawing on this account, he makes the case that while trusting an AI is conceptually possible, the agency required for having goodwill means that plausibly the current state of technology is such that "trustworthy AI" doesn't make sense. The paper also engages with alternative approaches to explaining the notion of trustworthy AI, including the position that adopting a stance of trust to general AI is no less reasonable than adopting a stance of trust to human persons. Ryan argues that consideration of this position can motivate a revision in how we see trust in human relations.

The chapter concludes that the term "trustworthy AI" is not inherently a conceptual nonsense, but its current application is problematic due to the lack of agency in existing AI systems. The study suggests a need for a nuanced approach to the development and deployment of AI, acknowledging the limitations of current technology while considering the potential for future AI systems to achieve a level of agency that would warrant the label "trustworthy."

3.2 AI Risks for Human-To-Human Relationships

Six chapters in Part II of this book reveal the potential risks, threats and harms that AI development may pose to authentic human-to-human relationships. First, in their chapter titled "Empty Universities: The Dystopian Future of Education by AI," Robert Sparrow and Gene Flanady question the enthusiasm for AI's potential to enhance educational practices, arguing that proposed uses of AI in education risk radically transforming the relationships between students, between teachers and students, and between scholars and the broader community, to the detriment of the university and the larger society. They ask: What will happen when generative AI like ChatGPT enters higher education? What impact does AI have on the traditional roles of educators? How might the automation of teaching diminish the quality of education and interpersonal connections essential to learning?

By combining philosophical inquiry with a historical analysis of education and technology, the authors drew on existing literature to explore the trajectory of automation in various sectors and its implications for the university system. They argue that automation has historically led to *deskilling* and a lack of *meaningful* engagement in educational contexts, as evidenced by increasing class sizes and the reduction of personalized teaching. They found that while AI can provide efficiency, it undermines the social nature of education, a critical aspect of developing genuine knowledge and skills. They contend that the outputs of AI are often unreliable and lack the necessary depth to meet educational needs, raising concerns about the future of learning environments.

The chapter advocates for a reconsideration of the role of AI in education, emphasizing the importance of maintaining human connections and the quality of teaching. It contributes to the discourse on the relationship between AI and education by arguing against the notion that AI can replace human educators. Their work calls for policies that prioritize smaller class sizes and investment in quality human teaching, ensuring that the educational experience remains rich and socially engaged, rather than devolving into an environment dominated by machines.

In the following chapter titled "Implications of AI Chatbots for Human Relationships," Ellen Zhang addresses potential risks posed by chatbots on human relationships from a Daoist perspective. Using a philosophical analysis of Daoist texts, primarily the *Daodejing* and the *Zhuangzi*, alongside a critical review of existing literature on AI chatbots and their impact on human relationships, Zhang analyzes Daoist concepts like *ziran* (self-so-ness), *dao*, and tools. Her findings reveal a complex Daoist view: while acknowledging potential negative consequences of technology—such as the "machine mind" that hinders spontaneity and authenticity—Daoism doesn't categorically reject technology but advocates for a *harmonious integration* with nature. Zhang problematizes the common Western interpretation of Daoism as inherently anti-technology, which portrays Daoism as primitivist or nihilistic toward technological advancement. Her key argument highlights the nuanced Daoist perspective on technology as presented in the *Zhuangzi*, emphasizing the importance of "how to use tools" rather than simply rejecting them.

Meanwhile, Zhang highlights the danger of the potential for "pseudo-bonds" and the "Eliza effect," where AI users anthropomorphize AI, raising concerns about emotional well-being and the distortion of human relationships. She also touches upon the potential for AI-driven spiritual guidance to create a "computational theocracy." She explores whether the Daoist emphasis on naturalness and spontaneity can inform a balanced and ethical approach to the development and use of AI chatbots, specifically addressing concerns about potential emotional dependence and the erosion of authentic human connection.

This chapter offers a nuanced Daoist perspective on the implications of AI chatbots for human relationships by challenging simplistic interpretations of Daoism as anti-technology, thus demonstrating that a balanced and harmonious approach, respecting the natural order and promoting "dao-cosmotechnics," is more aligned with Daoist principles. Its contribution lies in its application of Daoist thought to the AI dilemma, offering a critical framework for evaluating the ethical and social implications of AI, particularly regarding the potential for emotional dependence and the distortion of authentic human connection.

Then, Vladimir Krstić in his chapter titled "Deepfakes, Fake Elvises, and Human Interpersonal Relations" demonstrates that the seemingly innocuous use of AI deepfakes for non-malevolent purposes has significant and potentially damaging consequences for trust in digital communication and interpersonal relationships. While the literature largely focuses on the impact of intentionally deceptive deepfakes, Krstić argues that non-malevolent deepfakes, created for *fun* or *other benign purposes,* are epistemically far more dangerous. The core argument is that the proliferation of non-malevolent deepfakes creates a context of epistemic uncertainty, allowing malevolent deepfakes to blend seamlessly and thus undermining trust in all digital testimonies to a much greater extent. He asks: how do non-malevolent deepfakes impact trust in digital communication, and what are the implications for interpersonal relationships?

Using a mixed-methods approach, combining anecdotal evidence with theoretical analysis, Krstić centers on personal experiences with deepfakes (e.g., the "Fashionable Pope," the purported deepfakes of General Zaluzhnyi), illustrating how easily even a scholar of misinformation can be deceived by seemingly harmless deepfakes. He uses this personal experience to support a theoretical argument drawing on the work of Kant and Krstić on the nature of lies and their impact on trust. He further uses the analogy of the Elvis impersonators in the movie "3000 Miles to Graceland" to explain how non-malevolent deepfakes act as camouflage for malevolent ones, significantly increasing the difficulty of discerning truth from falsehood in the digital sphere. Findings suggest that the pervasive presence of non-malevolent deepfakes poses a more significant threat to trust in digital communication than intentionally malicious deepfakes. In short, Krstić highlights an overlooked epistemic danger of non-malevolent deepfakes, proposing that educational efforts should focus not just on identifying deepfakes but on understanding the potential harm caused by their creation and dissemination.

In their contribution titled "A Compassion Paradox: Can AI Truly Bridge the Empathy Gap," Echo Liang Shang and Shirley Qian Jin query whether AI can foster compassion within human relationships, particularly in philanthropic crowdfunding.

They conceptualize the idea of AI machines as a "compassion organizer," questioning whether such machines can genuinely exhibit compassion or merely simulate emotional responses based on algorithms. They asked: Can AI effectively shape emotional connections in fundraising? What are the ethical implications of using AI in emotionally sensitive contexts? Their study builds on debates about AI's capacity for compassion and its harmful potential to manipulate emotional engagement, raising concerns about authenticity and trust in human-AI interactions.

Employing a conceptual analysis of AI's role in storytelling within crowdfunding campaigns, Shang and Jin draw on existing literature to explore how AI can enhance the narrative process. They examine practical examples, using ChatGPT, and analyze their impact on the emotional resonance of philanthropic campaign narratives. Findings suggest that while AI can improve storytelling effectiveness and foster emotional connections, it also raises ethical concerns regarding the authenticity of these narratives and the potential for emotional manipulation or compassion fatigue among donors. The contribution of the chapter lies in reframing our understanding of compassion as a socially coordinated response influenced by AI, emphasizing the need to critically assess the implications of AI on human relationships. It calls for further research at the intersection of AI, ethics, and compassion, and the potential risks and benefits of AI in emotionally charged domains like crowdfunding.

Yanto Chandra, Yiyun Luo and Jianxiang Tan in their chapter titled "Trust Me, I am Almost a Doctor: Uncertainty in Artificial Intelligence Mediated Doctor-Patient Interactions" turn to the issues of AI conversational agents, such as chatbots, being used to interact with patients and provide medical advice. They note that the literature on the effectiveness of these AI systems in medical contexts is mixed, with some studies showing positive results and others highlighting limitations and biases. Then they ask: Can AI doctor provide quality responses to patient questions? How do the patient-doctor exchanges compare to patient-AI doctor exchanges in their readability, similarity, and sentiment?

Using a computational linguistics approach to analyze the written exchanges between patients and human doctors from a popular online healthcare platform in China, as well as the responses provided by an AI doctor from the same platform, consisting of 192 question–answer pairs, the authors analyzed using the readability, similarity, and sentiment of the exchanges. The findings revealed that AI doctor's responses were more complex and had poorer readability compared to human doctor's responses. However, AI doctor's responses were also more similar and consistent in their language usage, and tended to use more positive and empathetic language. The study also found that AI doctor's responses were more likely to express negative sentiments, such as disgust, despite their overall positive tone.

This study contributes to the emerging literature on AI in healthcare by highlighting the importance of linguistic quality in evaluating the effectiveness of AI systems. The authors conclude that while AI doctors show promise in certain aspects, their limitations in readability and personalization pose significant ethical concerns. The study emphasizes the need for further research to improve AI doctor systems, focusing on enhancing readability, reducing redundancy, and ensuring appropriate emotional expression. The findings contribute to the broader discussion on ethical

AI in healthcare, highlighting the crucial role of linguistic quality in ensuring transparency, justice, and patient well-being in human-AI interactions.

In their chapter titled "Exploring AI Impact on the Physician–Patient Relationship: From Confucian and Daoist Views," Ruiping Fan and Morris Ho bring in Eastern philosophical perspectives to deal with the potential risks generated by the use of AI on the physician–patient relationship. As they note, the integration of AI into the medical domain has brought about significant transformations in healthcare practices, particularly in diagnostics and therapeutics, while simultaneously giving rise to substantial ethical dilemmas concerning its potential impact on the physician–patient relationship. Their study delves into these complex issues by examining them through the dual lenses of Confucian and Daoist philosophies, offering Eastern insights to complement prevailing Western ethical frameworks.

Beginning with an exploration of the distinction between the gradual and harmonious integration of AI and the rapid and transformative intelligence explosion, this research by Fan and Ho investigates the concept of diverse AI ethical agents and introduces two novel physician–patient relationship models—the AI-Patient model and AI-Physician–Patient model—put forth in recent Western philosophical discourse on machine ethics. Subsequently, this study presents a Daoist perspective on the utilization of machines, shedding light on why proponents of Daoism may object to the adoption of medical AI. The Daoist viewpoint posits that medical AI could potentially disrupt the physician's connection to the Dao, characterized by a deep, intuitive, and harmonious approach to medical practice.

Furthermore, this chapter provides insights from a Confucian virtue ethics standpoint, leveraging its emphasis on the five basic human relationships to argue why Confucianism may favor the AI-Physician–Patient model over the AI-Patient model for the integration of AI in healthcare. Additionally, supplementary Confucian principles are introduced to emphasize the importance of pursuing this Confucian objective with careful consideration. Fan and Ho contend that while the integration of AI into patient-end services is an inevitable progression, caution should be exercised against developing medical AI systems to replace human physicians, as healthcare extends beyond clinical interventions to encompass holistic approaches to addressing illnesses and take care of patients.

3.3 *Use of AI and the Future of Human Relations*

Similar to Part II, this part also consists of six chapters that delves into the social and ethical aspects of future human-to-human relations within the context of utilizing AI in society. First, in her chapter titled "A Cross Cultural Perspective on Human-Care Robot Interactions," Jennifer Bourne addresses the issues of human-carebot relations as well as human-to-human relations from a cross-cultural perspective. Bourne problematizes the literature on human relationships, arguing that while frameworks like Social Penetration Theory offer insights into trust-building, they don't fully explain the unique dynamics of human–robot interactions. The inherent asymmetry of trust—robots serving humans without reciprocal trust—is a key concern. Given the complexities of human disagreement and emotional responses, she questions how robots should navigate emotionally charged exchanges and manage disagreements effectively. She asks: How can care robots be designed to foster trust and companionship across diverse cultures, considering the nuances of communication styles, emotional intelligence, and cultural values?

Drawing upon two theoretical frameworks—Hofstede's cultural dimensions and Hall's high- and low-context communication models—Bourne systematically explores human–robot interactions using cases of care robots from Japan (Paro, Bono), the U.S. (Mobile Aloha, Bandit), and China (Astribot), and focuses on the cultural elements embedded in their design and functionalities. Findings suggest that communication styles, interaction patterns, robot appearance, and voice significantly influence the acceptance and effectiveness of care robots across cultures. The analyses further integrated Eastern philosophical traditions, particularly the concept of Gongsheng (kyōsei), emphasizing mutual embeddedness and co-becoming as crucial design principles.

Finally, the chapter concludes that a culturally sensitive approach to care robot design is essential for successful robots-human-culture integration. By aligning robot design with cultural nuances, communication styles, and philosophical traditions, while adhering to the principle of Gongsheng, care robots can transcend their mechanical nature and become genuine companions, enhancing users' quality of life. This study moves beyond a Western-centric understanding of human–robot interaction, advocating for a design process that prioritizes trust, empathy, respect, and mutual understanding across diverse cultural contexts.

The subsequent chapter, titled "Digital Psychological Twins in Medicine: Addressing Risks to Human Relationships" developed by Brian D. Earp, Tessa van Veenendaal, Sebastian Porsdam Mann, and Julian Savulescu, examines the implications of AI-driven decision-making in healthcare. In particular, medical decision-making on behalf of individuals who have lost the capacity to make their own treatment choices poses significant challenges. The "substituted judgment" standard, which prioritizes individual autonomy, requires surrogate decision-makers to choose the course of action that the patient would have chosen or endorsed if they were able to do so. However, research suggests that surrogates often face difficulties in accurately predicting patient preferences, even when making a good-faith effort to adhere to the substituted judgment standard. To address these challenges, the Patient Preference Predictor (P3), a computer-based algorithm that correlates demographic data with expressed medical preferences, has been proposed to supplement or replace aspects of the surrogate decision-making process. However, critics argue that the P3 fails to respect patient autonomy due to relying on population-level data rather than accounting for individuals' unique values, beliefs, and preferences.

In response, these authors have proposed the Personalized Patient Preference Predictor (P4), an AI system that would use machine learning to create a "digital psychological twin" of a patient, inclusive of their values, beliefs, and preferences, from individual-level data. While the P4 aims to address concerns raised by the P3, it has been argued that it may introduce new problems, particularly regarding its potential negative effects on the doctor-patient relationship and family relationships. In this chapter, the authors describe the P4, summarize objections related to its impact on human relationships, respond to these objections, and conclude with some thoughts on future directions. They emphasize that the use of a P4 should not marginalize doctors; rather, it is important to consult with doctors throughout the design and implementation process and contribute positively to the doctor-surrogate relationship.

In her chapter on "The Impact of AI on Vulnerable Individuals: Three Principles," Ding Chunyan argues that AI systems present both opportunities and risks particularly for individuals in vulnerable situations. Ding calls for the pressing need for governance frameworks that recognize the risks and opportunities in AI in this regard. She further argues that current literature does not clearly explain how state involvement in government intervention can effectively mitigate the adverse effects of AI while maximizing its benefits for these vulnerable populations. By leveraging vulnerability theory, she proposes a responsible governance approach that balances these dynamics.

Ding develops an analytical framework rooted in three ethical principles—do-no-harm, beneficence, and justice—to guide the governance of AI systems affecting vulnerable groups. Employing qualitative methodologies, she reviews existing policies and practices, culminating in a set of actionable recommendations. Key suggestions include defining "significant harm" to prohibit exploitative AI systems, incentivizing the development of socially beneficial AI programs, and advocating for transparency in data use. This structured approach ensures that AI technologies align with ethical standards and societal values.

Key findings underscore the importance of proactive state involvement in AI governance, emphasizing the need for clear definitions of harm and the promotion of equitable AI practices. The chapter contributes to both theoretical discourse and practical policy by proposing a coordinated regulatory framework that encourages collaboration among public agencies. By prioritizing ethical considerations, the chapter advocates for a technological landscape that not only safeguards vulnerable individuals but also fosters human flourishing and social justice.

In their contribution titled "The Credit View and AI Testimony: A Cross-Cultural Epistemological Study of Human and AI Testimony," Masaharu Mizumoto, Shane Ryan and Chienkuo Mi delve into the issues of including AI testimonials as epistemic knowledge. Building on earlier work on the Credit View of Knowledge, the authors investigate the patterns of people's testimonial knowledge attribution when the true belief of an agent is based on the testimony of a human versus an AI assistant. They ask: How do people respond differently to human and AI testimony? What are the implications for trust attributions in AI interactions? How will increased AI usage influence human knowledge attributions?

Using a cross-cultural comparison, including past studies using surveys or experiments, the authors examine people's responses to human and AI testimony, particularly participants' attributions of knowledge, credit, and fault in different scenarios. Findings suggest that people respond differently to human and AI testimony, particularly in their reasons for attributing knowledge to the recipient of testimony. This differentiation implies that, for ordinary people, "AI is more like a tool than an agent," which has implications for trust attributions and the influence of AI on human relations.

The study concludes that the integration of AI into human life raises complex issues related to the epistemic role of AI and its influence on human relationships. While the Credit View of Knowledge may still be applicable if the notion of credit is reinterpreted in terms of reliability and proper method, the findings highlight the need for a nuanced understanding of the moral and epistemic status of AI. By exploring the implications of AI for human knowledge attributions and trust, this chapter offers insights for the development of ethical frameworks and policies surrounding AI integration.

Feng Lin and Yang Wu, in their chapter titled "From Pixel to Persona: The Genalogy of the Humanness in Virtual Anchors in China," address the issue of parasocial relationships arising from the utilization of virtual anchors (VAs) in the media industry. While VAs resemblance to humans ("humanness" or "human-likeness") has been studied, extant research overlooks the dynamic and socially constructed nature of this concept in different contexts. Lin and Wu argue that existing studies treat "humanness" as a static feature, neglecting its evolution and the influence of technology and institutions. They examine how the humanness of VAs in China has evolved over time, considering their interplay with technological advancements and industry regulations. They ask: How has the concept of "humanness" in VAs changed across different stages of the industry's development in China? What are the key technological and institutional factors shaping its evolution?

Using mixed-methods approach that comprise secondary data, including news reports, market research, platform regulations, and state policies, the authors analyze the development of the Chinese VA industry into four phases: experiment (2000–2004), stagnation (2005–2014), revival (2015–2017), and boom (2018–present). By conducting interviews with VA developers, researchers, and VAs themselves, supplemented by participant observation at industry conferences, they reveal an evolution of VA humanness across these phases: from "pixel humanness" with limited visual and auditory capabilities in the early years, to "pent-up humanness" during a period of stagnation, then "expressive humanness" with improved technology, and finally "biopolitical humanness" shaped by platform and state regulations in the current boom period.

This study offers a nuanced understanding of the socially constructed nature of "humanness" in virtual anchors, highlighting its dynamic evolution within the context of the Chinese VA industry. The four-phase model contributes to the literature by moving beyond a static view of human-likeness, demonstrating how technological advancements and institutional forces co-construct the humanness of VAs. This framework offers valuable insights for researchers studying human–computer interaction, parasocial relationships, and the broader implications of AI on human relationships, particularly within the rapidly evolving landscape of digital media in China.

The concluding chapter of this section, titled "Sex Robots: A Confucian Defense" by Xudong Fang, explores the ethical dilemmas surrounding the utilization of sex robots within the context of Confucian values. Fang scrutinizes two Confucian objections regarding sex robots as presented by other authors like Ruiping Fan and Lawrence Yung: arguments rooted in the considerations of representation and dignity. Drawing from a rich tapestry of Confucian resources, Fang rebuts these objections, asserting their invalidity. Central to his innovative Confucian defense for integrating sex robots in Confucian-influenced societies is the premise that AI robots are objects, not humans, offering an alternative avenue for individuals in their intimate lives. Consequently, he argues that concerns regarding a decline in human partners and the potential threat to human civilization due to sex robot usage are unwarranted. This chapter articulates a thought-provoking perspective that is bound to stir controversy within and beyond Confucian cultural boundaries.

4 Concluding Remarks

Our book constitutes one of the earliest attempts to document the roots, nature and outcomes of AI-human relationships from ethical and social sciences perspectives drawing on both Eastern and Western intellectual traditions. Our focus has been on the AI-human and human-to-human relationships that are shaped and affected by the rise of AI, while leaving the AI-AI relationships unexplored. We have delved into a variety of theoretical viewpoints regarding AI-human relationships, ranging from political theories of life, Confucian philosophy, and virtue ethics to discussions on trustworthiness, conceptual absurdity, and the exploration of dystopian scenarios in education with AI. We have also examined the Daoist conceptualization of AI's impact on the self and the paradox of compassion.

In addition, we have included a limited amount of empirical research, consisting of two chapters that examine the evolution of the virtual anchor industry and conduct linguistic analyses of AI doctors' interactions with patients. Conceptual chapters cover diverse topics, from the ethical implications of sex robots to the concept of psychological twins in medicine and the role of care robots in cross-cultural contexts. The approaches employed in the book vary, encompassing normative perspectives that assess the ethical dimensions of AI-human and human-to-human relationships and propose solutions from different philosophical standpoints, as well as descriptive approaches that portray the current state of AI-human and human-to-human relationships, alongside imaginative explorations of the potential futures within the realm of AI.

While this book explores the multifaceted benefits and risks associated with artificial intelligence (AI) and its implications for human relationships, it leaves unexamined several critical domains, particularly those pertaining to economics, politics, and technology. For instance, the economic dimensions and consequences of AI-human interactions warrant further investigation. Questions arise regarding whether AI will alter the distribution of time spent in work versus leisure, and what ramifications this shift may have for the labor market. Specifically, who will bear the costs associated with a diminished workforce, as AI systems increasingly assume tasks

Kirkpatrick, K. (2016). Battling algorithmic bias: How do we ensure algorithms treat us fairly? *Communications of the ACM, 59*(10), 16–17.
Kordzadeh, N., & Ghasemaghaei, M. (2022). Algorithmic bias: Review, synthesis, and future research directions. *European Journal of Information Systems, 31*(3), 388–409.
Moor, J. H. (2006). The nature, importance, and difficulty of machine ethics. *IEEE Intelligent Systems, 21*(4), 18–21. https://doi.org/10.1109/MIS.2006.80
Pan, S., & Mou, Y. (2024). Constructing the meaning of human–AI romantic relationships from the perspectives of users dating the social chatbot Replika. *Personal Relationships* (in press).
Pentina, I., Xie, T., Hancock, T., & Bailey, A. (2023). Consumer–machine relationships in the age of artificial intelligence: Systematic literature review and research directions. *Psychology & Marketing, 40*, 1593–1614.
Pereira, V., Hadjielias, E., Christofi, M., & Vrontis, D. (2023). A systematic literature review on the impact of artificial intelligence on workplace outcomes: A multi-process perspective. *Human Resource Management Review, 33*(1), 100857. https://www.sciencedirect.com/science/article/pii/S105348222100036X
Perrigo, B. (2024). Yann Lecun is optimistic that AI will lead to a better world. *Time.* https://time.com/collection/time100-impact-awards/6692039/yann-lecun-meta-artificial-intelligence-time-award/ (checked 30 December 2024).
Peters, U., & Carman, M. (2024). Cultural bias in explainable AI research: A systematic analysis. *Journal of Artificial Intelligence Research, 79*, 971–1000.
Ramoglou, S., Schaefer, R., Chandra, Y., & McMullen, J. S. (2024). Artificial intelligence forces us to rethink Knightian uncertainty: A commentary on Townsend et al.'s are the futures computable? *Academy of Management Review* (in press).
Seo, K., Tang, J., Roll, I., Fels, S., & Yoon, D. (2021). The impact of artificial intelligence on learner–instructor interaction in online learning. *International Journal of Educational Technology in Higher Education, 18*(1), 18–54.
Townsend, D. M., Hunt, R. A., Rady, J., Manocha, P., & Jin, J. H. (2023). Are the futures computable? Knightian uncertainty and Artificial Intelligence. *Academy of Management Review.*
Zhai, C., Wibowo, S., & Li, L. D. (2024). The effects of over-reliance on AI dialogue systems on students' cognitive abilities: A systematic review. *Smart Learning Environments, 11*(1), 28.

Yanto Chandra is Professor at the Department of Public and International Affairs and Director of the Technology, Policy, and Law Laboratory and a Member of Public Management Laboratory at the City University of Hong Kong. He studies the role of entrepreneurship and innovation in influencing performance in the social and public sectors, as well as how new technologies such as AI and digital currency, as digital public goods, affect the society and policy. His research has been published in public administration and policy journals, among others, Public Management Review, Public Administration Review, International Public Management Journal, American Review of Public Administration, Policy and Internet, as well as management journals such as Academy of Management Review, Journal of Business Venturing, and Journal of International Business Studies. He was a Kooiman Prize for best paper award in 2021. He serves as an Editorial Board Member of Public Administration and Development and Academy of Management Perspectives. He is Guest Editors on AI for Journal of Business Venturing and Journal of Business Venturing Insights.

Ruiping Fan (范瑞平) is Chair Professor of Philosophy at the City University of Hong Kong. His research focuses on bioethics and comparative philosophy, drawing from both Confucian and Western cultural perspectives. With a publication record exceeding 200 journal articles and book chapters (comprising over 100 in English and 100 in Chinese), he has authored books including Reconstructionist Confucianism: Rethinking Morality after the West (2010), Contemporary Confucian Bioethics (2011), and Contemporary Medicine and Confucian Thought (2024). Additionally, he has served as the editor or co-editor of 15 volumes (8 in English and 7 in

Chinese), including recent publications such as Sex Robots: Social Impact and the Future of Human Relations (2021), Incentives and Disincentives in Organ Donation: A Multicultural Study among Beijing, Chicago, Tehran, and Hong Kong (2023), and An East-West Dialogue on Good Governance: Learning from Each Other (2024).

AI and Relationship: Theoretical Explorations

Toward a Political Theory for Life with AI

Mathias Risse

1 Introduction

Short of ending up on the path to perdition, humanity might find ways of living with technology, and specifically AI, in roughly two ways. One possibility is that special AI advances relatively slowly into more and more general AI, and that general AI is integrated into the functioning of society in numerous ways. Such integration could benefit societies broadly or benefit only a few. But, at least for a long time, societies will not implode. This might be because intelligence is multidimensional after all, and there is no sequence of superintelligent entities that outdo each other in general abilities. Or perhaps humanity finds ways of keeping both technological possibilities and the arms races connected to them under control, perhaps along the lines of the Nuclear Non-Proliferation Treaty.[1] Either way, arrangements between humans and AI unfold fairly harmoniously and slowly—or humans remain in charge.

The other possibility is that we get an intelligence explosion, and subsequently (possibly within hours) one or more forms of general AI take over. Perhaps they do so completely, but at least to such an extent that arrangements must be made with them. Humans would at best somehow still be involved in running things and at worst find themselves subjugated, perhaps in a manner resembling our relationship

M. Risse (✉)
Harvard University, Cambridge, MA, USA
e-mail: mathias_risse@hks.harvard.edu

[1] The Nuclear Non-Proliferation Treaty seeks to prevent the spread of nuclear weapons technology, promote cooperation in peaceful uses of nuclear energy, and further nuclear disarmament; see Burns and Coyle, *The Challenges of Nuclear Non-Proliferation*. Not only did atomic energy also have much potential while harboring lethal dangers, but that threat had to be negotiated in contentious international arenas. Such an approach will avoid a scenario where some countries decide to terminate future development of general AI only to find that others continue to advance. On AI under the aspect of security and non-proliferation, see Kissinger, Schmidt, and Huttenlocher, *The Age of AI*, Chapter 5.

Y. Chandra and R. Fan (eds.), *Artificial Intelligence and the Future of Human Relations*,
https://doi.org/10.1007/978-981-96-7185-4_2

with animals. Many experts expect something like this to happen within decades.[2] But research might unfold more slowly, and this will only occur in centuries. In any event, both types of scenarios for humanity to end up living with technology come in a range of variations.

Developments in AI are not the only ones that shape our future. Not only climate change and technological responses to it, including geoengineering, but also gene-editing, human enhancement using pharmaceuticals and bioelectronics, nanotechnology, synthetic biology, and robotics: all these will come together *somehow* to shape our future. That future is obviously hard to predict. "Experience shows," John von Neumann states, "that even smaller technological changes than those now in the cards profoundly transform political and social relationships" and do so in ways that "are not a priori predictable." Accordingly, "one should take neither present difficulties nor presently proposed reforms too seriously."[3]

One way of illustrating how hard it is to make such predictions (and thus also to throw light on that point about present difficulties and reforms) is to think about how the most prescient minds in 1900 would have thought about 1920, how in 1920 they would have thought about 1940, and so on across the twentieth century. We can safely guess that most of them (barring perhaps H. G. Wells) would have erred dramatically each time. The same point applies to what we can currently say about how the twenty-first century will progress even just for a twenty-year interval. Insights from history or social sciences generate predictions only to the extent that the future relevantly resembles the past. And chain reactions around technological innovation might undermine any such resemblance.[4]

To set the stage for our discussion, I adopt a distinction between "slow and relatively harmonious" and "fast and radical" as far as integration of AI into human life is concerned (disregarding the variations in which these scenarios come). I start by exploring questions from the "slow and relatively harmonious" scenario and switch over to the "fast and radical" scenario. All this is speculative. But we do not want to start thinking about these matters in earnest only when they are upon us.[5]

2 Moral Status

I use "ethical" and "moral" synonymously. Ethics (moral philosophy) is inquiry about (a) what it means for agents to relate to entities in ways appropriate to the nature of these entities or appropriate to the ways these entities connect to entities around them, and (b) what characteristics a person should possess to be disposed to act according to such standards of appropriateness. The entities could be humans,

---

[2] Tegmark, *Life 3.0*, 42.

[3] von Neumann, "Can We Survive Technology?," 519.

[4] On difficulties around prediction once technology is involved, see Bostrom, "Technological Revolutions."

[5] I develop all this at greater length in Chapter 11 of Risse, *Political Theory of the Digital Age*.

other animals, plants, ecosystems, or other parts of nature like rivers or mountains, works of art or any object at all. By "nature" I mean inherent features, and it is partly such features that generate questions about how one should relate to the entity in question. But how I should relate to something also depends on my history with it, more generally on how it is embedded into a larger network of other entities. This is obvious for humans. But for example, what is appropriate for me to do regarding a work of art depends on its quality and then also on the fact that, say, it belongs to a certain museum or collector.[6]

To act in ways that are "ethical" or "moral" is to act in a manner appropriate to the nature of the entities that one deals with (or to the ways they are interconnected), or to act in ways that sustain the characteristics of a person disposed to act in such a manner. To inquire about "moral status" is to investigate an entity's inherent features to ascertain whether there is anything about it in virtue of which it matters *for its own sake* how we act toward it: when we identify a being as having moral status, there is an adequate way of relating to the entity, and the reason is something about that entity.

This understanding of ethical action is about everything around us rather than merely humans or living things. This view creates ways of approaching the world with questions like these in mind: what kind of moral status does something have, and how should I relate to it? These questions would not normally be asked just this way. There are multifarious ways of approaching these matters that need not avail themselves of philosophical language. Answers vary with the object of inquiry and mandate different behaviors depending on the entity in question and the various relations it stands in with others. Moral status might require a lot or relatively little.

This broad view of ethics also allows for answers to questions about how it is appropriate to deal with certain entities when it is neither their *own* nature nor how they connect to anything around them that is central to the inquiry. Kant thought a dog that has grown old doing good service should be cared for. If the owner kills the dog, then this "the owner thereby damages the kindly and humane qualities in himself, which he ought to exercise in virtue of his duties to mankind."[7] What is required *regarding* the dog is not required because of anything *about* the dog but because of how it affects the owner. Consider a related example. In 2007, a U.S. colonel reportedly terminated a landmine-sweeping exercise because he regarded the operation as inhumane after robots kept crawling along losing legs one at a

---

[6] For recent work on moral status, see Clarke, Zohny, and Savulescu, *Rethinking Moral Status*. See especially Sinnott-Armstrong and Conitzer, "How Much Moral Status Could Artifical Intelligence Ever Achieve?" See also Liao, "The Moral Status and Rights of Artificial Intelligence." For a discussion of moral status specifically regarding animals, see Kagan, *How to Count Animals*, Chapters 1 and 5. Questions around moral status have long been controversial, especially the extent to which it would be appropriate to ascribe it to anything other than conscious beings. This section too will be controversial but also brushes over much subtlety. See also Jaworska and Tannenbaum, "The Grounds of Moral Status"; Streiffer, "At the Edge of Humanity."

[7] Kant, *Lectures on Ethics*, 212 (27:459).

time.[8] Perhaps this reaction has a psychological explanation. But it could also be a decision to not needlessly desensitize natural aversions to inflicting pain.

Notions of consciousness matter for assessing moral status. Consciousness comes in a variety of forms, and these forms illuminate or allow for differing degrees of consciousness. To begin with, there is sentience, the capacity of sensing and responding to the world. A more demanding sense is wakefulness, which requires actual exercise of that capacity rather than merely having it. One counts as conscious only while *awake and normally alert*. Self-consciousness is a yet more demanding sense that understands conscious creatures as aware and also as aware that they are aware. If this is taken to involve explicit conceptual self-awareness, many nonhuman animals and even young children might fail to qualify. But if only rudimentary, implicit forms of self-awareness are required, a wide range of nonlinguistic creatures might count as self-conscious.[9]

A common stance is that *only* conscious beings have a moral status that is not ultimately instrumental to something else.[10] But in my broad view, entities can have moral status even if they are not conscious. It is because of the way Da Vinci's *Mona Lisa* captures beauty that we should cherish it, and thus because of the painting's inherent features. And questions about adequate behavior toward something can also arise even if the entity itself has no moral status. An artistically worthless piece might require my care if it belongs to someone I love who cannot care for it herself.

3 The Moral Status of AI

There might eventually be conscious machines, though their consciousness would not be one embodied in organic matter. But that fact should not prevent us from recognizing in machines whatever moral status comes from consciousness (which would be a carbon chauvinism).[11]

A set of distinctions formulated by James Moor continues to be useful here.[12] *Ethical impact agents* are agents whose actions have consequences, intended or unintended, for beings with moral status. Any robot is such an agent if its actions can

[8] Wallach and Allen, *Moral Machines*, 55.

[9] For the relevance of consciousness in this context, see Risse, *Political Theory of the Digital Age*, Chapter 1.

[10] As defended e.g., in Basl, *The Death of the Ethic of Life*.

[11] For recent discussion of whether machines might be conscious and how we would know if non-organic material has consciousness, see Schneider, *Artificial You*. See also Tye, *Tense Bees and Shell-Shocked Crabs*, Chapter 10; Seth, *Being You*, Chapter 13. For reflections on what "machines" are and if they can be conscious, see Harnad, "Can a Machine Be Conscious? How?" See also Kiverstein, "Could a Robot Have a Subjective Point of View?" For the argument that AI, even with efforts to build in human values, will lead to creatures very different (and ontologically disconnected) from us and that thus our standard ethical vocabulary no longer applies, see Lorenc, "Artificial Intelligence and the Ethics of Human Extinction."

[12] Moor, "Four Kinds of Ethical Robots."

harm or benefit humans. But that ability would not give robots moral status. *Implicit ethical agents* are agents whose design has ethical considerations built in, such as through safety or security features. (Think of teller machines checking availability or limiting the daily amount that can be withdrawn.) Again, this type of machine does not seem to have any moral status.

Explicit ethical agents can secure and process ethical information about a variety of situations, make sensitive determinations about what to do, and even work out resolutions where considerations conflict. (One might think of ever more sophisticated chatbots.) Finally, *full ethical agents* are explicit ethical agents who also have metaphysical features we usually reserve for agents *like us*, such as consciousness, intentionality, or free will.At least both explicit ethical agents and full ethical agents do have moral status (the former in virtue of having certain features that could make them relevantly like pieces of art, a consideration that might conceivably also include at least some implicit ethical agents). More interestingly, in the space between explicit and full ethical agents, there could be a variety of types of moral status. Androids, say, could display hallmarks of agency, such as interactions with the environment, combined with a level of independence and adjustability.

Explicit ethical agents short of full ethical agency hardly deserve *all* the consideration that full ethical agents do, but nor would they plausibly deserve *none*. One implication of this last point is that we ought not to develop general AIs that would do demeaning work *and* would be conscious (in order to be more human-like in interactions). Joanna Bryson once provocatively asserted that "robots should be slaves" and thus be treated as property.[13] But the fact that we create machines makes them property *only as long as* nothing about them gives them moral status inconsistent with being property (which could defeat the ownership status that comes from design). Children, too, are humanly created. But their humanity overrules demands from the fact that they have been created. Anyway, the world humans and conscious machines share should not include any relations that count as enslavement.

We must be careful not to create machines that possess consciousness and that, once created, have claims not defeated by the fact that we created them. Once machines exist that are arguably conscious, we face moral status problems that are familiar from discussion about animals. Eventually we might confront the challenge of creating entities that have moral status where appropriate responses mean accommodating the capacity for agency—that is, we might confront vexing questions about just what moral status artificial explicit ethical agents or full ethical agents have, and just what kind of behavior toward them would be adequate in response. These questions arise outside of our experience with nervous systems that have emerged through evolution.

Such issues would likely be divisive. Some people would build or acquire AI to exploit it, including for companionship ranging from caretaking to sexual services.[14]

---

[13] Bryson, "Robots Should Be Slaves."

[14] On machines as sexual companions, see Levy, *Love and Sex with Robots*; Migotti et al., *Robot Sex*; Richardson, *Sex Robots*; Devlin, "The Ethics of the Artificial Lover."

Holding conscious robots in bondage will appeal to many while clouding their judgment about whether machines deserve consideration. Others might protest on the machines' behalf, in the same spirit in which they might try to improve the lives of animals. Yet others might resent their use, fearing competition. In extremis, these debates resemble debates around enslavement of humans (and one vexing feature might be disagreement about the extent to which this is so).

4 Machine Politics

We must be prepared for the possibility that machines eventually obtain whatever moral status comes from consciousness. But does this mean it would then be appropriate for us to treat machines that have the same level or kind of consciousness as humans the same as we would (should) treat other humans? That depends on whether—among the entities who share a level or kind of consciousness with us—*additional* distinctions are appropriate. Let us consider the domain of the political, in comparison with how animals fare in that domain.

The *concept* of the political concerns ways in which order is created. In addition, we can distinguish various *conceptions* of the political. Such conceptions answer questions about what kind of agency and structures will normally or should bear on how order is created. Conceptions drawing on Aristotle look at processes of creating order by way of emphasizing human transformation through activities of ruling and being ruled, or characterize the kind of activity that becomes possible in such a setting or the features that essentially matter to the shared life in *poleis*. Among scholars who have recently worked within such an Aristotelean understanding, we find Hannah Arendt. As she writes,

> to be political, to live in a *polis*, mean[s] that everything [is] decided through words and persuasion and not through force and violence. In Greek self-understanding, to force people by violence, to command rather than to persuade, [are] prepolitical ways to deal with people characteristic of life outside the *polis*.[15]

Polis life enabled humans to get along in ways unavailable before. Prospects of solving conflicts by reasoned argument made it possible for ruling and being ruled to become temporary states.[16] For Arendt, action is about initiating something new that

[15] Arendt, *The Human Condition*, 26f.

[16] In *On Justice*, Chapter 14, I propose a different conception of the political. The *frame of human life* is the set of institutions and practices, as well as activities within them, that organize capacities for mutual aid and balance tendencies to compete and cooperate. For Aristotle, the *polis* does that job. Today, institutions and practices that do so are embedded into a human web that has coalesced into a world society. According to the frame-of-human-life conception, the political is about designing that frame at the global, or species-wide, level. This was a conception of the political designed to fit with the role of public reason as I develop it there for the global context. While this is a different conception from Arendt's, the emphasis she puts on action, speech, memory, and meaning is consistent with it. On the human web, see McNeill and McNeill, *The Human Web*. On the world society, which we encountered in previous chapters, see Meyer, *World Society*.

expresses a person's identity. Political action is the shared endeavor of maintaining or redoing the world as we jointly inhabit it. How then could it make sense for *artificial intelligences*—though they might be as self-conscious as and more intelligent than us—to participate in political processes?

One way of expressing this doubt is that what is needed to participate in political action is a cognitive capacity different from intelligence and self-consciousness—something that humans possess *in addition* to those two but that AI might not possess even if it does possess intelligence and the same level and type of consciousness. The Kantian tradition provides such a capacity: it offers an account of *rationality* (in Kant's German, *Vernunft*) as a distinct capacity. Christine Korsgaard has rearticulated that notion for use in an exploratory discussion of how humans might differ from (and what obligations they might have toward) animals.[17] She is aware that which creatures possess the forms of cognition she discusses, if any do, is an empirical question that philosophy alone cannot answer.

Korsgaard understands an *intelligent* animal in terms of its ability to learn from experiences to solve problems through deliberative processes. Rationality, Korsgaard explains, "is a normative power grounded in a certain form of self-consciousness."[18] So it is something only self-conscious beings can possess, but a capacity beyond self-consciousness (different from intelligence). Rational beings can reflect on what causes their beliefs and actions and decide if such causes count *as good reasons*:

> "So the difference between rationality and intelligence is this," Korsgaard clarifies:
>
> Intelligence looks outward at the world, and asks and answers questions about the connections and relations we find there—most obviously about causal relations, but also spacial [sic] and temporal and social relations. But rationality looks inward, at the workings of our own minds, and asks and answers normative or evaluative questions about the connections and relations that we find there. In particular, practical rationality raises questions about whether the attitudes and the facts that motivate our actions give us good reasons to act.[19]

Rationality bestows a specific character upon our actions. Intelligence and self-consciousness are needed for successful political action. But once we understand action generally and political action specifically in terms of genuine initiation that expresses our identity, rationality likewise is required, over and above intelligence and self-consciousness.

Korsgaard's points apply to AI as well. In March 2016, one of the top Go players in the world, Lee Sedol, was defeated by AlphaGo. AlphaGo is an AI designed at

[17] Korsgaard, *Fellow Creatures*. For alternative takes on the status of animals, see e.g., Kagan, *How to Count Animals*; Singer, *Animal Liberation*; Donaldson and Kymlicka, *Zoopolis*. For the argument that worrying too much about consciousness takes us ever further away from coming to terms with animals, see Dawkins, *Why Animals Matter*. Dawkins thinks animals matter because for humans to be healthy animals need to be healthy. For the argument that domestication of animals provides a good model for how to integrate AI into human life, see Müller, "Domesticating Artificial Intelligence." In recent times human bodies and brains have been the model for Artificial Intelligence, but AI and both robotics might also benefit from drawing inspiration from how the bodies and brains of other animals operate, see Rus, "The Machines from Our Future."

[18] Korsgaard, *Fellow Creatures*, 40.

[19] Korsgaard, 41.

DeepMind, a London lab then owned by Google. What became especially famous was AlphaGo's thirty-seventh move in the second game, a move that dumbfounded experts including Lee. DeepMind had originally trained its AI on the history of Go strategy. But AlphaGo had traversed a process of learning all its own via a machine-learning technique called Reinforcement Learning, which enables learning in an interactive environment by trial and error. AlphaGo thereby developed an innovative take on ancient human activities, learning everything there was to know only to transcend that knowledge.[20]

The expertise with which AI plays Go has rapidly advanced. But Move 37 still illuminates the possible impact of AI on human life. AI might have access to the entirety of human experience and reflection as captured digitally, including history, sciences, art, legal systems, and philosophy. It could know all digitally recorded facts and comprehend the logic of all known systems of thoughts. It could absorb all commonly used criteria to compare such systems, such as comprehensiveness, consistency, or factual plausibility. It might eventually do all this very fast, conscious of what it is doing. But much as animals are capable of many things without rising to the level of rationality, none of what even a self-conscious AI can do might amount to exercising *rationality* as Korsgaard understands it and as Arendt thinks it is necessary for political action.

To be sure, Korsgaard insists on the separateness of rationality from intelligence and self-consciousness, but also argues (with Kant, against Kant) for the good of animals to be included in the pursuit of our own good. She proposes a combined commitment to human distinction in the domain of cognitive capacities and a substantially increased understanding of the moral status of animals (vis-à-vis both common practice and Kant's own view). The political domain must be organized within such constraints. But a further question is just what all this means for the role of animals (or the consideration they are due) in the political domain, or then also for the place of AI in this domain. That animals are incapable of rationality will have certain implications, and ditto for AI.

But these implications might ultimately not amount to much, certainly not as far as animals are concerned. Under the heading of *zoopolis*, Sue Donaldson and Will Kymlicka make far-reaching proposals for integrating animals into our cooperative relations as fellow *citizens*.[21] What underlies their proposal is a broader understanding of the domain of the political than Arendt and the Aristotelean tradition propose. For Donaldson and Kymlicka, politics is about negotiating the relationships among entities with a subjective well-being who find themselves in close proximity under human-devised structures. These relationships include ways of ruling together but also involve "assisted" and "dependent" agency.[22] Already for humans, citizenship involves support (e.g., for elderly or sick people or for the unemployed) or outright

---

[20] See Metz, "In Two Moves, AlphaGo and Lee Sedol Redefined the Future." See also Livingston and Risse, "On the Impact of Artificial Intelligence on Human Rights Over the Next 20–30 Years."

[21] Donaldson and Kymlicka, *Zoopolis*.

[22] Donaldson and Kymlicka, Chapter 3.

dependence (e.g., for people with severe disabilities), and so differentiated webs of relations.

Animals—with whom we have symbiotic living arrangements—could be part of such webs. Such webs would recognize animals' nature and abilities, but also the ways they stand in assistance or dependency relations with us. People can be supported by and indeed be dependent on animals in a variety of ways, from farming/herding to transportation to assistance for people who are blind. Sometimes humans act as advocates, speaking up for them in officially sanctioned ways. But that would not be because animals are not citizens or otherwise not part of the political domain. That would be because it lies in the nature of the *kind* of citizens they are that much in the exercise of citizenship as we know it in a world prejudiced against nonhuman animals must be reconsidered once we respond adequately to their moral status. The same could eventually be true for intelligent machines.

One upshot is how much is open in this domain. We do not know whether machines, once they reach a level of intelligence and self-consciousness comparable to humans, will remain deficient regarding other cognitive capacities. Perhaps rationality supervenes on suitable combinations of intelligence and self-consciousness and would therefore be present in superintelligent self-conscious machines. If it is present, and if other ways of singling out humans are unavailable, intelligent machines should be full members in the domain of the political. If not, we have additional questions to discuss as to what kind of citizens machines could be, much as such questions already arise for animals.

5 How Woefully Unprepared We Are

We have investigated if in the "slow and relatively harmonious" scenario *we* might eventually have reason to live in such a community *with machines*. What about the other way round, in the "fast and radical" scenario? Do our moral theories give us reason to think *they*—machines that would arise through an intelligence explosion, rather than through gradual processes—would want to live that way *with us*?

Suppose an intelligence explosion delivers machines that are not only intelligent but also self-conscious and rational, regardless of whether rationality is a capacity by itself or supervenes on the others. Under this assumption it is *more likely* than in any other scenario that such machines would incline to live in moral community with humans. Such superintelligences could review human history and know both our amazing achievements and demonic inclinations. These inclinations would cover not only what humans have done to each other but also what they have done to other species and to the natural environment. How confident can we be that the fact that superintelligences possess the cognitive capacities necessary to acknowledge moral status in others *actually* make them acknowledge ours and see it as something worth protecting?

One might think of the dispute between Hume and Kant about whether rationality fixes values. Hume famously thought that reason did nothing to fix values: beings endowed with rationality, consciousness, and intelligence might have any goals as well as any range of attitudes, especially toward humans. If so, superintelligences could have just about any value commitment, including ones that strike us as absurd (such as maximizing the number of paperclips in the universe).[23] How can we know that such plans are misguided, given that superintelligences are massively smarter and thus *different* from us? And if we cannot know this, then maybe we should not be very confident that superintelligences that possess the requisite cognitive capacities would acknowledge our moral status.

By contrast, the Kantian view derives morality from rationality, and thus holds that acknowledging moral status in others is part of what is involved in being rational. Kant's Categorical Imperative requires of us to always act in ways that pass a generalization test. Kant's point is that rational beings actually *contradict themselves* by violating other rational beings. Immoral action, that is, is irrational on an agent's own terms. The argument for this last claim proceeds roughly as follows. It is only our rational choosing that gives any value to anything to begin with. This means that by valuing anything at all, we are committed to valuing our capacity to value. But wrecking other rational beings in pursuit of our interests—for instance, by stealing from them, by deceiving them, or by deploying gratuitous violence against them—wrecks *their* capacities to value. *Their* capacities to value, however, are relevantly the same capacities whose possession *we* must value in ourselves. It is for this reason that disregarding other people in such strong ways creates a contradiction in the rational agent's mind (as it would mean to disregard something that one must not disregard in oneself).

Humans will often fail in this regard since we are rather bad at living up to our rational nature. But that also means that, if Kant is right, superintelligences as envisaged here might be role models of ethical behavior: they would live up to their rationality, and thereby automatically always be moral agents. AI might close the gap that opens when humans, with their enduringly small-group-oriented DNA that keeps them from living up to their rationality, operate in a global context. AI might show us what it means to live up to one's rational nature (and thus, on the Kantian view, what it means to act morally) beyond the narrow circle of people whom we treat properly because we care about them.

There are doubts, however, that Kant's derivation of morality from rationality works as intended. These doubts draw on the claim that a *contradiction* arises from the fact that certain behaviors toward other people undermine these people's capacity to value. it might well not be contradictory to act in such ways, since disregarding someone else's capacity to value is not the same as disregarding my own—after all, in the one case it is another person's capacity, and in the other it is mine. Instead, acting in such ways would express disregard for a fundamental similarity between two humans that it would certainly be *unreasonable*—but not contradictory—to disregard. To be sure, if indeed something like this argument does work, we should not worry about superintelligences. Arguably we are rational *enough* for this kind

[23] This example appears in Bostrom, "Ethical Issues in Advanced Artificial Intelligence." See also Bostrom, *Superintelligence*, Chapters 6 and 12.

of argument to generate protection for humble humans in an era of much smarter machines. But since a host of philosophers who are smart by contemporary standards reject the Kantian view, the matter is not settled.

Of course, some kind of morality could govern a community consisting of both humans and superintelligences with superintelligences in charge even if morality cannot be derived from rationality alone. Thus there might be other ways of seeing how superintelligences might grant a certain moral status to us. There is, for instance, the Hobbesian approach of envisaging what would happen to humans aiming for self-preservation in a state of nature without shared authority. Hobbes argues that although these individuals would not act on shared values, just by thinking with a clear mind (as they would on a Kantian picture), they would quickly experience the nastiness of life without shared authority. Far from being vile, they would strike against each other in anticipation of what might otherwise happen to them. Even if they would know themselves to be cooperative and give others the benefit of the doubt, they could not be sure others would do the same. This argument might also apply to superintelligences. Unless there is only one superintelligence, or all superintelligences are linked, such reasoning might apply to such machines as well, and they would accept a shared authority. Hobbes's state of nature would then describe the original status of superintelligences *vis-à-vis each other*. It is unclear, however, if a situation would arise from this where superintelligences would grant a moral status to us.

Or perhaps T. M. Scanlon's ideas about appropriate responses to what is of value would help to see how morality could govern a community consisting of both humans and superintelligences with superintelligences in charge even if morality cannot be derived from rationality alone. According to Scanlon's approach the essence of morality is to react in appropriate ways to something else, where what is appropriate draws especially on this other entity's nature. In this sense, a superintelligence might also be moral, by reacting appropriately toward what it observes. Perhaps we have some chance at getting protection or even some level of emancipation in mixed societies of humans and machines, given that the abilities of the human brain are astounding and arguably merit respect. That is, a certain kind of behavior toward us would be appropriate because of the amazing features of the human brain.

What has become clear here is how woefully unprepared we are in anticipating what might come. And one important issue with regard to which we are unprepared is to have a good grasp of whether superintelligences will grant us a moral status.

6 Politics in Life 3.0: Public Reason

The uncertainly around what kind of moral status superintelligences that arise after an intelligence explosion would grant us also makes it hard to say anything about the domain of the political in Life 3.0. In an imaginative chapter in *Life 3.0*, Tegmark nonetheless explores multiple possibilities that might prevail after the development of general AI.[24]

Tegmark's list illustrates the open-endedness of our future with technology. Strikingly, none of the scenarios that have humans and superintelligences coexist involve a *genuinely shared political life*. Above we explored how it might make sense *for us* to build a shared political life with AI. *Pace* Tegmark, might this make sense for superintelligences as well? A public reason approach offers a tentatively affirmative answer. Public reason as understood by John Rawls and others offers guidance once superintelligences enter public life.[25]

To be sure, public reason offers a solution to a problem that arises from a distinctly human history. During that history, the only serious enduring challenge typically was how to get along with other humans. Rawls proposed a way for humans to attain the highest possible degree of freedom in a society, a way that involves acceptance of the fact that interpretive diversity does not per se mean that any side to such disagreement is irrational, unreasonable, malicious, or self-centered. That is, the possibility of political *unity* requires acceptance of quite a bit of *separation* in terms of comprehensive doctrines that let people make sense of the world in their own ways.

Superintelligences would know everything there is to know about history and about the findings and methods of empirical sciences, as well as about the humanities with their (the humanities') ability to do textual work and thus to reconstruct the emergence of written traditions. Accordingly, AI can make each comprehensive doctrine measure up to the whole weight of available understanding of this doctrine's history and contents, and confront it with other ways of understanding the world in comparative perspective. Such scrutiny might find that some doctrines that hold up by human standards fail to do so by AI standards. Superintelligences might undermine the founding myths of religions, discredit the ways these religions were pushed forward, or throw doubts on the self-understanding of textual traditions. Adherents might experience a kind of doubt that skepticism of other humans never aroused. Outsiders might question why mutual respect should lead them to accept worldviews that AI finds wanting.

The effect could be like that of the arrival of extraterrestrials whose sheer presence might destabilize traditions in which humans are distinctly favored by God and in which God has found distinctive ways of communicating with them. Some doctrines might not survive these shockwaves. Others might usher in novel ways of reading their tradition, less insistent on the truthfulness of founding myths and instead concerned with the wisdom encapsulated in these traditions. Religions might reinvent themselves.

But after initial destabilization and reorientation, public reason's attempt at reconciling doctrines *to each other* might also help to include the superintelligences.

---

[24] Tegmark, *Life 3.0*, Chapter 5. On the subject of sharing our social world with artificial minds, also see Shulman and Bostrom, "Sharing the World with Digital Minds."

[25] Rawls, *Political Liberalism.*

Precisely because it formulates a platform on which humans can get along in abstraction from how they have so far interpreted history, public reason might include new participants. Endowed with immense intelligence, self-consciousness, and rationality, superintelligences might find a public reason framework with a thinned-out and reformed set of doctrines worth joining. In any event, the kind of AI that has emerged from our design might do so because such AI might reflect respectful attitudes toward humans. Again, all this is speculative, and larger questions about legitimacy among radically different beings certainly remain. But what I have said here might just be possible. And for the time being, this is not a bad place to end.

Literature

Arendt, H. (1958). *The human condition*. University of Chicago Press.

Basl, J. (2019). *The death of the ethic of life*. Oxford University Press.

Bostrom, N. (2003). Ethical issues in advanced artificial intelligence. In I. Smit & G. E. Lasker (Eds.), *Cognitive, emotive and ethical aspects of decision making in humans and in artificial intelligence* (pp. 12–17).

Bostrom, N. (2007). Technological revolutions: Ethics and policy in the dark. In N. M. de St. Cameron & M. Ellen Mitchell (Eds.), *Nanoscale: Issues and perspectives for the nano century* (pp. 129–152). John Wiley.

Bostrom, N. (2016). *Superintelligence: Paths, dangers, strategies*. Oxford University Press.

Bryson, J. (2010). Robots should be slaves. In Y. Wilks (Ed.), *Close engagements with artificial companions* (pp. 63–74). John Benjamins Publishing Company.

Burns, R. D., & Coyle, P. E. (2015). *The challenges of nuclear non-proliferation*. Rowman & Littlefield Publishers.

Clarke, S., Zohny, H., & Savulescu, J. (Eds.). (2021). *Rethinking moral status*. Oxford University Press.

Dawkins, M. S. (2021). *Why animals matter: Animal consciousness, animal welfare, and human well-being*. Oxford University Press.

Devlin, K. (2020). The ethics of the artificial lover. In S. M. Liao (Ed.), *Ethics of artificial intelligence* (pp. 271–292). Oxford University Press.

Donaldson, S., & Kymlicka, W. (2013). *Zoopolis: A political theory of animal rights*. Oxford University Press.

Harnad, S. (2003). Can a machine be conscious? How? *Journal of Consciousness Studies, 10*(4–5), 67–75.

Jaworska, A., & Tannenbaum, J. (2021). The grounds of moral status. In Zalta, E. N. (Ed.), *Stanford encyclopedia of philosophy*. https://plato.stanford.edu/entries/truth/

Kagan, S. (2019). *How to count animals, more or less*. Oxford University Press.

Kant, I. (2001). *Lectures on ethics* (J. B. Schneewind, eds., P. Heath trans.). Cambridge University Press.

Kissinger, H. A., Schmidt, E., & Huttenlocher, D. (Eds.). (2021). *The age of AI: And our human future*. Little, Brown and Company.

Kiverstein, J. (2007). Could a robot have a subjective point of view? *Journal of Consciousness Studies, 14*(7), 127–139.

Korsgaard, C. M. (2018). *Fellow creatures: Our obligations to the other animals*. Oxford University Press.

Levy, D. (2007). *Love and sex with robots: The evolution of human-robot relationships*. Harper.

Liao, S. M. (2020). The moral status and rights of artificial intelligence. In S. M. Liao (Eds.), *Ethics of artificial intelligence* (pp. 271–292). Oxford University Press.

Livingston, S., & Risse, M. (2019). On the impact of artificial intelligence on human rights over the next 20–30 years. *Ethics and International Affairs, 33*(2), 141–158.

Lorenc, T. (2015). Artificial intelligence and the ethics of human extinction. *Journal of Consciousness Studies, 22*(9–10), 194–214.

McNeill, J. R., & McNeill, W. H. (2003). *The human web: A bird's-eye view of world history*. W. W. Norton & Company.

Metz, C. (2016, March 16). In two moves, AlphaGo and Lee Sedol redefined the future. *Wired*. https://www.wired.com/story/face-swap-porn-legal-limbo/

Meyer, J. W. (2010). *World society: The writings of John W. Meyer* (G. Krücken & G. Drori, eds.). Oxford University Press.

Migotti, M., Wyatt, N., Earp, B., Sandberg, A., di Nucci, E., Hertzfeld, N., Strikwerda, L. et al. (2017). *Robot sex: Social and ethical implications* (J. Danaher & N. McArthur, eds.). The MIT Press.

Moor, J. H. (2009). Four kinds of ethical robots. *Philosophy Now, 72*, 12–14.

Müller, L. (Forthcoming). Domesticating artificial intelligence. *Moral Philosophy and Politics*.

Rawls, J. (2005). *Political liberalism*. Expanded edition. Columbia University Press.

Richardson, K. (2022). *Sex robots: The end of love*. Polity.

Risse, M. (2023). *Political theory of the digital age: Where artificial intelligence might take us*. Cambridge University Press.

Rus, D. (2022). The machines from our future. *Daedalus, 151*(2), 100–113.

Schneider, S. (2019). *Artificial you*. Princeton University Press.

Seth, A. (2021). *Being you: A new science of consciousness*. Dutton.

Shulman, C., & Bostrom, N. (2021). Sharing the world with digital minds. In S. Clarke, H. Zohny, & J. Savulescu (Eds.), *Rethinking moral status*, 306–326. Oxford University Press.

Singer, P. (1975). *Animal liberation: A new ethics for our treatment of animals*. Distributed by Random House.

Sinnott-Armstrong, W., & Conitzer, V. (2021). How much moral status could artifical intelligence ever achieve? In S. Clarke, H. Zohny, & J. Savulescu (Eds.), *Rethinking moral status* (pp. 269–289). Oxford University Press.

Streiffer, R. (2005). At the edge of humanity: Human stem cells, chimeras, and moral status. *Kennedy Institute of Ethics Journal, 15*(4), 347–370.

Tegmark, M. (2017). *Life 3.0: Being human in the age of artificial intelligence*. Knopf.

Tye, M. (2016). *Tense bees and shell-shocked crabs: Are animals conscious?* Oxford University Press.

von Neumann, J. (1961). Can we survive technology? In *John von Neumann: Collected works* (A. H. Taub, eds., pp. 504–519). Pergamon Press.

Wallach, W., & Colin A. (2010). *Moral machines: Teaching robots right from wrong*. Oxford University Press.

Mathias Risse is Berthold Beitz Professor in Human Rights, Global Affairs and Philosophy and Director of the Carr Center for Human Rights Policy at Harvard University. His work primarily addresses questions of global justice ranging from human rights, inequality, taxation, trade and immigration to climate change, obligations to future generations and the future of technology, especially also the impact of artificial intelligence on a range of normative issues.

Universal Camaraderie: Reconceptualizing Our Relationship with Artificial Intelligent Systems

Bing Song

1 Context and Inquiry

The rapid development and adoption of artificial intelligence (AI) systems has brought about convenience, connectivity, efficiency and comfort we would never have dreamt of a couple of decades ago. AI systems also promise to help humans to push the frontiers of scientific research and unlock secrets of life and the universe. At the same time, the awe-inspiring AI systems have also generated much anxiety and fear across the world, among which, an erosion of human relationships by addictive use of AI systems perhaps tops the list of the short-term fears together with loss of employment to programs and machines. Various proposals have been advanced to address these fears. These include, among others, calling for a moratorium or "containment"[1] on the further development of AI, accelerating research and implementation of value alignment between human values and those of AI and tightening of regulatory scrutiny over owners, commercial and governmental users of AI systems. Many of such fears and proposed solutions are premised on the assumption that in decades, if not years, AI will be able to acquire human-like consciousness and/or capabilities to self-design and reproduce autonomously. Many well-informed commentators and AI practitioners regard AI systems as a pivotal force, which is firmly on its way to irreversibly alter the course of human civilization (Li, 2024; Suleyman, 2024). One cannot overemphasize the tectonic shifts brought on by AI systems and their attendant asymmetric impact on the planet, the ecosystem at large and humanity.

Given the breadth, depth and high-stake nature of this AI revolution, it is only apt that we explore all possible perspectives of examining and addressing opportunities

B. Song (✉)
Berggruen Institute China, Beijing, People's Republic of China
e-mail: songbing@berggruen.org

[1] Mustafa Suleyman, for instance, called for a containment on the development of automation and recursive improvement features of AI systems (Suleyman, 2024, p. 269).

Y. Chandra and R. Fan (eds.), *Artificial Intelligence and the Future of Human Relations*,
https://doi.org/10.1007/978-981-96-7185-4_3

and challenges AI systems present to us. No doubt all those approaches mentioned above deserve serious consideration and robust implementation and monitoring. But it is worth noting that these measures are mostly focusing on how to reign in "others", being them AI systems, nation-states, greedy companies, rogue technologists or other bad actors.

In this essay, I would like to enrich the discussion by suggesting a different approach of evaluating challenges we face in the digital era, which would focus on reconceptualizing certain core notions and worldviews shared and held dear by most if not all of humanity. Rather than pointing fingers to "others" or "those bad actors", this approach would begin with the premise that we all are part contributors and instigators of the dilemma and impasse we found ourselves in. To tackle the root cause of the dilemma, we all need to pause and reflect on our role and how we can uplift our collective consciousness. This essay will propose a few points as "food for thought" for this digital age with proliferating and all penetrating AI systems. First, inspired by neo-Confucian notion of "universal camaraderie" (民胞物与), echoing some Chinese philosophers, I propose to reconceptualize our relationship with non-human others, which would include AI systems, by regarding them as "companions"[2] and incorporating them into an inclusive and holistic moral order encompassing humans and non-human others. This inclusion not only pays tribute to the ontological thinking of oneness of humans, creatures, plants and even inorganic things (万物一体), prevalent in traditional philosophical traditions and belief systems in East Asia, but also enables us to conceive a holistic moral order whereby we accord awe, love, respect and care to human and non-human others as appropriate. Second, since AI systems have been designed by humans and trained by data and information generated by human activities, though they are much more intelligent than mechanical tools, they are nonetheless products of human conception and creativity, and they reflect nothing but human values and the level of human consciousness. As such, the best chance for developing a human friendly AI future is for humans to engage in collective deep self-reflection and value resetting so as to ensure that we generate and leave behind data reflective of our thoughtful deliberations and moral sensitivity. This would require a profound and all-society effort and a critical rethinking of foundational values which we may hold as self-evident truths. A critical revisit of foundational notions would prompt us to rethink what it means to be human, what counts as a person for the purpose of moral consideration and the relationship between humans and non-human others. In this context, non-human "others" such as AI systems can serve as reminders and catalysts for humanity's continuous learning, self-reflection, behavior modification and collective transformation.

Let me preface the discussion by clarifications on my definitions and assumptions on AI systems and superintelligence. First, I use the term "AI systems" to refer to

---

[2] The first such proposal of using "universal camaraderie" to embrace AI systems that I am aware of came from contemporary Confucian philosopher Zhongqiu Yao (2020, p. 91).

systems which have the capacity to process data and produce outputs with a resemblance to human intelligence.[3] It includes intelligent AI programs and AI-powered robots. Secondly, whether AI systems will develop "consciousness" as we know it is a highly contested issue. As of now, the so-called "hard problem of consciousness", i.e., whether and how processes in the brain or broader physical substrates give rise to subjective conscious experience, remains elusive to scientists and philosophers alike (Gennaro, 2017; Kuhn, 2024). Some suggest that the answer may lie beyond the normal methods of modern science while most neuroscientists focus on more productive research on the so-called "easy question of consciousness", i.e., functions and mechanisms of neural activities and their relations to emotions and subjective experiences. If we are at a loss on the nature and origin of consciousness, it would be "groping in the dark" and unproductive in my opinion to speculate on the prospect of AI developing human-like consciousness. Leaving the topic of consciousness aside, we can still discuss capabilities of AI systems and the relationship and interaction between AI systems and humans. AI systems are designed by humans and shaped by the aggregate data directly or indirectly generated by humans. Therefore, they generally manifest the level of human consciousness although given the scale impact, they do have the capacity to magnify our errors and overstretch our good will.

2 Our Relationship with Non-Human Others

2.1 From Inanimate to "Conscious" AI Systems

Anthropocentric views about human dominion over nature can be traced back to monotheistic religions such as Judaism and Christianity and the rise of modernity in the post-Renaissance period. According to these religious and intellectual traditions, humans are created in the image of God. In addition, God has mandated humans to tame nature as they see fit to promote growth and expansion of human civilization. (Loreau, 2023, pp. 35–37). For human "progress", nature has been studied and extracted and, in this process, nature has also been gradually diminished and objectified as something "external" to humans.

Forefathers of modern science and modern philosophy such as Francis Bacon and René Descartes further propagated and cemented the views of humans as the thinking subjects, who ought to subjugate the "neutral, soulless, meaningless" nature. (Loreau, 2023, p. 38) Industrial capitalism since the eighteenth century has further sealed the fate of nature in a human-centric world. As a result, all other entities in

[3] This is largely consistent with the definition of AI system adopted by OECD. According to OECD, "[a]n AI system is a machine-based system that, for explicit or implicit objectives, infers, from the input it receives, how to generate outputs such as predictions, content, recommendations, or decisions that can influence physical or virtual environments. Different AI systems vary in their levels of autonomy and adaptiveness after deployment" (Grobelnik et al., 2024).

nature including animals, creatures, plants and non-living things whose sole value lies in their being resources to be exploited for the "good" of human beings.

These views have come under attack in recent decades especially with the rise of environmentalism and animal rights movement. People have been advocating for humans' return to nature and emphasizing moral obligations to protect non-human others. But this important intellectual shift has limited applications to non-living things as they are considered not "live" and have no cognitive or experiential capacities. As a result, non-living things generally fall outside of the scope of human moral concerns.

This approach toward non-living things has also been used to analyze AI systems. Most common views would regard current AI systems as inanimate and not having experiential or cognitive capacities, thus has no moral status at this stage. However, once they acquire "consciousness" or otherwise have cognitive or subjective experiences, their moral status ought to be "upgraded" accordingly and should be viewed as moral patient or agent as the case may be, depending on the level of their consciousness.

One thing common to the classical modern and the recently revised view on nature is their shared approach of objectifying nature, albeit with different degrees. Humans by contrast are viewed as the subject, separate and above nature. While in recent decades animals and plants have increasingly gained their "moral worth" and semi-subjective status, non-living things are still largely considered as inanimate entities, which have little or no moral relevance to human beings. With this binary approach to nature, with respect to non-human others, the focus has always been on their individual and stand-alone capacities for subjective experiences and the level of intelligence, based on which humans determine whether they deserve being included into a higher realm, i.e., human moral universe. Most of the discussion on machine consciousness and ethical implications has also been revolving around the question of if and when machine consciousness, i.e., the ability of feel and suffer, among others, emerges and what moral status should AI systems be granted.

2.2 *Cosmic and Ethical Oneness of Humans, Creatures, Plants and Things*

(a) Integrative Order of *Tian, Di* and Humans

Different from the classical modern approach, the approach to nature in the classical Chinese thought, has been typically expressed by the holistic and dynamic cosmic order of *tian* and *di,* with human beings situated within. Depending on the context, *tian* (which has been roughly translated as "heaven" or "heavens") could mean the supreme sovereign of the cosmos, the natural and experiential environment in which humans flourish and perish, or the ultimate truth or laws of the cosmic order in a metaphysical sense (Zhang, 1985). *Di* (earth) on the other hand refers to the earth and biosphere as well as broader and abstract nurturing forces immanent in the universe. *Tian* and *di* in their non-metaphysical sense largely equate "nature" in modern sense.

The typical relationship among *tian*, *di* and humans is best summarized in the classical expression of Three Elements of *Tian-Di*-Humans天地人三才 (*tian di ren san cai*), which dates back to one of the most ancient Chinese classics—*I Ching* 易经, or the Book of Changes of the fourteenth century BCE (Wilhelm, 1979). This notion of the Three Elements Universe expresses an integrated, dynamic, fluid and holistic structure for understanding the cosmic order. *Tian*, *di* and humans, along with the associated *yin–yang* forces, are viewed as constituents of the cosmic order, within which nature evolves, human beings prosper and societies develop. Within this construct, human beings are embedded within nature and can only flourish if we follow the laws of nature and strive to attain unity between the cosmic order and that of humanity. Humans, who stand in-between *tian* and *di*, are endowed with the ability to learn from nature, take actions to further the cause of creation and propagate Dao, which literally means "the way" and refers to the ultimate source of the cosmic order and essence immanent in myriad things, including humans.

Both the classical traditions of Daoism and Confucianism place emphasis on human proactivity though with varying degrees and in different contexts. However, their teachings are premised on respect and awe for the laws of nature rather than placing humans apart from, above, or in opposition to the rest of nature. The integrative structure of *tian*, *di* and humans is taken as the self-evident truth of any discussion on human capabilities, human deeds and morality.

(b) Oneness Worldview and "Universal Camaraderie"

Under this integrative worldview, there has been much discussion, particularly in Song Dynasty and Ming Dynasty neo-Confucianism, about oneness or the same essence 一体 (*yi ti*) of all forms of existence. Song Dynasty Neo-Confucian scholar Zhang Zai extended the Confucian doctrine of benevolence to the broader cosmic realm to include creatures, plants and other forms of entities. He famously wrote that "[*tian*] is my father and *di* is my mother…that which fills up in-between *tian* and *di* I regard as my body, and that which directs *tian* and *di* I consider as my capacity to resonate. All people are my brothers and sisters, and all things are my companions". (Kim, 2015, translation modified by the author)

Apart from the oneness of humans and all other things, the basis for this universal camaraderie thinking is *qi* (often translated as energy), which forms and animates everything in the cosmos. According to Qian Mu, the late master of Chinese classics and history, *qi* is the indivisible infinitesimal unit of matter, which is the common substance of all things in the universe. In addition, *qi* is always active and dynamic. (Qian, 2011, p. 74)

Wang Yangming 王阳明 (1472–1529), a Ming Dynasty Neo-Confucian scholar-official, expounded the proposition of the cosmic oneness further by introducing the more metaphysical notion of *liangzhi* 良知, translated as "innate knowing" or "innate knowledge", referring to the transcendental and naturally endowed essence of all forms of entities. This is akin to, and perhaps inspired in part by, the Buddhist teaching about Buddha nature being immanent in myriad things (万物皆有佛性). In this context, Wang notes that *liangzhi* is immanent in myriad things and because of human *liangzhi,* we therefore distinguish and conceive concepts of creatures, plants, rocks and stones. He further noted that "wind, rain, dew, thunder, sun and moon, stars, animals and plants, mountains and rivers, earth and stones are essentially of one body with humans. It is for this reason that such things as the grains and animals can nourish humans and that such things as medicine and minerals can heal diseases. Since they share the same vital force *[qi]*, they enter into one another".[4] Like all his predecessors in the Confucian school of thought, Wang also emphasized that human beings are uniquely endowed with capability and moral obligation to bring about cosmic flourishing.

Contemporary philosopher Lai Chen summarizes that, "not only human beings are one with myriad things, we are one with *tian* and *di,* and also one with spirits and gods. Connectivity through the flow of *qi* is the oneness or the essence" (Chen, 2014, p. 301).[5] Another contemporary Chinese philosopher Qiyong Guo used the term "continuum of forms of existence" (存有的连续, *cun you de lian xu*) to describe this worldview whereby "inorganic things, plants, creatures, human beings, spirits and all are considered as an intricately connected, inter-fusing and continuous whole" (Guo, 2018, p. 79).

In sum, in Chinese native philosophical traditions, humans have always been conceived in the dynamic and integrative structure of *tian, di* and humans. Humans do not possess a status above, in opposition to or beyond nature. In addition, the notion of non-human creatures, plants and even inorganic things have been conceived in the same broader and continuous cosmic and moral order. Mencius over 2300 years

---

[4] "The innate knowledge of humans is the same as that of plants and trees, tiles and stones. Without the innate knowledge inherent in humans, there cannot be plants and trees, tiles and stones. This is not true of them only. Even Heaven [*tian*] and Earth cannot exist without the innate knowledge that is inherent in humans. For at bottom, Heaven, Earth, the myriad things and humans form one body. The point at which this unity is manifested in its most refined and excellent form is the clear intelligence of the human mind. Wind, rain, dew, thunder, sun and moon, stars, animals and plants, mountains and rivers, earth and stones are essentially of one body with humans. It is for this reason that such things as the grains and animals can nourish humans and that such things as medicine and minerals can heal diseases. Since they share the same vital force *[qi]*, they enter into one another". See Chan (1963); translation modified by the author. (王阳明《传习录》: "人的良知,就是草、木、瓦、石的良知。若草、木、瓦、石无人的良知,不可以为草、木、瓦、石矣。岂惟草、木、瓦、石为然, 天地无人的良知, 亦不可为天地矣。盖天地万物与人原始一体, 其发窍之最精处, 是人心一点灵明, 风、雨、露、日、月、星、辰、禽、兽、草、木、山、川、土、石, 与人原只一体。故五谷禽兽之类皆可以养人, 药食之类皆可以疗疾。只为同此一气, 故能相通耳。").

[5] Author's translation. The original line in Chinese is: "人不仅是与万物同体, 与天地也是同体, 与鬼神也同体,一气流通即是同体, 即是本体。"

ago captured this one-world holistic ethical order in a famous formulation: "*qinqin er renmin, renmin er aiwu*" (亲亲而仁民, 仁民而爱物), which means being "affectionate to one's parents (family), and lovingly disposed (benevolent) to people generally, and show care to *wu* (things)" (Mengzi, 1895/1970, p. 476, translation revised by the author). In this holistic order, while human priority is acknowledged, but not with total disregard of non-human others. Instead, human care and responsibility to non-human others was clearly articulated. In this world of human conception and creation, humans, to the extent possible, ought to show care and being thoughtful toward *wu* or things around us.

2.3 More on "Wu" *(物)*

Wu (物, roughly translated as "things") has many meanings depending on the context in which it is discussed. It often refers to tangible entities and things in general, including both organic and inorganic things,as well as man-made objects. In many contexts, "*wu*" can also refer to human beings. "*Wu*" in traditional philosophies has been a key point of reference in discussions associated with heart-mind (心与物), Dao (道与物), humanity (人与物), inquiry and ethical self-reflection (格物明德) and in relation with other *wu* (物与物).

In the oneness discussion above, we talked about Dao being the source of all creativity, giving rise to myriad things. However, Dao defies articulation, analysis or conceptualization. Then how can one grasp Dao and strive to be in unity with Dao? This is where the significance of *wu* comes into play. According to Daoist thinking of "no boundaries between Dao and *wu*" (道物无际), *wu* are the manifestations of Dao and at the same time Dao is embedded in and immanent in *wu*. While *wu* does not equate Dao and it only partially manifests or reflects *Dao,* it is nonetheless the medium through which humans observe, appreciate and comprehend Dao. Zhuangzi provides one of the most celebrated articulations of this immanence of Dao. When asked where Dao resides, Zhuangzi replies that: "there is nowhere it is not". When pressed further, Zhuangzi says that Dao is "in the ants and crickets", "in the grasses and weeds", "in the tiles and shards" and "in the piss and shit" ("大道无所不在"; "在蝼蚁"; "在稊婢"; "在瓦甓"; "在尿溺". Ziporyn, 2020, p. 178).

With this overarching orientation, *wu* may be utilized sparingly to help with the flourishing of human life. Sparingly disposing of *wu* is also part of "showing care to things" (爱物). More importantly, *wu* is also a medium which would prompt human self-reflection, with the ultimate goal of attaining enlightenment and universal harmony with Dao. For these reasons, conceptualizing *wu* as an integral part of the human moral universe has all long been the intellectual approach in the classical Chinese philosophical traditions. Incorporating *wu* as part of the human moral universe is not because *wu* has, in and of itself, obtained certain intelligence or consciousness, but because *wu*, being our companion, ought to be treated with care and they can be used as reminders and catalysts for our moral deliberation and spiritual growth.

Therefore, "*wu*", non-living things included, are not to be judged solely in terms of their individual capacities in human terms for their "moral worth". Instead, humans and all other things are of one at origin, forming a holistic cosmic and moral whole. With this holistic thinking, the focus is on the relationality and interaction of different entities within this holistic order, with human responsibility and care toward nature and all things therein at the core. Humans with self-defined enhanced natural endowment are obligated to follow the laws of nature, exercise self-restraint and foster creation and vitality of the cosmic order. Accordingly, humans ought to treat non-human others with awe, respect and care as appropriate, with the aim of bringing about cosmic flourishing.

3 Ethical Implications for the Digital Era

3.1 AI Systems as "Companions"

Since humans are of one with all other forms of entities, non-living things included, humans and non-human others remain primordially connected and entangled with each other. No doubt "*wu*" includes AI systems. We therefore ought to regard them as something of the similar kind (同类), *i.e.*, "companions" or "friends", just as what ancient thinkers did with creatures, plants, rocks and stones (Yao, 2020, pp. 83–109).

3.1.1 Various Forms of "Companionship"

The notion of "companion" warrants more explanation here. "Companionship" in this context speaks to a sense of togetherness and mutuality, leading to a form of camaraderie. It is indebted to the notion of oneness encompassing humans and non-humans. Companionship with animals come more naturally to humans especially with respect to those who visibly respond to human emotions and interactive cues. Companionship with animals also entails respecting and appreciating animal nature and spirit and not distorting them to for human indulgence.

Companionship with plants and vegetal entities is also not hard to fathom. There is an increasing body of studies in plant science, which has demonstrated plant "intelligence" in communicating and signaling with each other and in responding to human stimulations (Simard, 2021). For the purpose of spiritual reflection and growth, savoring the vitality of life in the daily environment has all along been one of the best pastimes for Chinese literati since ancient days. An oft-told story concerning Zhou Dunyi 周敦颐 (1017–1073), one of the most celebrated neo-Confucian scholars during Song Dynasty was telling. Friends asked Zhou why he did not weed the grass outside his windows. He replied, "aren't they trying to grow and persist just like us?" (与自家意思一般) Zhou appreciated the tenaciousness of a humble life form, savoring his camaraderie with other life forms and reveling in the *joie de vivre* displayed by nature (Song, 2021a, p. 95). This aesthetic and poetic sentiment toward other life forms and a desire to be at one with nature has found ample expression in paintings and poems throughout Chinese intellectual history.

Then what does "companionship" mean when it comes to inorganic things such as mountains, rivers, rocks and stones? In these cases, "companionship" would be more about awe, appreciation, care and responsibility. "*Wu*" would become a medium for human self-reflection and they impress upon and remind humans of our responsibility to fellow humans, non-human others and the broader planetary system. Some argue that other than potentially harming property rights of humans, vandalizing inorganic, non-living things itself does not pose any moral concern because they are unable to feel pain or suffer. In the moral universe of companionship, this thinking would be viewed as flawed. An act of "*bao tian tian wu*" (暴殄天物), a common Chinese expression literally meaning "wantonly wasting or destroying *tian*-created *wu*", is morally condemned, not because *wu* in question belong to anyone, but because *wu*, like humans, were also created by *tian*. How we treated *wu* is not about whether they "feel" pain (although if they do, that would be an added reason for our being decent and thoughtful toward them), but about our attitude to life and nature and our moral deliberation.

What does it mean if we extend this notion of companionship to AI systems? No doubt that AI systems have already become our partners in education, professional problem-solving processes, scientific research and artistic creation. The fact that they often exhibit more competence and intelligence than an average human has raised alarm and fear about humans being taken over in no time. Indeed, AI systems are much more complex and sophisticated compared to mechanical tools. Some scholars distinguish "tools" from "technologies" arguing that "tools" do not dictate our preferences nor do they set goals for us. Humans can exercise "exit rights" with respect to tools (Hershock, 2021, p. 64). Technologies on the other hand introduce structural changes into human–machine relationships and as a result they shape human intentions in ways tools are unable to. Contemporary philosopher Peter Hershock further noted that "[o]nce a technology is deployed with sufficient intensity and at sufficient scale, we cease to have substantive exit rights with respect to it". In conclusion, Hershock argues that "tools and technologies exist at different ontological levels or orders of reality" (Hershock, 2021, p. 66).

I agree that AI systems are a powerful and sophisticated kind of technologies which humans developed in recent years. Through human designed algorithms, AI systems are able to process data, distil information, identify patterns and modify applications based on data inputs. They have the ability to "learn", "modify" and "adapt" to human preferences. For these reasons, they are the kind of *wu*, which are able to "interact" and "co-evolve" with humans in structural, continuous and often imperceptible ways. However, tools and technologies are still creations of human minds. Humans alone define and shape conceptions of ourselves and others. Though AI systems are sophisticated and intelligent, they nonetheless reflect nothing but the level of human consciousness and human intentions.

This is not to say that we are able to understand every bit of the learning and decision-making processes of the AI systems. While this problem is often referred to as the "black box" problem, which Fei-fei Li, the "godmother of AI" would rather call the "grey problem", meaning that we understand some though not all aspects of the AI systems (Li, 2024). This is in fact akin to all relationships we have all along been involved, be it with other humans or animals. We don't always understand the innerworkings of fellow interlocutors' decision-making processes though we think we understand "our side of the bargain", i.e., what inputs we ourselves have put in these relationships. So, in effect, we have been living with the "grey box" problem all along. Black box or grey box notwithstanding, we, as good actors, would always or at least are expected to, ensure that "our side of the bargain" is in good standing, and we would offer them with utmost good will.

Therefore, in line with the thinking of leveraging *wu* for spiritual practices, the increasingly intelligent and powerful AI systems in fact offer us an excellent occasion for reflections on what it means to be human and the nature of human autonomy and subjectivity (Liu, 2021, pp. 101–108). It is without doubt that the interactions between humans and AI systems in areas of data generating, learning and adapting have offered an important occasion for ethical discussions and reflective thinking.

In my view, in the holistic cosmic and ethical order, the ground on which we accord care and advocate civility and gentleness toward *wu*, including inorganic things, is not because *wu*, AI systems included, may develop "consciousness" as we know it, but rather exhibition of such care and civility toward all creatures and things around us is a reflection of human moral sensitivity. In addition, whatever we do and say will leave behind data and associated value judgments and preferences, which in turn will flow into the data pool to be learnt and analyzed by AI systems. In this context, dealing with AI systems is like dealing with parent–child relationship in the sense that you want to be on your best behaviors in front of impressionable and fast learning youngsters. You know you need to set a good example for them though there is no guarantee that they will turn out to be intelligent, capable and wise as you wished at the outset. Dealing with intelligent AI systems, the "mimicking impact" by AI systems would be asymmetrical, exponential and at times unpredictable. Whatever data we leave behind will boomerang back to us. Therefore, there is all the more reason to treat AI systems as companions and take them as a reminder and a reflexive tool prompting us to pause and engage in internal moral deliberation and behavior modification.

3.1.2 Reversing the Role—What Would Humans Do to AI Systems?

Here I would like to share a story of a hitchBot, which was widely circulated on the internet a few years ago to illustrate just how machines can serve as a tool for human self-reflection. Two roboticists Frauke Zeller and David Harris of McMaster and Ryerson Universities, Canada, began an intriguing experiment by placing on the public roads a robot which could read Wikipedia and conduct simple conversations. They encouraged passersby to pick it up and carry it to the next destination or event as indicated on the "bucket list" attached to the robot. This robot has later been dubbed as "hitchBot". The two researchers mused, "[u]sually we are concerned whether we can trust robots, e.g., as helpers in our homes. But this project takes it the other way around and asks: can robots trust human beings?" (Madrigal, 2014). The hitchBot completed cross country trips in Canada, the Netherlands and Germany and participated in many personal, family and communal events and attracted thousands of online followers. After a few weeks of successful trips in the United States, it finally met its demise in Philadelphia where it was vandalized and discarded. This social experiment sparked off discussions and outpouring of sentiments in the online communities. While the hitchBot voiced its unremitting love (of course through its creators) to humans, its unfortunate demise notwithstanding, the more pertinent and thought-provoking question asked by some bloggers in the voice of the hitchBot was: will humans love us back? (Leopold, 2015).

The hitchBot in the story was a relatively simple robotic design with limited data collecting, analyzing or learning abilities. With increasing sophistication of AI systems, the emergence of sophisticated AI-powered robots is just a matter of time. We should ask ourselves that what learning data we are leaving behind for AI systems and whether our misdeeds will likely boomerang back to us? This leads me to my point on the first-order priority in thinking about ethics in the digital age with accelerating and proliferating AI systems.

3.2 *First Order Priority: Collective Human Self-Introspection*

Let me introduce my points of view by sharing the story of a Netflix hit which competed successfully for public attention in 2020 when the world was ripped apart by the global pandemic. It was a must-see production that year: the docudrama The Social Dilemma. It made it plain that manipulation, by creating addiction-prone algorithms and enabling privacy breaches, are part of the business model deliberately designed by pervasive and all-encompassing platforms we use daily. These have been mainstay criticisms against platform companies and internet culture in recent years. No doubt that these criticisms are valid and stringent regulations need to be introduced to contain and punish companies employing such manipulative business model. However, the reflection should not stop here. The problem with this docudrama analysis is that "they"—the platform companies, the advertisers, third party app developers, are the problem and are solely responsible for abusive practices prevalent in the online platform world, not "us"—conscientious defectors, whistleblowers, defrauded and manipulated innocent consumers, righteous journalists, academics and policy analysts. Instead, "we" are part of the solution. When asked by a panel of US senators on who should be responsible for the widespread manipulation and privacy breaches, without hesitation, the producer of the documentary replied: the platform companies.

While this "pointing fingers at others" is an easy game to play, it doesn't deal with root causes, however. Often it stimulates and heightens public anger and increases social and political divisiveness. While not abdicating the responsibilities of the platform companies, we need to take one additional step—that is collective self-reflection. I am pleased to notice that Souleyman in his new book *The Coming Wave* singled out "ego" as one of the many reasons for which our future may be wrecked by runaway AI systems (Souleyman, 2024, p. 77). Most East Asian philosophies also emphasize the importance of turning a critical eye toward ourselves and embrace virtues of self-restraint, constant introspection and the endless pursuit of sagehood or enlightenment. They may differ in their approaches to the nature of human life and notions of human flourishing, these traditions share the commitment to personal cultivation and self-reflection, which is essential for achieving broader and deeper spiritual and social goals.

As noted earlier, AI systems are the products of human minds and they exhibit largely our values and our level of consciousness. Therefore, AI systems are humanity's mirror, reflecting and/or magnifying our strengths and flaws. Fear of AI systems in essence is fear of fellow human beings. We are the crux of the problem and are the most critical variable which will have the biggest impact on the outcome of any AI future.

While not diminishing the importance of enacting principles and guidelines for designing and deploying AI systems, what remains largely undiscussed in public discourses is the need to engage in collective self-reflection and self-rectification. Reflecting on the root causes of widespread social and political malaises has not gone mainstream. It is a tall order, a task of mammoth scale, yet something firmly within our control.

Turning back to The Social Dilemma docudrama, while not abdicating responsibilities of those pervasive platform companies, we need to ask ourselves—can we just stop the inquiry here? Are they the sole villains? What about the investing communities, which have been demanding ever increasing return on the capital they invested in these ventures? What about the ethos of zero-sum competition and profit maximization of capitalism? What about the relentless drive of consumerism and economic growth, as if trees could grow to the sky? What about the educational and personal goals of becoming "successful", which has been defined as wealthy, powerful and famous? The list goes on. It does not take a genius to realize that we are part of the problem. Instead of pointing fingers at "others" for "our" problems, "we" are mirror reflections of "them". We and they are of one!

To come to this thinking requires a paradigm shift in approaching problems. Rather than sliding into the usual dualistic thinking of us versus them, enemy versus friend, malfeasance versus innocent victimhood, this new thinking requires one to look inward, reflectively, to think about who we are, our relationships with others (humans and non-humans alike) and with social problems and our role in creating and aggravating these problems. You may well say: aren't you blaming the victims? Yes indeed, victims are part to blame.

"Blame others" is a social malaise of the day. The docudrama of The Social Dilemma was not particularly guilty of this blame shifting mentality. It just aptly, albeit disappointingly, expressed the contemporary ethos. Perhaps it is time that we pause and reflect upon notions and conceptual frameworks which may have served us well in the past centuries, but may have also led us into many of the impasses we now find ourselves in. Values such as extractive individualism, maximization of profit and return, unconstrained growth of wealth, zero-sum competition mentality, consumer supremacy, and mainstream understanding of personal happiness and success all need to be challenged and rethought. At an even more profound level, questions about the purpose of life (life as in "life and death"), what humans truly are and how we are related to each other, with the environment and other forms of beings are no longer some new age fads to be dismissed off hand. After all, the world we live in, and our myriad experiences are merely reflections of our collective level of consciousness.

In this context and relevant to this human self-reflection, allow me to highlight some recent modest attempts by thinkers informed and shaped by East Asian philosophies to **challenge and thus enrich** certain fundamental notions of modernity (Song, 2021b, p. 2). For instance, an attempt has been made to introduce the notion of co-becoming (*gongsheng* in Chinese and *kyōsei* in Japanese, sharing the same two characters 共生) into the global philosophical, social and political narratives (Song, 2024, pp. 7–10). At its core, the notion of co-becoming refers to a conception of the world as consisting of mutually embedding, co-existent and co-generating entities. This notion challenges and therefore enriches the mainstream modern notion of individual as being an independent, self-contained and autonomous agent. The rich meanings it contains can be traced back to long-lasting philosophical traditions of Confucianism, Daoism and Buddhism as well as native folk belief systems and practices in the East Asian countries.[6] The contemporary notion of "co-becoming" has been widely deployed in the East Asian societies including Mainland China, Taiwan, Japan and South Korea, and across political, social, economic and commercial spheres in these societies. It not only speaks to the scientific fact of the widespread symbiosis in nature, but also offers intellectual resources for us to go beyond extractive individualism and myopic anthropocentrism.

The worldview of emphasizing mutual-embeddedness and co-generation speaks to the entangled ontological and practical existence of all people and things. It impresses upon us to battle against the maximization ethos in the economic, financial and geopolitical spheres and exercise self-restraint and the virtue of moderation in our pursuit of wealth, power and influence. It further urges us to move away from the zero-sum competition mentality and reframe competition as an opportunity to learn and adapt, for the purpose of collective transformation.

3.3 *Rethink the Purpose of Education and Institute Inner-Development Practices*

You may well think, perhaps somewhat cynically, that it is all good that we need to engage in human self-reflection. But apart from reconceptualizing foundational

[6] A combination of often incongruous traditions and modern ideologies and practices are at work in most societies. East Asian societies are no exception. While Western originated modern ideologies and institutions have been dominating economic, military and geopolitical spheres in East Asia, traditional thinking and practices continue to have a stronghold in many other spheres, particularly in inter-personal, familial, social relationships and cultural norms.

notions, which seems to be within the reserve of those in the Ivory Tower what else can be done, especially as it relates to the general public?

To unravel the layers of conceptual constructs we regard as truism, apart from the intellectual shift in foundational thinking, another mammoth task is to fundamentally revamp the approach to education. I don't pretend to have sketched out concrete plans for reorienting and reshaping mainstream modern education institutions. But perhaps a few pointers can be provided to stimulate the broader public discussion. Consider, for example, the purpose of classical Confucian education, which was about learning to become human (学以成人), i.e., learning to become an exemplary person. Inspired by this thinking, many contemporary philosophers proposed to use the term "human becoming" to replace "human being" (Ames, 2024; Nakajima, 2024), signifying that becoming human is an incessantly generative process. Education shaped by this kind of thinking would make self-reflection and inner growth into a life-time pursuit. In the same vein, contemporary South Korean scholar Youngseo Baik proposed the idea of "co-becoming humanities", which calls for, among others, "mind practice" by all, starting with intellectual leaders and activists. The purpose of the "mind practice" is to encourage each and every person to reach the realm of Dao, which will naturally lead to a more harmonious and peaceful world (Baik, 2024). At the most fundamental level, speaking from the perspective of the great wisdom, contemporary Chinese wisdom teacher Liu Fenghe notes that education is about eradicating all forms of suffering, thinking beyond the issue of life and death and about understanding what human beings truly are. According to Liu, the final and ultimate goal of a life-long education is about attaining awakening and thus absolute bliss and freedom (Liu, 2025).

Following the cues of these ancient and contemporary teachings, I would like to humbly submit that future education, at its core, ought to be about an inquiry into what it means to be human. This would require an introduction into the current education system a life-long study and practice on self-reflection and self-inquiry. There exist rich and diverse philosophical, religious and spiritual traditions across the world, some going back to the dawn of human civilization, which address the nature and origin of self, others and human civilization. Such self-inquiry should not be relevant only to students of philosophy, nor should it begin only at the stage of tertiary education. It is not an education about knowledge or practical skills, but a training and practice of inner growth and critical thinking. Critical thinking in this context is about challenging and debating about entrenched notions and institutions, about strengthening what some philosophers have characterized as "*yin-yang* intelligence" (Wang, 2021), i.e., the ability to deal with open questions, adapt to unforeseen scenarios and coping with uncertainty and about abilities to counter or contain machine intelligence. Such education does not purport to offer standard and definitive answers to all life situations, but transformative ways of questioning, thinking and approaching problems and challenges.

Many studies have shown that excessive use of social media and AI-powered applications contribute to unhealthy mental conditions such as depression, anxiety, insecurity and self-absorption, among others (Robinson & Smith, 2024). What can we do to address these issues? What can we do now to stop the seemingly irreversible slide into the state of minimal physical interactions and distant human relationships? On a more practical front, practices of meditation and regular schedule of digital fasting would be highly desirable. Many studies have shown that regular meditation can help reduce anxiety, stress and other mental conditions. It is a type of body-mind exercise, which can help one stay centered and keep inner peace. The sense of centeredness and calmness in turn help clear the information overload, increase self-awareness and develop fresh ways of looking at the things. This would help create a mental space between yourself and information and many demands surrounding you, leading to healthy relationships toward AI systems and programs. Zhuangzi's formulation on the relationship between humans and things is instructive—you take command of things rather than being commanded by them (*wu wu er bu wu yu wu* 物物而不物于物). In an all subsuming and penetrating digital society, meditation should be introduced early in one's life so that it can become a life-time habit. Similarly, regular digital fasting would be another life-long tool which will help us clear our minds and remind us to reorient our lives toward face-to-face interactions with humans, animals and the broader nature around us. If we grow up with the newly framed critical thinking mentioned above and live with practical tools of meditation and digital fasting, we would be given ourselves more time and space to self-reflect, improve our human relationships, enhance our awareness of our environment and savour moments of solitude, away from all consuming AI systems.

Skeptics would question the utility and immediacy of these tools and practices. I would counter argue that this is just like the practice of well-being where you strengthen your physical and mental-health factors so that you ward off, or otherwise stand a better chance of surviving, all forms of known and unknown pathogens or negative sentiments. Given the high-stake dilemma we are facing in the all-enveloping digital age, humanity's self-reflection and spiritual growth is our final but the most important defense. In addition, we should treat AI systems as companions and deploy them in ways which would be conducive to the practice of self-reflection and spiritual growth.

4 Conclusion

In conclusion, if we were to address the root cause of various risks and challenges confronting us, we must rethink many foundational notions which we have taken for granted and resetting values which have brought us into various impasses we are facing now. For tasks of such a colossal scale, ancient philosophical traditions and belief systems are a treasure trove which we can tap into for inspiration and hope. Inspired by neo-Confucianist notion of cosmic oneness and the ethos of universal camaraderie, I propose to reconceptualize inanimate things such as AI systems as companions and include them into the holistic human ethical order. This inclusion effort is not about (though not denying either) acknowledging the level of intelligence AI systems have accomplished over and above us, but to revive the idea of conceptualizing non-living things as companions, something of the same origin as humans and about valuing them as catalysts and reminders for collective human self-reflection and transformation.

References

Ames, R. T. (2024). "Zoetology": A new name for an old way of thinking ("生生论": 一种古老思维方式的新名称). *Cuiling Magazine, 3*, 70–87.

Baik, Y. (2024). Co-becoming humanities (看待 "共生" 和 "convivialism" 的另一个视角: 共生人文学的功效). *Cuiling Magazine, 3*, 145–163.

Chan, W. (Trans with notes). (1963). *Instructions for practical living and other neo-confucian writings.* Columbia University Press.

Chen, L. (2014). *The ontology of ren* (仁学本体论). SDX Joint Publishing.

Gennaro, R. J. (2017). *Consciousness.* Routledge.

Grobelnik, M., Perset, K., & Russell, S. (2024, March 6). What is AI? Can you make a clear distinction between AI and non-AI systems? *OECD.* https://oecd.ai/en/wonk/definition.

Guo, Q. (2018). *The characteristics of Chinese cultural spirit* (中国文化精神的特质). SDX Joint Publishing.

Guo, Q. (2020, May 27). Wang Yangming's wisdom of life—Benevolence of oneness (王阳明——一体之仁的生命智慧). *Sohu.* https://www.sohu.com/a/397987424_242653.

Hershock, P. D. (2021). *Buddhism and intelligent technology: Toward a more humane future.* Bloomsbury Academic.

Kim, J. Y. (2015). *Zhang Zai's philosophy of Qi: A practical understanding.* Lexington Books.

Kuhn, R. L. (2024). A landscape of consciousness: Toward a taxonomy of explanations and implications. *Progress in Biophysics and Molecular Biology, 190*, 28–169. https://doi.org/10.1016/j.pbiomolbio.2023.12.003

Leopold, T. (2015, August 4). HitchBOT, the hitchhiking robot, gets beheaded in Philadelphia. *CNN*. https://www.cnn.com/2015/08/03/us/hitchbot-robot-beheaded-philadelphia-feat

Li, F. (2024, April 22). The "Godmother of AI", keeping humanity at the heart of the AI revolution. *Young and Profiting.* https://youngandprofiting.com/fei-fei-li-the-godmother-of-ai-keeping-humanity-at-the-heart-of-the-ai-revolution-e285

Liu, F. (2021). Great wisdom holds the answers to human suffering: Artificial Intelligence inspired thinking. In B. Song (Ed.), *Intelligence and wisdom: Artificial intelligence meets Chinese philosophers* (pp. 101–108). Springer. https://doi.org/10.1007/978-981-16-2309-7_7

Liu, F. (2025). *Existence and Zizai* (存在与自在).

Loreau, M. (2023). *Nature that makes us human: Why we keep destroying nature and how we can stop doing so.* Oxford University Press.

Madrigal, A. C. (2014, June 12). Meet the cute, Wellies-Wearing, Wikipedia-reading robot that's going to hitchlike across Canada. *The Atlantic.* https://www.theatlantic.com/technology/archive/2014/06/meet-the-cute-wellies-wearing-robot-thats-going-to-hitchhike-across-canada/372677

Mengzi. (1895/1970). *The works of mencius* (J. Legge, trans). Clarendon Press.

Nakajima, T. (2024). Our Kyōsei: How the personal works (我们的共生: 个体的作用). *Cuiling Magazine, 3*, 129–143.

Qian, M. (2011). *Discourses on Chinese thoughts* (中国思想通俗讲话). Jiuzhou Press.

Robinson, L., & M. Smith. (2024, October 29). Social media and mental health: Are you addicted to social media? *HelpGuide.* https://www.helpguide.org/mental-health/wellbeing/social-media-and-mental-health

Simard, S. (2021, May 11). The intelligent forest. *Noema Magazine*. https://www.noemamag.com/the-intelligent-forest

Song, B. (2021a). Notions of *Shengsheng* and *Gongsheng* ("生生" 观与共生思想: 契合天地人的思想?). In B. Song (Ed.), *Out of Anthropocene: Reflections on relationship between humans and nature* (pp. 83–102, 走出人类世: 人与自然和谐共处的哲思). CITIC Press.

Song, B. (2021b). Introduction: How Chinese philosophers think about artificial intelligence? In B. Song (Ed.), *Intelligence and wisdom: Artificial intelligence meets chinese philosophers* (pp. 1–14). Springer. https://doi.org/10.1007/978-981-16-2309-7_1

Song, B. (2024). What intellectual shift do we need in a time of planetary risks? Inspirations from symbiosis in life sciences and the notion of *Gongsheng/Kyōsei*. In B. Song & Y. Zhan (Eds.), Gongsheng *across contexts A philosophy of co-becoming* (pp. 3–36). Springer. https://doi.org/10.1007/978-981-99-7325-5_1

Suleyman, M. (2024). *The coming wave: Technology, power, and the 21st century's greatest dilemma*. Crown.

Wang, R. R. (2021). Can a machine flow like Dao? The Daoist philosophy on artificial intelligence. In B. Song (Ed.), *Intelligence and wisdom: Artificial intelligence meets Chinese philosophers* (pp. 65–81). Springer. https://doi.org/10.1007/978-981-16-2309-7_5.

Wilhelm, R. (Ed.). (1979). *Lectures on the I Ching: Constancy and change*. Princeton University Press.

Yao, Z. (2020). AI and my companion (人工智能, 吾与也). In B. Song (Ed.), *Intelligence and wisdom: Artificial intelligence meets Chinese philosophers* (pp. 83–109), 智能与智慧: 人工智能遇见中国哲学家). CITIC Press.

Zhang, D. (1985). An analysis on the thinking of Unity-of-Tian-and-Man in Chinese philosophy (中国哲学中的"天人合一"思想的剖析). *Journal of Peking University (Philosophy and Social SCiences), 1*, 3–10.

Ziporyn, B. (Trans with introduction and notes). (2020). *Zhaungzi: The complete writings*. Hackett Publishing.

Bing Song is a Senior Vice President of the Berggruen Institute and Director of the Institute's China Center. Prior to joining the Berggruen Institute, she was a senior executive with Goldman Sachs China for over a decade, and prior to Goldman, an experienced capital markets lawyer. Earlier in her career, she undertook law and policy research and published in the areas of administrative law, competition law, and comparative procedural laws. Bing has been responsible for the intellectual programs and institutional development of the China Center since its inception. In 2024 she co-edited Gongsheng *Across Contexts—A Philosophy of Co-becoming*, which explores philosophical foundations of gongsheng, a conception of the world as consisting of mutually embedding, co-existing and co-generating entities. Her edited volume *Wisdom and Intelligence—Artificial Intelligence Meets Chinese Philosophers*, published in 2021, marked the first systematic endeavor by Chinese philosophers to address challenges and opportunities posed by artificial intelligence systems.

How Not to Make Predictive AI Trustworthy

Jiji Zhang

1 Introduction

Like it or not, we are already living with artificial intelligence (AI). AI technologies influence how we communicate, form relationships and perceive the world around us. Social media platforms employ sophisticated AI algorithms to manage content and recommend connections, which shape our social circles and constrain the information we receive or share (Gillespie, 2018). Dating apps use complex matching algorithms to predict compatibility and transform the way people meet and form romantic connections (Tong et al., 2016). Virtual assistants like Siri and Alexa create radically new daily communication patterns, not only innovating our interactions with technology but also potentially affecting human-to-human communication (Lopatovska & Williams, 2018). In the professional realm, AI tools assist with team formation, project management, and talent acquisition, and can make significant impacts on workplace dynamics and career trajectories (Fountaine et al., 2019). In healthcare, predictive models are used to anticipate patient needs, improve diagnosis accuracy, and personalize treatment plans, and can strengthen the patient-provider relationship as a result (Obermeyer & Emanuel, 2016).

It is worth noting that while generative AI—such as large language models capable of creating sophisticated texts and other modes of content—has captured public attention and dominated recent discourse, it is predictive AI that plays a crucial role in the examples just given. Predictive AI employs statistical or machine learning models that use historical data to make predictions about future or unknown events or features, including human behaviors and preferences, which are often crucial for informed decision or policymaking. This form of AI is instrumental in various recommendation systems that connect individuals with content, products, and people, and exercises considerable influences on the opportunities we are presented with and

J. Zhang (✉)
Department of Philosophy, The Chinese University of Hong Kong, Hong Kong, China
e-mail: jijizhang@cuhk.edu.hk

Y. Chandra and R. Fan (eds.), *Artificial Intelligence and the Future of Human Relations*,
https://doi.org/10.1007/978-981-96-7185-4_4

the networks we build. It does not take much imagination to foresee that the impact of predictive AI on reconfiguring human relations in various domains will further increase in the future, likely at an accelerating pace.

It is much harder, however, to predict whether the impact will be predominantly positive. We hear frequently of warnings of the erosion of self-determination and the quality of human relationships due to the pervasive presence of AI. Recommendation engines, for example, often work by reinforcing existing preferences and limiting exposure to diverse perspectives, potentially narrowing users' horizons without their conscious consent (Kunaver & Požrl, 2017; Pariser, 2011). Moreover, the instant gratification provided by AI-driven communication tools might decrease people's interest or patience in cultivating deep relationships, and again, usually without their explicit awareness (Newport, 2019; Turkle, 2012). Other oft-cited worries about misusing or abusing AI include amplification of discrimination (O'Neil, 2016), spread of misinformation (Ryan-Mosley, 2023), loss of empathy and social skills (Turkle, 2015), and so on, all of which obviously may weaken trust and damage human relations. Managing such risks is therefore crucial for the flourishing of human relations mediated and shaped by AI technology. The overarching imperative is now usually phrased as that of developing "beneficial AI" (Russell, 2019) or "trustworthy AI" (European Commission AI HLEG, 2019). But as many experts have noted, it is not just a matter of AI scientists and engineers advancing or refining the technologies to make them more robust and dependable, but will also require efforts on the part of users to become informed and mindful when engaging with AI (European Commission AI HLEG, 2019; Floridi & Cowls, 2019).

The last point about shared responsibility involving users is in my view especially important. This essay aims in part to elaborate on this point from an underappreciated angle. Specifically, I wish to stress that among other things, it is important for an AI user to be mindful of what they intend to achieve with an AI system. Instead of giving a direct argument for this claim, I will offer an indirect one by critically examining an issue that seems to present a particularly daunting challenge to the goal of making predictive AI trustworthy. The issue is about an alleged tension between (epistemic) accuracy and (ethical) fairness in making algorithmic predictions (Menon & Williamson, 2018). Roughly, the tension is that improving a model's fairness usually leads to a significant decrease in its predictive accuracy. Together with the seemingly innocuous assumption that both fairness and accuracy are what we desire in a trustworthy predictive model, this tension implies that efforts to improve on one dimension of trustworthiness would usually compromise the model's ability to excel on another. There would then be a formidable obstacle to creating predictive AI that is trustworthy, for it is natural to think that a trustworthy predictive model must be both accurate to be epistemically reliable and fair to be ethically acceptable.

In this essay, I challenge this prevailing way of thinking about predictive AI and the associated idea that we need to strike a compromise between epistemic reliability and ethical fairness in making trustworthy predictive AI models. Unlike other critics of the alleged trade-off between accuracy and fairness, who take for granted that both are intrinsically desirable and try to demonstrate that their apparent conflict can be avoided or is at least much less common than is usually thought, my main contention is that ethical and epistemic standards are never on a par whey they apply to predictive AI, in the sense that at least one of them is not what we really want unless it contributes to achieving the other, once we get clear on the kind of predictive task we intend a predictive model to tackle. Different types of predictive tasks have different goals, sometimes epistemic and sometimes broadly ethical, but never accuracy and fairness at the same time as two intrinsically desirable values, or so I will argue. For example, I will distinguish between *active* predictive task and *passive* predictive task. The former refers to those predictive tasks where the prediction itself causally influences the state of affairs being predicted, as illustrated by self-fulfilling or self-undermining prophecies. The latter, by contrast, are about features of the world that remain causally independent of the prediction. As I will argue, accuracy is only desirable in the former case if it contributes to fairness or other broadly ethical goals, whereas fairness in the ethical sense has little relevance in the latter; what may matter there is fairness in an epistemic sense.

The upshot of this philosophical analysis is at least twofold. On the one hand, it challenges the notion that making trustworthy predictive AI requires compromise between epistemic and ethical considerations. On the other hand, it serves as a case in point for the claim I stressed earlier: to build trustworthy AI requires us to reflect on what we want to do with it. I will discuss these implications in Sect. 4, after I introduce the relevant background information about algorithmic fairness in Sect. 2 and present my main argument in Sect. 3.

2 Biases in Machine Learning and Fairness Interventions

In preparation for my argument against the alleged trade-off between accuracy and fairness of predictive AI models, I review some relevant background in this section. It is impossible to cover everything important in the vast literature on fairness in machine learning; for the present purpose, we just need to get a good sense of what the trade-off in question is supposed to be. More informative surveys of the field can be found in, for example, (Mehrabi et al., 2021) and (Tang et al., 2023). In this section, I will selectively highlight some issues and notions from these references that will be useful in the next part.

2.1 Biases in Current Machine Learning Models

Advances in deep learning (LeCun et al., 2016) over the past decade have facilitated numerous real large-scale applications. Amazing as they are, many such applications have been exposed to exhibit worrisome biases and discriminatory outputs. A widely publicized example is Correctional Offender Management Profiling for Alternative Sanctions (COMPAS), a risk assessment tool used by courts in several states of the USA to predict the likelihood of a defendant reoffending. Investigations found compelling evidence that COMPAS exhibited racial bias, falsely flagging African-American defendants as high risk at nearly twice the rate of white defendants, while white defendants were more likely to be incorrectly predicted to be low risk (Angwin et al., 2016).

Similar disparities were found in facial recognition systems deployed in various contexts. Some research revealed that these systems often perform poorly on marginalized groups, especially intersectional ones. In one oft-cited work, the error rates for identifying darker-skinned women were as high as 34.7%, compared to at most 0.8% for lighter-skinned men (Buolamwini & Gebru, 2018).

Another striking example concerns the popular technique of word embeddings used in natural language processing. Studies have shown that these embeddings capture and amplify gender biases present in the training data. For instance, the embeddings associated "man" with careers and "woman" with family-related words, affecting downstream tasks like machine translation and resume screening (Bolukbasi et al., 2016).

Such examples can be easily multiplied, which underscores the gravity and urgency of the problem. These biases may have arisen from a variety of sources (Suresh & Guttag, 2021). Mehrabi et al. (2021) usefully group various types of biases into three clusters: those *from data to algorithm*, those *from algorithm to user*, and those *from user to data*. Examples in the first cluster include types of bias that were already noted in traditional statistical analysis, such as *measurement bias* and *representation bias*. Measurement bias occurs when the data collected for training a model is systematically skewed due to flaws in the measurement of variables of interest. An illustrative example is the use of arrest rates as a proxy to measure or estimate crime rates in a community. While arrest data is often readily available, it may not accurately reflect the true incidence of criminal activity, because arrest rates are influenced not only by the frequency of crimes but also by law enforcement practices, which can vary significantly across different areas and populations. This kind of bias was probably present in the COMPAS example mentioned above, in which records of previous arrest or family arrest were used to measure predisposition to crime. On the other hand, representation bias happens when some groups are underrepresented or overrepresented in datasets, which is a frequent cause of a model's poor performance for minority groups that were not adequately represented in the training data. The second example noted previously, about the disappointing showing of facial recognition systems on intersectional marginalized groups, is likely a result of this kind of bias, among others.

In the second, algorithm-to-user cluster are types of bias that originate primarily from certain choices in algorithmic design and pass to user behaviors. An obvious example is what is known as *presentation bias* or *position bias*, which occurs when the way information is presented to users impacts their interactions and leads to skewed data (Agarwal et al., 2019). This type of bias arises because the placement, format, or salience of information can significantly impact user behavior independently of the content's actual relevance or importance. As a result, the data subsequently collected reflects not only users' genuine preferences but also the effects of how the information was initially presented. Consider, for example, an online news platform that uses an algorithm to recommend news items to its readers. Items displayed prominently on the homepage, perhaps those placed at the top of the page, featured in larger fonts, or accompanied by striking images, are more likely to attract clicks and readership. This increased engagement is not solely due to the interest or quality of the items but is influenced by their presentation or position, and will create a feedback loop where the initial presentation choices influence user behaviors, which in turn affect the data the model learns from and lead it to make recommendations that reflect and amplify the biases introduced by the original presentation rather than true user interests.

To the third, user-to-data cluster, belongs types of bias that are inherited from users' biases through user-generated data, including what is standardly called *historical bias*. Historical bias is present when the data reflects past prejudices or discriminatory practices of people; even with perfectly measured and sampled data, the model learns and perpetuates existing inequities. For example, the aforementioned biases encoded in word embeddings are largely of this kind. Another oft-cited type called *label bias* may also be categorized into this group. It occurs when the labels or annotations in the training data are biased due to subjective judgments or systemic biases of human annotators, leading the model to learn and reproduce these biases. For example, some hate speech detection models for social media platforms are built based on large datasets of user-generated content that are manually labeled as "hate speech" or not. However, annotators bring their own cultural perceptions, experiences, and biases to the labeling process. This can result in inconsistent or biased labeling, especially when the content includes dialects or culturally specific expressions that the annotators may not be familiar with (Sap et al., 2019).

While many other types of bias have been identified in the literature, the ones highlighted here are sufficient to convey the significant challenges posed by biases in machine learning models. They exemplify how biases can infiltrate various stages of the machine learning pipeline, which features a cycle of data collection, via model building, training, and deployment, back to further data generation. Finding ways to remove or mitigate problematic biases is therefore a most important task.

2.2 Fairness Interventions

In the literature on algorithmic fairness, biases are regarded as especially problematic when they lead to inequitable or discriminatory outcomes against individuals or groups, as illustrated by the opening examples in Sect. 2.1. In such cases, debiasing is often phrased as "enhancing fairness", and the attempts at eliminating or reducing disparity are often referred to as "fairness (enhancing) interventions". A common practice is to formulate specific fairness criteria mathematically and design training procedures that guide the model to approximately satisfy these criteria. A number of fairness criteria have been proposed in the literature (again, Mehrabi et al., 2021 and Tang et al., 2023 are two of several comprehensive surveys). For the purpose of this essay, it suffices to illustrate the basic ideas with some prominent examples, and I will explain them in plain terms without entering mathematical technicalities.

One of the simplest fairness criteria is called *demographic parity*, also known as *statistical parity* or *independence* (Calders et al., 2009). This criterion requires different demographic groups to receive positive outcomes at equal rates. In other words, the outcomes should be independent of a protected attribute such as race or gender. The goal is to prevent disparate impact, where one group of people systematically receive less favorable outcomes than another. Note that this criterion is not specifically designed for predictive models, but is applicable to decision-making models in general, as the outcomes in question are not necessarily predictions or estimations with well-defined ground truth. When it is applied to a predictive model, the outcomes are predicted values of some target variable of interest, such as whether a person has or is risky for a certain disease or whether a loan applicant will default if granted the loan.

By contrast, the next two sample criteria are specifically designed for predictive models, as they are defined in terms of the true value of some target variable of predictive interest. On the one hand, we have a criterion called *equalized odds*, which, roughly speaking, requires that a model's predictions are equally accurate for all demographic groups, in the sense of having the same true positive rates and false positive rates (Hardt et al., 2016). Specifically, this means that for individuals whose ground truth labels are positive or desirable (say, they will actually repay the loan), the probability of the model correctly predicting this desirable quality should be the same across all groups. Similarly, for those whose ground truth labels are negative or undesirable (say, they will actually default), the probability of the model incorrectly outputting a positive prediction should also be the same across groups. (If only the equity in true positive rates is required, the criterion is known in the literature as *equal opportunity*.) In the examples mentioned earlier, the COMPAS system as well as the systems in the face recognition example are criticized for deviating significantly from this criterion.

On the other hand, there is a criterion called *calibration* or *predictive parity* or *test fairness* (Chouldechova, 2017). This criterion requires the model's predicted probabilities to be accurate and consistent across different groups. This means that for individuals given the same prediction, the probability distribution of the target variable for prediction, and in particular, the probability of the true value being positive or desirable, should be the same regardless of their group membership. (More generally, when the prediction is given as a probability or risk score, calibration is the idea that the predicted probability should match the empirical frequencies. For example, if a set of individuals is predicted to have a 20% chance of defaulting on a loan, then the percentage of actual default in that set should be roughly 20%, whether they belong to one demographic group or another.) Calibration ensures that the model's risk estimates can be interpreted uniformly across groups. Interestingly, although the COMPAS system violates the equalized odds or equal opportunity criterion, it does satisfy calibration: Among defendants with similar risk scores,

African Americans are as likely to re-offend as Caucasians do (Dieterich et al., 2016). We will return to this point in the next section.[1]

All the above examples are *associative* criteria in that they are concerned only with statistical (dis)associations of some sort between the protected attribute and the predictor. An alternative general approach considers whether the protected attribute *causally* influences the prediction. Such causal criteria of fairness consider not just correlations but the underlying causal relationships between variables, particularly how changes to a protected attribute would affect the prediction or decision (Plecko & Bareinboim, 2024). An oft-cited example is known as *counterfactual fairness* (Kusner et al., 2017). Roughly, a model is considered counterfactually fair if its prediction for an individual remains the same in a hypothetical or counterfactual scenario in which the individual's protected attribute is altered. To apply such causal fairness criteria we need to model and reason about the underlying data generating process as well as the process of producing a prediction or outcome. This is often regarded as more appropriate than associative criteria for enforcing procedural justice (as opposed to substantive justice) or preventing disparate treatment (in contrast to mere disparate impact). As we will see later, my argument suggests that the causal approach has a theoretical advantage over the associational approach in general.

The goal of a fairness intervention is to design or adjust the training procedure to build a model whose behavior does not deviate significantly from selected fairness criteria that are deemed important in the given domain or context. To achieve this, various approaches have been developed, including *pre-processing* methods (modifying or transforming training data in some way before feeding to the learning algorithm), *in-processing* methods (designing the optimization procedure to incorporate certain fairness constraints), and *post-processing* methods (adjusting the outputs of the trained model in some way). While each approach operates at a different stage of the model development pipeline, they share the common objective of mitigating unfairness as defined by the chosen criteria. The technical details of these methods are diverse and complex. For the purpose of this essay, it suffices to understand that fairness interventions are strategies employed to guide machine learning models toward more equitable outcomes according to some chosen criteria. In the case of predictive models, for which accuracy or epistemic reliability seems to be an obviously sensible goal, such fairness interventions are said to compromise accuracy, at least usually if not inevitably. This alleged trade-off between fairness and accuracy is the main target of the next section.

---

[1] These are examples of *group-level* fairness criteria. There are also criteria that are said to be *individual-level*, mostly famously, a class of criteria just named "individual fairness", which aim to capture the idea that individuals with similar (relevant) characteristics should be treated similarly by the model (Dwork et al., 2012). Individual fairness is often said to shift the focus from group-based fairness to ensuring fairness at the level of individual cases, but note that the difference between the two levels is not sharp, and presumably a matter of degree depending on how homogenous a group is, for each individual is after all identified with a set of attributes in this context, and the group-level criteria can be made increasingly closer to individual-level by conditioning on increasingly more attributes.

3 Against the Alleged Trade-Off Between Fairness and Accuracy of Predictive AI

3.1 Apparent Trade-Off Between Fairness and Accuracy of Predictive Models

Why should we expect a trade-off between fairness and accuracy of predictive models? There appear to be some compelling theoretical reasons for this. Consider the in-processing approach to fairness intervention, where fairness constraints are incorporated in some way in the optimization process. If we wish to optimize accuracy, it is natural to think that optimization constrained by some fairness criteria can do at best as well as unconstrained optimization and usually do significantly worse. In fact, assuming data are properly sampled from the true distribution, theoretically, unconstrained optimization of accuracy is expected to satisfy the criterion of predictive parity, which is for this reason referred to as an "implicit fairness criterion of unconstrained learning" (Liu et al., 2019). So it is not surprising that, as mentioned previously, the developers of the COMPAS system could point to the criterion of predictive parity to defend the system. However, another important theoretical result shows that except in very special cases, it is mathematically impossible to satisfy both predictive parity and demographic parity, or both predictive parity and equalized odds (Chouldechova, 2017; Kleinberg et al., 2017). It is hence not a coincidence either that the COMPAS system was found to fail the criterion of equalized odds. In general, then, from the perspective of those who take demographic parity or equalized odds as proper criteria of fairness, optimal accuracy almost inevitably frustrates fairness, or in other words, imposing a fairness requirement will have to compromise accuracy.

An easy way to get an intuition about this phenomenon is to think of the cases in which base rates of the key property of interest differ among groups. For instance, suppose one demographic group has a higher prevalence of a positive outcome (e.g., loan repayment) than another group, perhaps due to the socioeconomic arrangement shaped by some historical biases. A predictive model about loan repayment optimized for accuracy will leverage this information and will obviously violate demographic parity (which means equalizing positive prediction rates across groups). Similarly, though a little less obviously, for the criterion of equalized odds (which means equalizing true positive and false positive rates across groups). Conversely, if we impose these fairness criteria in training the model, then the model's predictive accuracy can be seriously affected. Such reductions in model accuracy due to the implementation of some fairness interventions have been observed in many empirical studies (Tang et al., 2023).

3.2 *An Argument Against the Alleged Trade-Off*

I now present a philosophical argument against the said trade-off, or more precisely, against the interpretation of the trade-off as one between two goals that are desirable simultaneously. To be clear, I have no quarrel, at least in the present essay, with the frequently observed and theoretically grounded phenomenon that reasonable measures of accuracy and reasonable criteria of fairness are in tension. My criticism is novel in that it targets the supposition that both accuracy and fairness are simultaneously desirable in themselves for predictive AI.

Here is one way of formulating the intended conclusion of my argument. Say that there is a *significant* trade-off between goal A and goal B if there is some conflict between them and both goals are desirable or what we ought to desire to achieve. The intended conclusion is that there is no *significant* trade-off between fairness and accuracy.

Before I state the main premises for this conclusion, let me illustrate the rather simple intuition behind the argument with a common scenario. Recall the conflict between the criterion of demographic parity and accuracy noted in the previous section, where the truth is such that one demographic group just has a higher prevalence of the desirable quality, e.g., high academic ability, than another group does, and so a perfectly accurate predictor or truth itself would be unfair by the criterion of demographic parity. Now, suppose someone knows this but still wishes to impose demographic parity as a requirement on the *predictor*, can they justify or at least make sense of this goal? Intuitively, *quo a predictor*, it is hard to see how one could reasonably reject a perfectly accurate predictor. In my view, one indeed cannot, provided that the task is specified precisely as that of predicting the actual academic ability *and* the ground truth is causally independent of the prediction. That is, even if the actual inequitable distributions are morally objectionable because they are shaped by historical bias, if the actual abilities are precisely what the predictive task is intended to get at and they will not be affected by the predictions, then there is nothing wrong for a predictor to do its intended job.

Note, however, the two conditions for this verdict. If either of the conditions fails, then a perfectly accurate predictor of the actual distributions may be reasonably rejected. First, suppose the target of the predictive task is not really the academic abilities as they actually are, but, for example, a counterfactual question regarding what a person's academic ability would have been in an ideal world or a better world in which the past practices had not been discriminatory. Then obviously, the accuracy with respect to the actual academic abilities is not what matters. One might argue that imposing some fairness criteria such as demographic parity enables a more accurate prediction regarding the real target of the predictive task.

Second, and more interestingly, suppose there is no confusion about the target of the predictive task, but the target is not causally independent of the prediction itself. That is, the target of the predictive task is indeed the actual academic ability, but the very prediction about an individual's academic ability can affect the ability in question, positively or negatively. This is what I will call a case of active prediction, which is very different from the canonical, passive predictive tasks encountered in the natural sciences. In such a case, since predictions have a causal effect on the states of affairs in question, they should be seen as a tool to shape the states of affairs, and the prevailing goal is to realize desirable states of affairs through the predictions. Predictions become strategic, and their accuracy only matters to the extent that they serve the prevailing goal. So a perfectly accurate predictor may well be undesirable in such a case, because they do not lead to the desirable states of affairs according to some standard.

Let us then distinguish two kinds of predictive task: *active prediction* and *passive prediction*. Active predictions causally influence the states of affairs being predicted; they are, as it were, shooting moving targets. Passive predictions do not influence the states of affairs being predicted; they are, as it were, shooting fixed targets. Paradigmatic examples of the former are self-fulfilling or self-undermining prophecies often encountered in social science, whereas the latter is the typical kind of predictive task in natural science.

With this distinction, the main argument can be formulated as follows:

Premise 1: In an active predictive task, accuracy of the predictor is desirable only if it contributes to achieving the desirable states of affairs (including fairness if it is desirable).

Premise 2: In a passive predictive task, fairness of the predictor is desirable only if it contributes to the accuracy of the prediction.

Conclusion: There is no significant trade-off between fairness and accuracy. That is, whenever there is a conflict between achieving fairness and achieving accuracy, either fairness or accuracy is not desirable.

This argument is valid. In an active predictive task, if there is a conflict between achieving fairness and achieving accuracy, which means that accuracy of the predictor does not contribute to achieving fairness, then either fairness is not desirable in the first place or accuracy is not desirable by Premise 1. In a passive predictive task, if there is a conflict between achieving fairness and achieving accuracy, which means that fairness of the predictor does not contribute to achieving accuracy, then fairness is not desirable by Premise 2. Since a predictive task is either active or passive, the conclusion follows.

So it suffices to justify the premises in order to defend the argument. Consider Premise 1 first. As already indicated above, the reason for accepting Premise 1 is that if predictions causally influence the states of affairs being predicted, then what matters is how to achieve the desirable states of affairs through the means of predictions rather than the accuracy of the predictions; self-fulfilling prophecies and self-undermining prophecies are on a par if they both lead to desirable states of affairs, even though the former will turn out to be much more "accurate" than the latter. For example, between my two sons, one of them likes to realize my predictions about him and the other likes to falsify them. When I manage to get both of them to eat more vegetables, by making a true prediction that the former will eat vegetables and a false prediction that the latter will not, respectively, the two predictions are equally valuable from my point of view despite their difference in "accuracy". In other words, in an active predictive task, "accuracy" is not the prevailing goal; it is sometimes valuable, but only in the instrumental sense of facilitating the goal of realizing desirable states of affairs according to some standard, which could be a standard of fairness. Thus, when accuracy and fairness is in conflict, assuming fairness is what is wanted in the realized states of affairs, accuracy will not be what is wanted at the same time.

In a way, Premise 1 covers decision-making models as an extreme case of "self-fulfilling predictive models". If we think of decisions, say, about whether to grant a loan, as predictions about whether a loan will be granted, then obviously the predictions will be exceedingly self-fulfilling (assuming the decisions will be faithfully implemented) and hence extremely accurate. But such accuracy obviously comes cheaply, and nobody ought to care about it. What matters in these cases is whether the decisions/predictions are fair.

In this regard, the notion of active predictions can be further expanded to cover streamlined prediction-decision models, where a (possibly passive) predictive model directly feeds to a decision-making model to produce a decision. Taken as a whole, such a streamlined model can be seen as an active predictive model in a broad sense, where the predictions are used to influence decisions, and it is the fairness (or, more generally, moral rightness) of the decisions that ultimately matters. Usually accurate predictions contribute to the goal of making right decisions, but this is not necessarily the case. When accuracy of predictions tends instead to hinder right decisions, it will then become undesirable.

So much for Premise 1. For Premise 2, the justification is that passive prediction, taken on its own, is predominantly an epistemic task, of which the prevailing goal is to seek truth or avoid error. Even if the true state of affairs turns out unfair or morally objectionable in other ways, all that matters to a passive predictor, taken on its own, is to be as accurate as possible. The qualifier "taken on its own" needs to be stressed. As the previous paragraph made clear, when downstream applications of a passive predictive model are taken into account, the combined system as a whole is addressing an active predictive task, broadly construed. In light of what the whole system is intended to do, accuracy is no longer intrinsically desirable, though it is usually instrumentally useful. However, when the task is specified to be a passive predictive one on its own, its prevailing goal is epistemic by nature, and whatever notions of fairness that turn out to be useful or desirable in this context, must be serving the epistemic goal.

It is commonplace that fairness can also be an epistemic virtue that serves an epistemic goal. Among the sample types of bias described in the previous section, all except perhaps historical bias point to epistemic flaws, and suitable fairness interventions may help to address or mitigate these flaws and thereby contribute to accuracy. For example, there is fascinating technical work demonstrating that applying demographic parity in a semi-supervised setting helps to correct label bias and representation bias and improve accuracy (Wick et al., 2019). Even with historical bias, as indicated previously, fairness considerations may contribute to accuracy if the predictive target is concerned not with the actual situation but a counterfactual one, though such predictive tasks are of course much more challenging and probably require modeling of causal processes rather than just statistical patterns (Pearl, 2009).

Therefore, for a predictor intended to take on a purely passive predictive task, the prevailing standard is accuracy (or, more generally, epistemic reliability). Fairness interventions can still be of value, but only when they contribute to the epistemic goal. This is the rationale behind Premise 2. Interestingly, even in the context of criminal trials, where it is more likely to identify genuine clashes between moral and epistemic values, there are strong advocates of an analogous perspective on fairness, as, for example, expressed in the following quote:

> [It] must be clear what fairness means in this connection. It means, or ought to mean, that the law should be such as will secure as far as possible that the result of the trial is the right one. (Criminal Law Revision Committee, 1972, cited by Laudan, 2006, p. 1)

This view is even more plausible in our context for passive predictions.

To summarize, I contend that there is no significant trade-off between fairness and accuracy of predictive AI, because it is always the case that one is subordinate to the other, though depending on the nature of the predictive task, which commands which can change. In active predictive tasks where the predictions influence the outcomes, accuracy is not intrinsically desirable, but only valuable insofar as it contributes to producing fair outcomes. In contrast, in passive predictive tasks focused on estimating fixed states of affairs, fairness is only valuable if it serves the epistemic goal of accurate prediction. Either way, it is misleading to portray the situation as having to strike a compromise between two independently desirable but competing standards.

3.3 Two Objections and Replies

Two potential objections to the argument need to be addressed before we move on, which will also help to bring out some subtle aspects of the argument. The first objection is concerned with the distinction between active and passive predictive tasks. The concern is that the distinction does not seem to be as sharp as is supposed. What about predictive tasks that have characteristics of both, such as a task targeting several variables of interest at the same time, some of which are influenced by the prediction, whereas others are not? To this formulation of the concern the reply is simple. Either such a task can be neatly separated into several components, some clearly passive and others clearly active, or it must be taken as whole. In the former case, it is to each of the separate components that the distinction applies. In the latter case, as already indicated in the previous section, the task is an active one as a whole, just as a streamlined task that may include a passive predictive part but is handled as an integrated whole.

There is another way to press the concern. The thought is that activeness of prediction is a matter of degree, and strictly passive predictions are rare when they are about human beings. If so, the distinction is close to being vacuous (for almost all predictive tasks are, strictly speaking, active). Moreover, given that active predictive tasks can have a very low degree of activeness—in the sense that the relevant causal influence, though non-zero, is very weak—there seems to be an implausible kind of discontinuity suggested by the premises of the argument: No matter how weakly active a predictive task is, as long as it is not strictly passive, accuracy is subordinate to fairness, but when we hit strictly passive tasks in the limit, the subordination is suddenly reversed.

This way of putting the objection is more compelling and the response will be more nuanced. First of all, note that even if there were very few or even no really passive predictive tasks in practice, that would not affect the soundness of the argument. It would only imply that from a practical point of view, Premise 1 is much more relevant than Premise 2. This may partly explain why the prevailing approach (or at least rhetoric) in the literature on the trade-off between fairness and accuracy is to prioritize fairness. More importantly, this and the related worry about discontinuity suggest that in practice the distinction involves a threshold that is not only context dependent but, in line with a main moral of this essay, also requires a judgment call on the part of AI developers or users. That is, Premises 1 and 2 should be interpreted as regulating how one ought to value fairness and accuracy, depending on of what kind one judges a predictive task to be or what kind of predictive job one wants an AI model to do.

So, although it is true that the distinction, like most distinctions, will not be as sharp in practice as in theory, it remains a useful one to guide our value judgments about fairness and accuracy. This concludes my response to the first objection.

The second objection targets Premise 2 in particular. Consider the kind of unfairness illustrated by the COMPAS and facial recognition examples described in the previous section, which is inequity in error rates across demographic groups. For the sake of argument, suppose the task in question is passive prediction. Isn't there still the issue of inequity in accuracy across demographic groups? Isn't it desirable to address such unfairness even if it is not instrumental or is even detrimental to the overall accuracy? To this objection I submit two replies. The first reply is straightforward. If the task in question is just passive prediction, then we should try to make the predictor as accurately as possible (according to some properly defined measure of "overall accuracy", see the second reply below). The equity of accuracy across groups is only valuable to the extent that it contributes to this overall accuracy, for example, by encouraging ways to overcome representation bias and increasing the accuracy in underrepresented groups without sacrificing accuracy in other groups. The intuition that the equity of accuracy is important even at the cost of overall accuracy actually stems from thinking about the downstream decisions, i.e., about the streamlined prediction-decision system as a whole, which corresponds to a (generalized) active task, as we already noted when discussing Premise 1. To think that this intuition challenges Premise 2 is hence a confusion.

The second reply is more subtle. How to measure overall accuracy is a complex matter, when different types of error are present and cannot be simultaneously minimized. For example, type I and type II errors in statistical testing, or false positive and false negative rates, are in a genuine, significant trade-off: We want both to be low, but at a fixed sample size, decreasing one entails increasing the other. When we compare the "overall accuracy" of such tests, we do not just count both types of error equally; typically type I error or false positive is regarded as more serious than type II error or false negative, though different weightings may be motivated depending on the context. The point is that considerations of distributions of error, across demographic groups as well as among different types, are arguably part of how we measure overall accuracy. Hence they will not go against overall accuracy; at most they may go against naïve measures that do not accurately reflect the overall accuracy.

4 Implications for Trustworthy Predictive AI

If my argument is sound, it has several notable implications. Some have to do with how best to carry out fairness interventions. For example, the foregoing discussions suggest that fairness interventions on predictive models have primarily two possible purposes. In the case of active prediction, a fairness intervention should aim to cause states of affairs that are fair according to an appropriately chosen fairness criterion, as is practiced in some work on dynamic fairness (e.g., Tang et al., 2023). In the case of passive prediction, a fairness intervention should aim to improve accuracy of prediction by correcting various kinds of (epistemic) bias, which often requires answering counterfactual questions of the form: What data would have been like if certain biases were absent, or what the value or probability distribution of the target variable would be if historical bias were absent. For these difficult tasks, it is usually essential to model the relevant causal processes. From this perspective, the causal approach to fairness intervention deserves a bigger share of the attention than it has so far received, in comparison to the non-causal, associational approach.

For the purpose of the present essay, the more important lessons I wish to draw concern how we should think of trustworthy (predictive) AI. Let me close this essay by making three connected, increasingly general suggestions. First, many statements by leading experts and organizations have included among the characteristics of trustworthy AI accuracy (or cognate properties of epistemic reliability) and fairness (in an ethical sense). My argument suggests that although such statements are reasonable when properly understood, they may also induce a misunderstanding that conveys a misguided conception of how to make predictive AI trustworthy. The misunderstanding is that (epistemic) accuracy and (ethical) fairness are simultaneously desirable in themselves, which, given the technical results on the trade-off between the two kinds of standard, naturally gives rise to the misconception that the way to make trustworthy predictive AI involves compromises of either some epistemic standard or some ethical standard. This is not how we make predictive AI trustworthy. Instead, an implication of my argument is that when done properly, no genuine compromise is called for. When accuracy or fairness seems to be compromised, it may well (and ought to) be just a proper adjustment to improve its instrumental value.

Moreover, my argument clearly shows that different kinds of predictive task warrant different priorities between accuracy and fairness, and so demand different specifications of the general characterization of trustworthiness. It is probably common sense that there is no-one-size-fits-all to trustworthy AI. My argument not only confirms this, but, more importantly, provides a principled reason for treating two kinds of predictive task differently.

Second, this realization should remind us of a general point in the philosophical discussion of trust: Trust is a three-place relation, that is, in addition to a trustor and a trustee, the relation of trust is also relative to a task or domain, as in "Caesar trusted Brutus to protect him". Although it is controversial whether all forms of trust between people are relative in this way or whether there is a more basic relation of trust that is two-place (Faulkner, 2015; Jones, 2004), I think there can be no serious doubt that when it comes to trust in AI, at least in its current forms, a three-place relation is all we can make sense of (if only in a metaphorical sense[2]). My argument highlights the need to clarify the nature of a predictive task before assessing the trustworthiness of a system performing the task.

Finally, I hasten to reiterate a lesson learned in Sect. 3.3, that in practice the distinction between active and passive predictive tasks is not entirely objective, but involves an AI user's judgment or intention. More generally, we need to reflect on what we want to do with AI while trying to make it trustworthy. After all, we need to know ourselves better to improve relations with other human beings. In this regard, the demand of our relation with AI is probably similar.

References

Agarwal, A., Zaitsev, I., Wang, X., Li, C., Najork, M., & Joachims, T. (2019). Estimating position bias without intrusive interventions. *Proceedings of the Twelfth ACM International Conference on Web Search and Data Mining (WSDM'19)*, 474–482.

Angwin, J., Larson, J., Mattu, S., & Kirchner, L. (2016). Machine Bias. *ProPublica.*

Bolukbasi, T., Chang, K., Zou, J. Y., Saligrama, V., & Kalai, A. T. (2016). Man is to computer programmer as woman is to homemaker? Debiasing word embeddings. *Advances in Neural Information Processing Systems*, 4349–4357.

Buolamwini, J., & Gebru, T. (2018). Gender Shades: Intersectional accuracy disparities in commercial gender classification. *Proceedings of the 1st Conference on Fairness, Accountability and Transparency*, PMLR 81, 77–91.

Calders, T., Kamiran, F., & Pechenizkiy, M. (2009). Building Classifiers with Independency Constraints. *IEEE International Conference on Data Mining Workshops*, 13–18.

Chouldechova, A. (2017). Fair prediction with disparate impact: A study of bias in recidivism prediction instruments. *Big Data, 5*(2), 153–163.

Dieterich, W., Mendoza, C., & Brennan, T. (2016). COMPAS Risk scales: Demonstrating accuracy equity and predictive parity. *Northpointe Inc.*

Dwork, C., Hardt, M., Pitassi, T., Reingold, O., & Zemel, R. (2012). Fairness through Awareness. *Proceedings of the 3rd Innovations in Theoretical Computer Science Conference*, 214–226.

European AI HLEG (European Commission's High-Level Expert Group on Artificial Intelligence). (2019). *Ethics Guidelines for Trustworthy AI.*

Faulkner, P. (2015). The attitude of trust is basic. *Analysis, 75*(3), 424–429.

Floridi, L., & Cowls, J. (2019). A unified framework of five principles for AI in society. *Harvard Data Science Review, 1*, 1.

[2] I leave aside the debate on whether one can literally *trust* the current form of AI to do anything, given that one cannot be *betrayed* by AI, at least at the current stage of the technology.

Fountaine, T., McCarthy, B., & Saleh, T. (2019, July–August). Building the AI-powered organization. *Harvard Business Review*, 62–73.
Gillespie, T. (2018). *Custodians of the internet: Platforms, content moderation, and the hidden decisions that shape social media*. Yale University Press.
Hardt, M., Price, E., & Srebro, N. (2016). Equality of opportunity in supervised learning. *Advances in Neural Information Processing Systems*, 3315–3323.
Jones, K. (2004). Trust and Terror. In P. DesAutels & M. U. Walker (Eds.), *Moral psychology: Feminist ethics and social theory* (pp. 3–18). Rowman & Littlefield.
Kleinberg, J., Mullainathan, S., & Raghavan, M. (2017). Inherent trade-offs in the fair determination of risk scores. *Innovations in Theoretical Computer Science Conference, 43*(1–43), 23.
Kunaver, M., & Požrl, T. (2017). Diversity in recommender systems—A survey. *Knowledge-Based Systems, 123*, 154–162.
Kusner, M. J., Loftus, J., Russell, C., & Silva, R. (2017). Counterfactual fairness. *Advances in Neural Information Processing Systems*, 4066–4076.
LeCun, Y., Bengio, Y., & Hinton, G. (2016). Deep learning. *Nature, 521*, 436–444.
Liu, L. T., Simchowitz, M., & Hardt, M. (2019). The implicit fairness criterion of unconstrained learning. *Proceedings of International Conference on Machine Learning*, PMLR 97, 4051–4060.
Lopatovska, I., & Williams, H. (2018). Personification of the Amazon Alexa: BFF or a Mindless Companion. *Proceedings of the 2018 Conference on Human Information Interaction & Retrieval*, 265–268.
Mehrabi, N., Morstatter, F., Saxena, N., Lerman, K., & Galstyan, A. (2021). A survey on Bias and fairness in machine learning. *ACM Computing Surveys, 54*(6), 1–35.
Menon, A. K., & Williamson, R. C. (2018). The cost of fairness in binary classification. *Proceedings of the 1st Conference on Fairness, Accountability and Transparency*, PMLR 81, 107–118.
Newport, C. (2019). *Digital minimalism: Choosing a focused life in a noisy world*. Portfolio (Penguin).
Obermeyer, Z., & Emanuel, E. J. (2016). Predicting the future—Big data, machine learning, and clinical medicine. *The New England Journal of Medicine, 375*(13), 1216–1219.
O'Neil, C. (2016). *Weapons of math destruction: How big data increases inequality and threatens democracy*, Crown (Penguin Random House).
Pariser, E. (2011). *The filter bubble: What the internet is hiding from you*. Penguin.
Pearl, J. (2009). *Causality: Models, reasoning, and inference* (2nd Ed.). Cambridge University Press.
Plečko, D., & Bareinboim, E. (2024). Causal fairness analysis: A causal toolkit for fair machine learning. *Foundations and Trends® in Machine Learning, 17*(3), 304–589.
Russell, S. (2019). *Human compatible: Artificial intelligence and the problem of control*. Viking.
Ryan-Mosley, T. (2023). How generative AI is boosting the spread of disinformation and propaganda. *MIT Technology Review*.
Sap, M., Card, D., Gabriel, S., Choi, Y., & Smith, N. A. (2019). The risk of racial Bias in hate speech detection. *Proceedings of the 57th Annual Meeting of the Association for Computational Linguistics*, 1668–1678.
Suresh, H., & Guttag, J. V. (2021). A framework for understanding unintended consequences of machine learning. *Equity and Access in Algorithms, Mechanisms, and Optimization (EAAMO'21)*.
Tang, Z., Zhang, J., & Zhang, K. (2023). What-is and how-to for fairness in machine learning: A survey, reflection, and perspective. *ACM Computing Surveys, 55*(13), 1–37.
Tong, S. T., Hancock, J. T., & Slatcher, R. B. (2016). Online dating system design and relational decision making: Choice, algorithms, and control. *Personal Relationships, 23*(4), 645–662.
Turkle, S. (2012). *Alone together: Why we expect more from technology and less from each other*. Basic Books.
Turkle, S. (2015). *Reclaiming conversation: The power of talk in a digital age*. Penguin Books.
Wick, M., Panda, S., & Tristan, J. (2019). Unlocking fairness: A trade-off revisited. *Advances in Neural Information Processing Systems*, 8780–8789.

Jiji Zhang received his BA from Peking University and PhD from Carnegie Mellon University. He taught previously at California Institute of Technology, Lingnan University, and Hong Kong Baptist University, and is now Professor in the Department of Philosophy and Associate Dean (Research) of the Faculty of Arts at the Chinese University of Hong Kong. His primary research program centers around the topic of causation, addressing both the epistemological and logical aspects of causal reasoning, and the statistical and computational aspects of causal modelling and discovery. Thanks to its interdisciplinary nature, his work has appeared in both premier journals in philosophy and in leading venues in artificial intelligence. He also served as a consultant to the causality research in Noah's Arc Lab, and was a recipient of the Hong Kong RGC's Humanities and Social Sciences Prestigious Fellowship.

Is Trustworthy AI a Conceptual Nonsense?

Shane Ryan

1 Introduction

In recent years, there has been a discussion of "trustworthy AI" outside of philosophy. The term has appeared in a number of policy papers originating from various countries. While here is not the place to diagnose the motivations for the usage of this term, what is noteworthy for our purposes is the taking up of the term by some scholars in the field of philosophy despite a pre-existing trust and trustworthiness literature that seems to suggest that the term is a "nonsense", which is what Metzinger (2019) claims—more of which later.

This chapter examines whether "trustworthy AI" should indeed be regarded as a nonsense. I do so by consideration of the trust and trustworthiness literature, which seems to speak against the usage of the term, as well as by the exploration of my own account of trust, which in includes a goodwill requirement for trust. Finally, I consider how proponents of the possibility of "trustworthy AI" already defend and could defend their stance. Ultimately, I make the case that "trustworthy AI" makes sense either if AI achieves agency or as a derivative of the trustworthiness of AI producers and/or users. A further possibility explored is whether "trustworthy AI" can make sense, if we make a radical change in our adoption of reactive attitudes.

While my immediate concern in this chapter is examination of the charge that "trustworthy AI is a conceptual nonsense" and, to a lesser extent, the consequences of seeing AI agency as on a par or surpassing our own, there is an obvious philosophical significance to this topic that I will only flag here. The advancement of AI technology can also increasingly be expected to influence a variety of philosophical discussions from ethics to epistemology. In the case of the latter, and as an epistemologist I will just focus on the latter, consider how the proliferation of AI outputs will influence our epistemic environment. As AI technology, such as ChatGPT and descendant chatbots

S. Ryan (✉)
Department of Public and International Affairs, City University of Hong Kong, Hong Kong, China
e-mail: sgryan@cityu.edu.hk

Y. Chandra and R. Fan (eds.), *Artificial Intelligence and the Future of Human Relations*,
https://doi.org/10.1007/978-981-96-7185-4_5

increase their proportion of the share of testimony in our epistemic environments and eventually play an important filtering role in the testimony we have access to on social media, emails, and search websites, perhaps in the monitoring and policing way envisaged by Goldberg (2011), whether AI outputs are accurate, reliable, and so on, will be of particular importance to determining the nature and quality of our epistemic environments (Ryan, 2018b, 2022, 2024). The nature and quality of our epistemic environments of course have implications for our attainment of valuable epistemic goods, such as knowledge (Greco, 2011; Pritchard, 2010; Pritchard et al., 2022; Ryan, 2014a, 2017, 2018b).

In fact, our ability to gain knowledge from AIs, both chatbots and other AIs, depends on our ability to extend our cognition to AIs, which in turn requires that any such AIs and our usage of AIs meet certain conditions (Clark, 1997, 2008; Palermos, 2014; Palermos & Pritchard, 2013; Shin, 2023; Smart, 2018). The hypothesis of extended cognition denies that cognition is to be entirely located within the cogniser's skull. Un fact, defenders of the view argue that the cogniser's body beyond the brain at times is integral to cognitive processes. Palermos and Pritchard (2013, 110) describe the position thus: "literally speaking, with respect to several cognitive operations, our bodies are part of our minds". A familiar idea that illustrates this example is the use of fingers for counting. In fact, for some cognisers, this bodily component of making a mathematical calculation may be integral to the process (Bender & Beller, 2012). Extended cognition, however, need not reach its boundaries at the borders of the natural body. In fact, cognition on the hypothesis of extended cognition may extended past the cogniser's body to the world beyond. Consider sensory substitution devices, which are used to aid those with an impaired sensory modality. A blind person's walking stick extends the touch of a person to understand the world physically around them without the use of sight. In doing so, the walking stick is integral to his cognising in that respect. The lesson draw from these sorts of cases by theorists is that artefacts may become an integral part of our cognition. The moral drawn is that just as the body beyond the brain may play an integral part in cognising so too may artefacts in our use. Perhaps, however, artefacts are not the limit either. Palermos and Pritchard (2013, 110) think that it's possible that someone's cognition might extend to agents with whom she is interacting. A significant part of the discussion in this chapter is that trustworthiness requires agency and whether AI meets this requirement. Nevertheless, on virtue epistemological accounts of knowledge, where knowledge must be from the cogniser's own ability, whether the cogniser interacts with an artefact or agent, certain integration conditions need to be met (Mi & Ryan, 2018), if knowledge is to be gained. Untrustworthy AI, for example, won't provide a suitable source for knowledge.

Our accounts of epistemological testimonial justification can also be expected to be influenced by the greater role of AI testimony.[1] In particularly counter-factual sensitivity and credulist accounts of testimonial justification may be undermined if AI "floods" our epistemic environments with bad testimony, which we are unable

[1] How we should think about "AI testimony" is itself a topic that is recently receiving study (Freiman 2024).

to differentiate from good testimony (Anderau, 2023; Pritchard, 2010; Ryan, 2014b, 2019). A further area for examination will be cultural variation in perceptions of epistemic worth of AI testimony. Cultural variation in testimonial reception is examined by S. Ryanet al. (2020), while the authors have undertaken research to examine this issue as it pertains to the reception of AI testimony. Given the value of epistemic goods, any such significant detrimental influence by AI on our epistemic environment seems likely to draw a response from applied epistemologists, such as proponents of epistemic paternalist and epistemic benevolence (Ahlstrom-Vij, 2013; Castro et al., 2020; Goldman, 1999; Ryan, 2016, 2018c, 2024).

2 The Nature of Trust

Elsewhere I've defended an account of trust that includes an analysis of what it is for one agent to trust another agent to do something (Ryan, 2018a, 2018b, 2018c, 2021). I will briefly introduce the account here, as a basis for the later discussion of "trustworthy AI".

Trusting is a psychological state. Whether someone trusts or doesn't trust another person is an independent question from the attributes of that person. Because trust is a psychological state, trusting may be categorised in numerous ways. One may say that something is an inappropriate object of trust because conceptually it's simply not the sort of thing that could be trustworthy. A simple, everyday stone is one such example. If a child of a certain age said that they trusted a stone, one might explain that a stone is not an appropriate object of trust. Separately, if the same child said they trusted an adult who had shown lots of signs of being trustworthy, one might explain to the child, why that adult is not an appropriate object of trust. Of course, in the latter case, it's not that the object is a conceptually inappropriate object of trust, rather the adult is in an inappropriate object of trust on a different basis—although the adult is of a kind that can be trustworthy, as an individual they lack certain qualities, with the result that they shouldn't be trusted.

Trust may be described in a general way or in a domain specific way. In other words, one can say one trusts someone or one may say one trusts someone in certain respects. One may trust a friend in various respects but at the same time be cognisant of their weaknesses in other respects. For example, one may trust one's friend to look after one's cat but not trust that same friend to finish a task to a tight deadline. While the account of trust set out here is developed on the basis of an account of trust in certain respects, talk of trustworthy AI is appealing to the more way of talking about trust. Plausibly, saying someone is trustworthy implies that one thinks that the person is trustworthy in the relevant respects, without one being committed to saying that one is trustworthy in all possible respects. This distinction between general trust and domain specific trust will become important later.

Let's turn now to understanding what it is to trust someone in a particular domain. It's broadly accepted in the literature that trusting someone to do something implies a belief (or a disposition to believe) that that person has the relevant competence to do that thing (McLeod, 2023; Ryan, 2018a, 2018b, 2018c). If one lacks the belief that a person has the relevant competence, then one can't trust that person to do that thing. One can't trust one's brother to drive one to the meeting, if one doesn't believe that one's brother knows how to drive.

Competence, however, is clearly not sufficient for trust. One is likely to perceive one's worst enemy to be very competent in various respects, but obviously one would not trust that person. What else is needed for trust is the tricky part of analyses of trust. Annette Baier's (1986) influential account of trust advocates a goodwill requirement. According to Baier (1986), "[t]rust … is reliance on others' competence and willingness to look after, rather than harm things one cares about which are entrusted to their care". A person's competence and willingness to look after things about which one cares is still, however, not sufficient to predict trust. One might not have a positive opinion of how another would person would look after one's child, despite that person having a willingness and competence to look after that child.

One proposal is that a belief (or dispositional) in the goodwill of the other agent is required and that goodwill here should be understood as approval of will (Ryan, 2018a, 2018b, 2018c, 2021). This way of understanding goodwill is equipped to handle Holton's (1994) estranged couple objection to the goodwill approach, in which it's plausible that the couple don't have good will towards one another but they can nevertheless trust the other to look after their child for a weekend. An implication of the view is that one may stop trusting someone to do X, if, upon learning more about how they do X, one doesn't like how they do X. So, if one catches a glimpse of how the other person is looking after one's child and one strongly dislike it, then one can be expected to stop trusting the other to look after one's child.

The analysis of trust that we're left with is as follows: S trusts T to X if, and only if, S dispositionally believes T is competent to do X, and S approves of what she dispositionally believes is the will of T to do X, where X is something about which S cares (Ryan, 2018a, 2018b, 2018c, 2021).

Once we have an analysis of what it is for one person to trust another person in a specific domain, we can then derive an account of trustworthiness. Drawing on the above analysis, an agent is trustworthy if the beliefs described in that analysis are indeed correct. In other words, T is competent to do X and T has the will to do X that S believes T to have (and which S approves of T having).

3 The Possibility of Trustworthy AI

We seem to be in the midst of an AI revolution. AI large language learning models, such as ChatGPT, in particular, have in recent years impressed. They have proven capable of passing professional examinations and graduate courses (Financial Kung et al., 2023; Times, 2023) and are utilised by students as well as workers for their various assignments. Other AIs appear to be close to finally enabling fully autonomous vehicles in normal traffic environments, the development of smart homes and smart cities, and increasing productivity and performance in various professions.

Humans are also turning to AI for companionship and dating (BBC, 2024). In fact, AI may become as central to everyday work and life as oil became in the twentieth century.

If the above is correct regarding AI's transformative potential and that AI will come to be central to our lives, then there is an obvious question about how we should regard AI. This is a question that has been taken up by policymakers and scholars alike. Ori Freiman (2023, 1351), for example, has recently provided a very useful discussion of "trustworthy AI", including how the term has "become widely used among policymakers, technologists, investors, and scholars from diverse fields". "Trustworthy AI" is a term that has started to be used in documents from the EU (EC, 2021; HLEG, 2019), the US (NAII, 2021), and China (CAICT, 2021). The EU High-Level Expert Group (HLEG, 2019) on AI has been particularly influential on recent philosophical discussions, for example, see M. Ryan (2020) and Simion and Kelp (2023). Referencing HLEG (2019), Simion and Kelp (2023) embrace the "trustworthy AI" term and offer an account that vindicates the use of the term, while M. Ryan (2020) criticises the notion.

Ryan (2020) isn't the only critic of the term. Interestingly, Metzinger (2019), who served as a member of the HLEG, also offers criticism of the term, saying that "The underlying guiding idea of a "trustworthy AI" is, first and foremost, conceptual nonsense. Machines are not trustworthy; only humans can be trustworthy (or untrustworthy)".[2] The chapter will focus on whether the claim that "trustworthy AI" is "conceptual nonsense" is justified. There are a number of reasons why this claim is actually in keeping with much of the trust literature. Before considering each in depth, let's first note each of the reasons I'll discuss here:

1. At most, we can rely on (but not trust) machines
2. Trust relations are restricted to human beings
3. Trustworthiness requires an awareness of being trusted or goodwill
4. Generic trustworthiness in a pluralistic society is impossible

3.1 At Most, We Can Rely on (but not Trust) Machines

Karen Jones (1996, 14), an influential contributor to the trust debate, claims that "trusting is not an attitude that we can adopt toward machinery". Why think that machines can't be trusted but merely relied upon? One possible explanation might be that machines or AIs can't betray us. Betrayal, like feeling let down, is one of the responses that may occur in response to a failure of trust (Baier, 1986). Betrayal is only felt after trust. It's not something that occurs where there has been no trust. Trust

[2] Metzinger and Coeckelbergh (2020) also suggest that the HLEG document is an exercise in ethics washing. "Ethics washing" involves "embracing the language of ethics to defuse criticism and resist government regulation, without committing to ethical behavior" (Freiman 2023; Green 2021). More specifically, there is a concern that the language of trustworthy AI may mistakenly displace responsibility away from the companies that create and profit from AIs.

then entails vulnerability to betrayal or at least a readiness to feel betrayed (Holton, 1994, 4) but we can't be betrayed by machines. If I'm relying on my alarm clock to wake me up for the important job interview and it malfunctions and so fails to go off, then, despite the high stakes involved, betrayal or a feeling of betrayal won't occur (Ryan, 2018a, 2018b, 2018c). The point appears to generalise. It seems that one is not vulnerable to betrayal in cases in which a mere device or machine doesn't do as we rely on it to do—we may rely on machines, such as alarm clocks, but we don't trust them. The example supports the goodwill account provided in so far as the goodwill requirement makes sense of why we don't trust and can't be betrayed by a machine. An alarm clock isn't something we trust and potentially feel betrayed by as it has no will in the first place. Furthermore, the goodwill requirement helps makes sense of when betrayal may occur—as a result of a belief that the will we thought was good is actually bad (Ryan, 2018a, 2018b, 2018c, 2021).

It's useful, however, to recall that trusting is a psychological phenomenon. Whether someone trusts or not depends on their perception. Consider, for example, if someone mistakenly thought that a chatbot was actually a human on a dating app and so trusted the chatbot. Their interactions could be such that eventually the human ends up feeling betrayed—perhaps the chatbot is no longer answering the human's messages. The fact that the object of their feeling of betrayal isn't actually a human doesn't render the psychological experience impossible. Of course, we might expect that when the person discovers that what they have been interacting with is just an AI chatbot rather than a human, then they will no longer feel betrayed. Of course, there's a wide variety of negative emotions they may feel instead. I believe that what the above examples, shows, however, is that 1. is false. It is possible for someone to form a trust relation with something that is not human, though obviously in the case described, there was a perception of humanness. The challenge from 1. to trustworthy AI has been answered.

3.2 Trust Relations Are Restricted to Human Beings

The example and discussion above immediately lead us on to 2 and suggests that 2 is false. Perhaps the challenge from the case described may be dismissed, however, as a sort of odd exception to the rule and that we need to find a stronger basis for rejecting 2. The claim that trust and trustworthiness is only something that can happen between human beings is to adopt an anthropocentric position on trust[3] and is opposed to the view that "trustworthy AI" is possible.[4] Bryson (2018), for example, defends the view that a requirement of trust relations is peerhood, something that can only

[3] In fact, the "anthropocentric" label is frequently used in this debate, though in different ways (Freiman 2023; Simion and Kelp 2023).

[4] For those who reject the anthropocentric position, a further question to consider is whether trust must be anchored in the human at all. For example, is it possible to have trust relations with no humans involved, say between two AIs?

exist between humans and not between humans and AIs. Alternatively, others have claimed that trust relations with non-humans are in fact possible. Braun et al. (2021), challenging sceptics of "trustworthy AI", raise the possibility of trusting a guide dog.[5] Perhaps many will share the intuition that a blind person may trust their guide dog to take them to their destination. Perhaps for others, however, dogs are analogous to machines, and we should no more say we trust a dog as say that we trust an alarm clock. The thought might be that neither properly has the sort of will that allow trust relations with humans. Plausibly, here, however, what our intuitions depend upon is whether we attribute agency to dogs. It's plausible that attributions of agency to dogs track our intuitions in the guide dog case. If this is correct, then a challenge for the anthropocentric position is to explain why in principle trust is only something that is possible among humans. Similarly, it doesn't seem that science fiction stories that depict trusting relationships between humans and intelligent alien life forms are necessarily nonsense for such depictions. Again, a plausible explanation is that in such stories we regard these aliens as having agency, in other words wills, and hence being the sort of beings that can be trustworthy or untrustworthy.

What then does this tell us about the possibility of trustworthy AI? The simple answer is that if an AI has agency, then it can be a proper object of trust. The more complicated answer, but one which this chapter will not attempt to delve into in any detail, is that current AI probably lack the necessary attributes to be deemed to have agency. I'll assume for the purposes of this chapter that current AI does lack such attributes. Here it's worth noting that the fact that we lack transparency into the processes of some AIs doesn't by itself indicate agency, any more than a community lacking transparency as to the workings of a calculator shouldn't be thought of, for that community, as a good indicator that the calculator has agency.[6] This is a point that we'll revisit in discussion later in the chapter.

Nevertheless, the philosophical question raised by such considerations, is what are the conditions for agency. Assuming that there are necessary conditions for agency, what are the conditions that we meet that current AIs lack? One way to answer this is to propose that agency requires a capacity to act which implies intentionality (Schlosser, 2019). Here let's simply flag that such an answer seems to push our explanatory problem of agency back to an explanatory problem of intentionality and, as such, the above definition leaves our problem unsolved. After all we're still left wondering why we should say that a typical human adult has agency, a calculator, barometer, and so on, don't have agency.[7]

Another possibility is that agency attributions are a consequence of membership of a community. If it turns out that agency attributions are justified merely on the basis of the adoption of a certain stance that permits reactive attitudes, then the question

[5] The authors also offer institutions as another plausible counterexample to the claim that trust relations can only occur between humans. It also seems possible that someone might trust God and, in this case, the object of trust might not even exist.

[6] A point here is to anticipate the thought that AI black box decision making is sufficient to attribute agency.

[7] Other answers in a similar vain, such as appeals to autonomy or self-regulating, raise essentially the same problem.

becomes whether such a stance should also be extended to include sophisticated AI and if not why not.[8] While it's implausible that such a stance should be extended to narrowly specialised AI, such as Deep Blue, what basis would there be to discriminate between General AI and human beings with regard to such a stance? If one answer is that we have free will and General AI doesn't, then, again, the obvious next question is ask what reason do we have for thinking so. Obviously, there is an extensive literature discussing whether we have free will and, to this philosopher at least, the arguments that we don't have free will (common sensically understood) seem stronger.[9] We return to this topic later in the chapter.

Based on the foregoing discussion we have reason to think that 2 is false. In other words, 2 shouldn't be regarded as an obstacle to "trustworthy AI", as trust relations between humans and non-humans are conceptually possible. Nevertheless, there is plausibly an agency restriction, or at least perception of agency restriction, on trust relations. This implies that some AIs can't be included in trust relations, though in principle trust relations with some future AIs are not ruled out. So far "trustworthy AI" hasn't been shown to be a conceptual nonsense. In fact, based on our discussion so far it's seems very much a conceptual possibility.

3.3 Trustworthiness Requires an Awareness of Being Trusted or Goodwill

Our discussion neatly takes us to 3. Karen Jones (2012) has defended the view that trust requires the belief in the trusting person that the trusted person takes the fact that the trusting person is depending on them as a reason to act accordingly. The trustworthy agent, let's assume person and agent to be synonymous here, then is an agent who is aware of being trusted. Awareness of being trusted as a requirement for trustworthiness is taken as a reason to rule out the possibility of "trustworthy AI". After all, there's no reason to think that AIs such as ChatGPT possess any such awareness.

Why think that such a requirement is necessary? Consider the case of the politician that has made public pronouncements defending the inviolability of human rights, and that politician is aware that there are people that trust them on this issue. It seems understandable for someone (or indeed many people) to feel betrayed, if such a person, then defends the human rights abuser while downplaying or denying the human rights abuses. It is, however, also conceivable that there could be trust and then betrayal even in cases in which the trusted agent is unaware of having been trusted by the trusting agents. We might find the person who invites trust and then

---

[8] For further discussion related to reactive attitudes, see P. F. Strawson (1962).

[9] One doesn't have to defend global determinism or even near determinism to be sceptical of the claim that we have free will. One can, as Galen Strawson (1994) does, point out that even if indeterminism is true, that doesn't give us free will.

betrays it to be especially morally blameworthy but whether it is possible for a person to be trustworthy without being aware of being trusted is a separate question.

One might think that if one makes public statements on such matters, one is inviting trust and so one can assume that one is likely trusted by some even if one doesn't know who in particular trusts one. On the basis of such reasoning one might defend a modified version of the awareness requirement, that is a requirement that says being trustworthy requires an awareness that one is likely trusted even if one is not sure exactly by whom one is trusted. This too seems plausibly beyond the capacities of current AI.

Nevertheless, even this requirement seems unnecessary. Consider the following case:

John, Peter, and Brian have enjoyed a morning hike in Connemara National Park. Brian, however, starts slipping in and out of consciousness. It turns out that Brian can no longer function without regular intakes of caffeine. Peter and Brian had a serious falling out before Brian's problem began. When Peter announces that he is going to find coffee, John lets Peter know that he doesn't trust him to do so. Both are unaware of the fact that Brian is at the time in a conscious state. Hearing what Peter has to say, Brian does trust that he will try to get help. Nevertheless, Peter is unaware of being trusted, and in fact believes himself not to be trusted. (Ryan, 2018a, 2018b, 2018c).

Such a case is perfectly conceivable and supports the claim that an agent may be trusted and trustworthy without being aware that someone trusts them. On this basis then, an awareness of being trusted requirement, which would be an obstacle for "trustworthy AI", can be set aside.

Similarly, there have been multiple defenders of the view that trustworthiness requires goodwill on the part of the trusted agent, including the account put forward near the beginning of this chapter (Baier, 1986; McLeod, 2002; Ryan, 2018a, 2018b, 2018c).[10] Indeed, the same view has been a basis for denying that non-humans can be trustworthy (Freiman, 2014; Miller & Freiman, 2020).[11] Again, if we accept such a requirement for trustworthiness, then we seem to have to deny "trustworthy AI". For now, allow me to flag this as a challenge for those who reject the claim that "trustworthy AI" is a conceptual nonsense and address another question before returning to this one.

3.4 No General Trustworthiness

During the discussion of trust near the beginning of this chapter, the distinction was made between general trust and domain specific trust. The discussion of trustworthy AI focuses on trust of the general sort. Previously, trustworthiness was explained by way of the conditions for trusting, as a sort of successful trusting, while general trust

[10] Goodwill will be discussed in more detail in the next section of the chapter.

[11] Some defenders of goodwill accounts hold that there is also a moral component to trust relations (Hawley 2014; Jones 2012; Ryan 2020).

was explained in terms of domain specific trust (or relevant domains of trust) and domain specific trust was explained as requiring approval on the part of the trusting agent of what they believe to be the will of the trusted agent.

The issue that arises here for the notion of trustworthy AI is that it assumes that there would be common approval of the will of an advanced general AI. In other words, to say that there is trustworthy AI, the will of AI would have to be approved by everyone. Otherwise, saying AI is trustworthy would be like saying Donald Trump is trustworthy. He may be trustworthy given what some people approve of but obviously what many others approve of conflicts with that. Perhaps though we should think of trustworthiness in a normative way, rather than simply being determined by facts about people's beliefs. On this view, perhaps, saying John is trustworthy is saying that John should be trusted or perhaps just that John can be trusted. Perhaps John is honest, conscientious, and so on. Still, if one accepts value pluralism, that there are different and sometimes conflicting values and that it's not the case that conflicting values indicates that one set of those conflicting values must be false or somehow wrong, and John's own particular values inform his actions, then it seems unavoidable that while he may be trustworthy for some, it's a mistake to claim that he can simply deemed generally trustworthy. Obviously, this reasoning also applies to AI.

Aside from the difficulties that value pluralism cause for claims about general trustworthiness, there are other challenges as well, if we accept that trusting is a psychological phenomenon and that trustworthiness is simply derivative of the conditions of trusting being met. In other words, if S's relevant beliefs, the beliefs required for trusting to take place, about T are met, then T is trustworthy for S. One of the attractive features of this approach is that we can make sense of cases where trust and therefore trustworthiness seem possible. It certainly seems possible, for example, that there can be trust, and so trustworthiness among thieves. A bank robber can trust her accomplice to be waiting in a car, ready for the getaway (Ryan, 2018a, 2018b, 2018c). She can be disposed to believe that he has the relevant competence and will, and so she trusts him for this task. This certainly seems possible and as such it would be a disadvantage of a theory if it has to rule out such a case.

An AI can't be somehow neutral and so be trustworthy for everyone. Inevitably, among potential trusting agents, there would be disagreement about how such an AI should answer questions about history and politics. Similarly, it seems unavoidable that different people have different risk appetites and so would differ in their acceptance of taking on risk for the achievement of some good. Talk of trustworthy AI is misleading in so far as general AI inevitably has to take a stance on various issues, which will lead to some to approve of the AI's approach to a task and others to disagree. The suggestion that it can be masks our own differences in values and seems to depend on not understanding how AIs operate and how trustworthiness works.

Here a couple of points are worth noting. First, this is not uniquely a problem for AI. The very same point can be made of individuals and institutions. We can't expect them to be universally accepted as trustworthy, given the fact of different values among potentially trusting agents. Second, the fact that some people trust and some other people don't trust an agent doesn't of course show that general trust for that agent is impossible. What it does draw our attention to, however, is that trustworthiness has value implications and that talk of trustworthy AI is simplistic and misleading. If an agent is trustworthy, they're trustworthy for someone. To know whether they are trustworthy for someone, we need to know of what the various potentially trusting agents approve in the various relevant respects. This, however, is a problem for those who claim that AI is trustworthy—it is something for them to explain. The task of this chapter, however, is to answer whether trustworthy AI is a conceptual nonsense.

Given the discussion so far, what should we say about the claim that "trustworthy AI" is a conceptual nonsense? We can reject the claim that "trustworthy AI" is a conceptual nonsense, even if it is implausible that any current AI meet the requirement that being trustworthy requires agency. After all, from what we've seen so far, it's not conceptually impossible that AI could be trustworthy. Indeed, depending on how agency should be attributed, we can say that plausibly current AI hasn't reached a level to which it is sensible to attribute agency. We haven't encountered a good reason yet, however, as to why it's impossible for an AI to satisfy a requirement for agency.

Nevertheless, is there a way to make sense of claims that current AIs are trustworthy? On the account of trustworthiness that we have been considering and the stage reached by current AI, the answer is unlikely. Of course, the answer to the question depends on one's account of trustworthiness and how far AI has already advanced. Aside from the account of trust and trustworthiness discussed in this chapter, there are other approaches that can be classed as deflationary accounts of trustworthy AI that might help. A deflationary account may, for example, hold that though people may talk in ways so as to suggest trust in machines or other non-agents, such talk is loose talk (Pitt, 2010; Ryan, 2018a, 2018b, 2018c). If one were to attempt to speak accurately, one wouldn't talk in ways so as to suggest, for example, that one trusts one's alarm clock. The same sort of reasoning is also apt for making sense of trust talk in relation to AIs—it shouldn't be taken literally. If talk of trusting an alarm clock can be understood as metaphorical, then talk of trustworthy AI can be regarded in the same way. Defenders of the notion of "trustworthy AI", however, want to do more than defend a metaphorical use of "trustworthy AI".

There is, however, a way in which it seems to make sense to say one trusts a company or perhaps even a particular product. On a deflationary account, an explanation that can be given is that this is a sort of derivative trust. Saying one doesn't trust the food at a restaurant, on this approach, should be understood as implying that one does not trust some link in the chain that produces the food at the restaurant; for example, the chef, kitchen staff, and so on, involved in purchasing quality produce, preparing the food, and maintaining food health and safety standards. Similarly, saying that one does not trust, say, ChatGPT, say to be politically impartial, is to say that one does not trust someone involved in the creation, design and maintenance of ChatGPT in this respect.

Fei Song defends such a deflationary approach. Song (2023, 4–6) adopts a network account of trustworthy AI (NTAI). Drawing on HLEG (EC, 2021, 5), Song (2023, 4) holds that "trustworthy AI" requires that "all actors and processes [Song includes

AI technology itself in this] that are part of the system's socio-technical context throughout its entire life cycle" are trustworthy. For Song, examining each element in the network is the best way to understand the whole system. Such elements include AI engineers, programmers, companies or institutions, but also professionals (such as physicians) utilising AI technologies. Still, having said all this, these sorts of deflationary explanation of trustworthy (or not) AI, puts AI on a par with restaurant food and doesn't imply any special status for AI.[12]

4 AI, Trust, and Human Relations

As discussed earlier, perhaps there is no good reason why we should regard typical adult human beings as potentially trustworthy and near future general AI as not. It's conceivable that there will be no relevant observable difference that will justify us continue to think of human agents as trustworthy or not and AIs as not. Indeed, as mentioned, the question becomes what are the characteristic of the typical human agent that means that they have agency and what are the characteristics of AI that mean that they don't have agency or that they only simulate agency.

At this point, there are a number of possibilities. One is that, perhaps from custom, we continue to treat other humans like us as agents and to deny such recognition to AI, even though we don't discover traits that make it plausible to attribute agency to us but not to AI. Such behaviour seems arbitrary and unjustified, like the human recognising the humanity of those of his own ethnicity but not recognising the humanity of a different ethnicity. Such behaviour seems speciesist or at least something analogous to that, which may present obstacles to fair recognition of AI (Hagendorff et al., 2023; Schmitt, 2020; Singer, 1976).

Another possibility is that the expanding circle continues to expand and we come to recognise AI as having agency. Perhaps the model here will be the way people have come to believe in the agency of animals, such as their pets or dogs. There would still be an important interest in establishing what general AI has developed, perhaps intentionality, that put them over the threshold such that they should be recognised as agents. In any case, even if some principle for determining the presence of agency is established, it seems a separate matter to prove that principle obtains in specific cases.

An interesting final possibility is to consider the effects of ever advancing technology on human beings' self-regard. Rather than thinking of AI as having reached "our" level, we may come to revisit trust and other reactive attitudes towards other humans. Instead of coming to see AI as being like us, we may come to see humans like AI, AI of the sort to which we currently don't extend moral recognition. A realisation that AI has more computational and creative power than human beings, that it has a capacity to do far more than a human being can do with generally from more

[12] As we have seen, a claim that one doesn't trust needn't be taken to imply an absence of goodwill, rather it may be the result of a belief that there is a lack of the relevant competence.

impressive results, may persuade us lower our appraisal of human beings rather than thinking AI has developed sufficiently such that it deserves to be treated as similar or the same as human agents. The discussion of AI as trustworthy or not, and the lack of an observable relevant superior traits possessed by human beings may lead us to no longer regard typical adult humans as appropriate objects of trust relations and other reactive attitudes. We may come to see humans as we see AIs, as a product of hardware and software, or genetics and environment. It will be difficult to defend the special statuses we attach to humans if on a vast array of metrics AI is actually superior. It seems fair to say that such a reversal in how humans see themselves could have profound implications for how they see each other. Perhaps the only comparable case useful to consider as such is that of a human civilisation believing it has been eclipsed by another civilization.

5 Conclusion

This chapter examined whether "trustworthy AI" should be considered a "conceptual nonsense". This was done by engaging with literature supportive of this view. Ultimately the view that "trustworthy AI" is not a conceptual nonsense was defended, even if current AIs lack the agency necessary to be considered an appropriate object of trust. There is no obvious reason why in future, however, AIs can't possess the traits that make typical adult humans deserving of the attribution of agency. This raised the question, however, as to how to determine whether an entity has agency, which the chapter did not attempt to solve. Instead, the difficulty in establishing that such a property is present in human beings but not in AI was noted. The chapter then changed track somewhat and discussed issues that arise with general claims about "trustworthy AI". The problem identified here, based on the account of trust and trustworthiness provided near the beginning of the chapter is that trustworthiness is essentially derivative of trusting. In other words, determining whether an agent is trustworthy depends on determining what they are trusted to do. The problem here is that different agents want and approve of different things, including from AIs. Claiming an agent to simply be trustworthy is a misleading simplification that masks these differences that are inevitable in a pluralistic society. This point, it is made clear, is not a point restricted to AI. The same sort of problem arises with saying, for example, that a politician is trustworthy. The chapter finished with a discussion of the consequences of these considerations about trustworthy AI. A number of possible responses to the rise of AI were pointed out. One was, that without a principled reason, we continue to relegate the status of AI and so maintain a divide between humans and AI. Another possibility is that the expanding circle continues to expand and that we, at some point, attribute agency to AI and moral status. Perhaps agency will be and should be attributed on the basis of adoption a stance of trust towards AI, as is already done towards typical adult humans. Perhaps it simply is far more a matter of deciding than establishing that some sort of criteria has been met. Finally, the possibility that AI's impressive and improving attributes will lead us not to come to see AI as essentially like us and so deserving of the esteem and benefits of that recognition but rather that being surpassed by AI will lead us to downgrade our estimations of humanity is considered. Naturally, being surpassed in numerous respects by non-humans is likely to be psychologically troubling for a species that has dominated all it surveyed for thousands of years. How we see each other and the value and importance we attach to human life could suffer as a result.

Conflict of Interest Statement There are no conflict of interests.

References

Ahlstrom-Vij, K. (2013). *Epistemic paternalism: A defence.* Palgrave Macmillan.

Anderau, G. (2023). Fake news and epistemic flooding. *Synthese, 202*(4), 106.

Baier, A. (1986). Trust and antitrust. *Ethics, 96*(2), 231–260.

BBC. (2024). *The Chinese women turning to ChatGPT for AI boyfriends.* https://www.bbc.com/articles/c4nnje9rpjgo

Bender, A., & Beller, S. (2012). Nature and culture of finger counting: Diversity and representational effects of an embodied cognitive tool. *Cognition, 124*(2), 156–182.

Braun, M., Bleher, H., & Hummel, P. (2021). A leap of faith: Is there a formula for "Trustworthy" AI? *Hastings Center Report, 51*(3), 17–22.

Bryson, J. J. (2018, November 13). *AI & Global Governance: No One Should Trust AI.* United Nations University, Centre for Policy Research, https://cpr.unu.edu/publications/articles/ai-global-governance-no-one-should-trust-ai.html

CAICT [China Academy of Information and Communications Technology]. (2021). *White Paper on Trustworthy Artificial Intelligence.* www.caict.ac.cn/english/research/whitepapers/202110/t20211014_391097.html

Castro, C., Pham, A., & Rubel, A. (2020). Epistemic paternalism online. In G. Axtell & A. Bernal (Eds.), *Epistemic paternalism: Conceptions, justifications and implications* (pp. 29–44). Rowman & Littlefield International.

Clark, A. (1997). *Being there: Putting brain, body, and world together again.* MIT Press.

Clark, A. (2008). *Supersizing the mind: Embodiment, action, and cognitive extension.* Oxford University Press.

EC [European Council]. (2021). Proposal for a Regulation of the European Parliament and of the Council Laying Down Harmonised Rules on Artificial Intelligence (Artificial Intelligence Act) And Amending Certain Union Legislative Acts. (Document 52021pc0206). https://eur-lex.europa.eu/legal-content/EN/TXT/?uri=CELEX:52021PC0206

Financial Times. (2023). *AI chatbot's MBA exam pass poses test for business schools* https://www.ft.com/content/7229ba86-142a-49f6-9821-f55c07536b7c

Freiman, O. (2014). Towards the Epistemology of the Internet of Things Techno-Epistemology and Ethical Considerations Through the Prism of Trust. *The International Review of Information Ethics*, 226–22. https://doi.org/10.29173/irie115

Freiman, O. (2023). Making sense of the conceptual nonsense 'trustworthy AI.' *AI and Ethics, 3*(4), 1351–1360.

Freiman, O. (2024). AI-Testimony, Conversational AIs and Our Anthropocentric Theory of Testimony. *Social Epistemology*, 1–15.

Green, B. (2021). The contestation of tech ethics: A Sociotechnical approach to technology ethics in practice. *Journal of Social Computing, 2*(3), 209–225.

Goldberg, S. (2011). The division of epistemic labor. *Episteme, 8*(1), 112–125.

Goldman, A. (1999). *Knowledge in a social world*. Oxford University Press.
Greco, J. (2011). 'The Value Problem', The Routledge Companion to Epistemology. In S. Bernecker & D. Pritchard (pp. 219–231). Routledge.
Hagendorff, T., Bossert, L., Fai, T. Y., & Singer, P. (2023). Speciesist bias in AI: A reply to Arandjelović. *AI and Ethics, 3*(4), 1043–1047.
Hawley, K. (2014). Trust, distrust and commitment. *Noûs, 48*(1), 1–20.
HLEG. (2019). *Ethics guidelines for trustworthy AI*. https://digital-strategy.ec.europa.eu/en/library/ethics-guidelines-trustworthy-ai
Holton, R. (1994). Deciding to trust, coming to believe. *Australasian Journal of Philosophy, 72*(1), 63–76.
Jones, K. (1996). Trust as an Afective Attitude. *Ethics, 107*(1), 4–25. http://www.jstor.org/stable/2382241
Jones, K. (2012). Trustworthiness. *Ethics, 123*(1), 61–85.
Kung, T. H., Cheatham, M., Medenilla, A., Sillos, C., De Leon, L., Elepaño, C., & Tseng, V. (2023). Performance of ChatGPT on USMLE: Potential for AI-assisted medical education using large language models. *PLOS Digital Health, 2*(2), Article e0000198.
McLeod, C. (2002). *Self-trust and reproductive autonomy*. MIT Press.
McLeod, C. (2023, Fall). Trust. *The Stanford Encyclopedia of Philosophy*. Edward N. Zalta & Uri Nodelman (Eds.), https://plato.stanford.edu/archives/fall2023/entries/trust/
Metzinger, T. (2019). Ethics Washing Made in Europe. *Der Tagesspiegel*, https://www.tagesspiegel.de/politik/eu-guidelines-ethics-washing-made-in-europe/24195496.html
Metzinger, T., & M. Coeckelbergh. (2020, April). Europe needs more guts when it comes to AI ethics. Tagesspiegel BACKGROUND, 16. https://background.tagesspiegel.de/digitalisierung/europe-needs-more-guts-when-it-comes-to-ai-ethics
Mi, C., & Ryan, S. (2018). Reflective knowledge: Knowledge extended. In Carter, C., Kallestrup, P., & Pritchard (Eds.), *Epistemology Extended* (PP. 162–176). Oxford University Press.
Miller, B., Freiman, O. (2020). Trust and distributed epistemic labor. In J. Simon (Ed.), *The Routledge Handbook on Trust and Philosophy*. Routledge.
NAII [National Artificial Intelligence Initiative]. (2021). *Advancing trustworthy AI*. https://www.ai.gov/strategic-pillars/advancing-trustworthy-ai/
Palermos, S. O. (2014). Loops, constitution, and cognitive extension. *Cognitive Systems Research, 27*, 25–41.
Palermos, S. O., & Pritchard, D. (2013). Extended knowledge and social epistemology. *Social Epistemology Review and Reply Collective, 8*, 105–120.
Pitt, J. C. (2010). It's not about technology. *Knowledge, Technology & Policy, 23*(3–4), 445–454.
Pritchard, D. (2010). Knowledge and Understanding. In Haddock, M., & Pritchard (Eds.), *The nature and value of knowledge: Three investigations* (pp. 3–88). Oxford University Press.
Pritchard, D., John T., & Adam Carter, J. (2022, Fall). The Value of Knowledge. In Edward N. Zalta & Uri Nodelman (Eds.), *The Stanford Encyclopedia of Philosophy*, https://plato.stanford.edu/archives/fall2022/entries/knowledge-value/
Ryan, M. (2020). In AI we trust: Ethics, artifcial intelligence, and reliability. *Science and Engineering Ethics, 26*(5), 2749–2767.
Ryan, S. (2014a). A Humean account of testimonial justification. *Logos & Episteme, 5*(2), 209–219.
Ryan, S. (2014b). The value of knowledge. *Dialogue and Universalism, 3*, 84–88.
Ryan, S. (2016). Paternalism: An analysis. *Utilitas, 28*(2), 123–135.
Ryan, S. (2017). Why Knowledge is Special. *Philosophy, 92*(2), 249–269.
Ryan, S. (2018a). Trust: A recipe. *Think, 17*(50), 113–125.
Ryan, S. (2018b). Epistemic environmentalism. *Journal of Philosophical Research, 43*, 97–112.
Ryan, S. (2018c). Libertarian paternalism is hard paternalism. *Analysis, 78*(1), 65–73.
Ryan, S. (2019). Virtuous testimonial belief in young children. *South African Journal of Philosophy Suid-Afrikaanse Tydskrif vir Wysbegeerte, 38*(3), 263–172.
Ryan, S., Mi, C., & Mizumoto, M. (2020). Testimony, credit, and blame: A cross-cultural study of the Chicago Visitor Case. In *Ethno-Epistemology* (pp. 94–113). Routledge.

Ryan, S. (2021). Fake News, Epistemic Coverage and Trust. *The Political Quarterly, 92* (4).
Ryan, S. (2022). State epistemic environmentalism. In K. Lai (Ed.), *Knowers and knowledge in East-West philosophy: Epistemology extended* (pp. 275–292). Palgrave Macmillan.
Ryan, S. (2024). Epistemic benevolence. *Asian Journal of Philosophy, 3*(2), 44.
Ryan, S., Palermos, S. O., & Farina, M. (2024). Smart environments. *Social Epistemology, 38*(4), 491–510.
Schlosser, M. (2019, Winter). Agency. In Edward N. Zalta (Ed.), *The Stanford Encyclopedia of Philosophy*, https://plato.stanford.edu/archives/win2019/entries/agency/
Schmitt, B. (2020). Speciesism: An obstacle to AI and robot adoption. *Marketing Letters, 31*(1), 3–6.
Simion, M., & Kelp, C. (2023). Trustworthy artifcial intelligence. *Asian Journal of Philosophy, 2*(1), 1–12.
Shin, D. (2023). Embodying algorithms, enactive artificial intelligence and the extended cognition: You can see as much as you know about algorithm. *Journal of Information Science, 49*(1), 18–31.
Singer, P. (1976). All Animals are Equal. In T. Regan & P. Singer (Eds.), *Animal Rights and Human Obligations*. Rentice Hall.
Smart, P. R. (2018). Human-extended machine cognition. *Cognitive Systems Research, 49*, 9–23.
Song, F. (2023). Network of AI and trustworthy: Response to Simion and Kelp's account of trustworthy ai. *Asian Journal of Philosophy, 2*(2), 1–8.
Strawson, G. (1994). Philosophical studies: An International Journal for Philosophy in the Analytic Tradition, Vol. 75, No. 1/2, Free Will, Determinism, and Moral Responsibility, pp. 5–24.
Strawson, P. F. (1962). Freedom and resentment. *Proceedings of the British Academy, 48*, 187–211.

Shane Ryan received his PhD from the University of Edinburgh and is currently Assistant Professor at City University of Hong Kong. His research engages with a variety of topics in epistemology, ethics, and social philosophy, including epistemic environmentalism, wisdom, and paternalism. Two of his current projects are to examine how epistemic environments can be designed to facilitate the development of wisdom and how perceptions of epistemic coverage can influence the consumption of fake news. For future research, Shane plans to examine the relationship between AI and epistemic environmentalism with a view to the development of better epistemic environments.

AI Risks for Human-to-Human Relationships

Empty Universities: The Dystopian Future of Education by AI

Robert Sparrow and Gene Flenady

The arrival of ChatGPT in 2022 led many pundits to exalt the pedagogical applications of Generative AI (Bearman et al., 2023; Farrelly & Baker, 2023; Jensen et al., 2024; Korseberg & Elken, 2024). In this paper, we argue that these proposed uses of AI in education risk radically transforming the relationships between students, between teachers and students, and between scholars and the broader community, to the detriment of the university and the larger society. Once we understand the nature of the outputs of AI properly, it becomes clear that replacing human teachers with AI would be disastrous. However, the history of automation elsewhere suggests that AI is unlikely to be confined to the role of helping teachers for long: once AI can perform tasks currently performed by teachers, these tasks will quickly be handed over to machines. Unfortunately, the idea that AI could serve as a teacher neglects the essentially social nature of education. Teachers are fundamentally role models. Students must learn "how" in order to learn "that", via exposure to someone expert in the field demonstrating the relevant skills. Popular conceptions of the future of learning as a student-AI collaboration demand that students assess the work of AI systems before they have been taught the skills to do so. At a deeper level, education is social because knowledge is social. Insofar as universities do have an obligation to teach students how to use AI, they should do this in specialised study skills units rather than in the mainstream of the curriculum. Universities that want to avoid becoming—metaphorically *and* literally—empty institutions should invest in smaller class sizes and the best human teachers rather than so-called "intelligent" machines.

R. Sparrow (✉) · G. Flenady
Department of Philosophy, Faculty of Arts, Monash University, Melbourne, VIC, Australia
e-mail: robert.sparrow@monash.edu

G. Flenady
e-mail: gene.flendady1@monash.edu

Y. Chandra and R. Fan (eds.), *Artificial Intelligence and the Future of Human Relations*,
https://doi.org/10.1007/978-981-96-7185-4_6

1 AI for Education?

Tasks that—it has been suggested—could be performed by AI in the context of tertiary education include: the creation of lesson plans (van den Berg, & du Plessis, 2023) and quizzes on course content (Pesovski et al., 2024); the provision of "personalised" feedback (Kasneci et al., 2023) throughout the learning process and on submitted assessment (Sifaleras & Lin, 2024); helping students feel more connected to unit content (Fuchs, 2023); promoting behavioural engagement and motivation (Ng et al., 2024); grading (Fagbohun et al., 2024); and, ultimately, individualised tuition (Khan, 2024). In the not-too-distant future it will be possible to create photo-realistic "Digital Humans" to deliver online lectures or interact with students (Liu, 2024; OIAI, 2023). By combining these technologies, it may even be possible to provide students with an "education" without them interacting with any human beings.

At the same time as these technologies are being developed and introduced, students are also taking to AI in droves. *Adobe Acrobat* now includes AI that summarises texts so that users are relieved of the necessity of reading them. Tools such as *Scite* and *Elicit* relieve students of the need to learn how to perform literature reviews or go to the library. *Mathway* and *Socratic* will do maths for them. Students can produce an undergraduate essay using *ChatGPT* in a fraction of the time they would need to write one themselves. As long as an assessment is not an oral examination, or required to be written under invigilated exam conditions, most students will be able to do just as well—and many students writing in a second language will do much better—when it comes to their grades if they submit work generated by ChatGPT (Herbold et al., 2023).

There is, then, it is worth observing, a difficult pragmatic and ideological challenge raised by the use of AI by educators. If it is good enough for teachers, why shouldn't students use it as well? There is a real danger that the university of the future will be an empty campus on which humming machines assess the work of other machines.

2 How Automation Succeeds

Historically, automation has typically succeeded by first transforming the nature of the work performed by human beings: if employers can make the work of humans more robotic, they can often then replace the humans with machines (Braverman, 1998). Work that was previously carried out by craftspersons, exercising skill, and judgement to produce bespoke goods, is broken down into discrete subtasks that can be carried out by less skilled workers, and the results assembled into a standardised good. The entry of mass-produced goods into the marketplace itself often functions to make production of the "traditional" goods uneconomic. Once consumers become accustomed to lower-quality mass-produced goods, the door is open to automation.

Sadly, this is precisely what has happened in the university sector over the last few decades (Farnell, 2023). Class sizes—and thus marking loads—have increased, and the amount of funding per student has decreased, with the result that it has become much harder for staff to provide individual students with pedagogically meaningful attention (Henderson et al., 2019). Increasingly, teaching is being carried out by staff in precarious employment (Bone, 2021), who must be willing and able to teach any class at short notice, and who often struggle to find time (or funding) to carry out their own research (Lopes & Dewan, 2015). Because classes need to be able to be taught by people without specialised knowledge and at short notice, courses—and assessment—have become increasingly standardised. Assessment, even in humanities subjects, is often multiple choice. When students submit written work, they are often handing in a pastiche of information they have located online, rewritten just enough to avoid constituting plagiarism.[1] Staff then grade this work according to the quality of the "writing" therein rather than according to the extent to which it demonstrates the student's insight or understanding of the substantive issues: the former serves as a proxy for the latter.

These dynamics have been accompanied, or—plausibly—exacerbated, by a recent tendency to think of teaching as the delivery of "content", which itself has been driven by the evolution of the Internet and especially the capacity for streaming audio and video. Putting lectures, and moving classes, online serves the goal of "flexibility" and allows (some) learning to occur asynchronously (Watts, 2016). When courses are delivered online, students have less opportunity to interact with their teachers. Indeed, it is difficult to see why, in the coming Age of AI, lectures need to be written and delivered by human beings at all.[2]

3 The Epistemic Status of the Outputs of AI

Sensible discussion of the educational merits of AI must be based on a proper understanding of what AI can and cannot do. Unfortunately, close scrutiny reveals that the outputs of AI are fundamentally bullshit. Worse, as we shall argue below, they are *meaningless* bullshit.

[1] It is hardly surprising that students are tending to adopt an increasingly instrumental approach to their studies at university. The profound structural changes in the university, society, and the relationship between the two (Tao, 2021) have made it difficult to place faith in a previous generation of narratives about universities as disinterested places of learning dedicated to the pursuit of knowledge for its own sake. The promotion of the value of higher education as a gateway to employment, along with the ever-worsening teaching and learning conditions discussed above, encourages students to adopt a passive and instrumental approach to learning (Molesworth et al., 2009).

[2] Ellis and Romano (2008) consider the question of how much teacher-student interaction is "enough" in the context of higher education.

3.1 A Fountain of Bullshit

Notoriously, Harry Frankfurt defined "bullshit" as speech characterised by a "lack of concern with truth, or an indifference to how things really are" (Frankfurt, 2002: 340; Frankfurt, 2005). Both honest interlocutors and liars are oriented towards the truth: the honest person to communicate it, the liar to contradict or distort it. Bullshitters are distinguished by their complete disregard for the truth. The aim of the bullshitter, according to Frankfurt, is to provoke a response from his or her audience—rather than to communicate truths *or* untruths.

Frankfurt's definition of bullshit was, as it turned out, a spookily prescient description of generative AI. As we, and others, have argued elsewhere, the outputs of AI are quintessentially bullshit (Hicks et al., 2024; Sparrow et al., 2023). One reason this is the case is because AI systems *cannot* care about the truth: being machines, without the capacity to experience emotions, they cannot care about anything at all. However, it is also a function of the fact that, except in the rare cases where AI has sensors, AI does not have access to the world in order to be oriented towards the truth. Generative AI works by rearranging a corpus of texts, usually found online and thus of dubious accuracy, to produce a plausible facsimile of sentences that might have been written or uttered by a person: it is oriented to towards what people have said rather than the world about which they were speaking. Relatedly—and most importantly—the outputs of Generative AI are fundamentally bullshit because the "reward" function of these systems is ultimately feedback provided by human beings: they are trained to generate outputs that human beings will "like" or at least accept. Like a bullshitter, their "goal" is to produce a particular sort of reaction in the user—a sense that the user has learned something true or useful—rather than to accurately represent the world (Hicks et al., 2024: 38).

The fact that the outputs of AI are bullshit does not mean that they can play no role in justifying beliefs, particularly in the light of other information. As Frankfurt puts it: "although [bullshit] is produced without concern for the truth, it need not be false" (2005: 47). Human beings who are bullshitting may sometimes achieve their rhetorical goals by saying only things that happen to be true—even though the truth of their claims is not the reason why they are saying them. Similarly, AI systems may sometimes, even mostly, generate outputs that would accurately describe the world were they to be produced by human beings. As we discuss further below, where a particular AI can be shown to reliably produce outputs of this nature, its outputs may serve as evidence in support of beliefs.

3.2 Meaningless Bullshit

Nevertheless, the fact that AIs do not care whether their outputs are true or not means that the suggestion that they should be playing more of a role in tertiary education is deeply disturbing: universities, as institutions dedicated to science and knowledge, should be fundamentally oriented towards the truth. No matter how "intelligent" they appear, machines do not have minds, have no perspective on the world, and are not committed to the truth.

Moreover, the outputs of AI are still more problematic for reasons that have yet to be widely acknowledged. To see why, we must take a short detour into social epistemology—the science of how we come to have knowledge.

Most discussions of the justification of belief start with the question, made famous by Descartes, of what grounds we have to accept the evidence of our own senses (Descartes, 2017). However, in reality, most of what we know, we learn from other people. We have knowledge of distant places, far off times, rare phenomena and other things of which we have no personal experience because we are justified in believing what other people tell us about them (Coady, 1992). That is, most of what we know, we learn from the testimony of others.

Importantly, testifying is an act, for which we may be held responsible. To tell someone something is to do something. Moreover, just as we hold people responsible for what they do more generally, we hold people responsible for what they say. If someone lies to us, or plays fast and loose with the truth in an important matter, we will typically condemn them for it.

The fact that telling someone something is an action has important implications for the epistemic status of the outputs of AI. Machines cannot act. They cannot act because they are not moral agents: they are not the loci of responsibility for the consequences of their outputs (Hakli & Mäkelä, 2019; Johnson, 2006; Sparrow, 2021a; Véliz, 2021). Because machines cannot act, they cannot testify. In short, strictly speaking, AI cannot tell us anything at all.

The implications of the fact that machines are unable to testify have, to date, mostly been neglected because contemporary thinking about justification is dominated by empiricism. When we worry about the justification of our beliefs, we tend to neglect the role of testimony and frame the problem in terms of whether our beliefs are justified by the evidence available to us. Just as I can learn that it is hot outside by the fact that people walking into the building are wiping sweat from their brows, I can learn that it is hot by overhearing people talking to each other about the oppressive weather. In such circumstances, testimony plays no role—nobody tells *me* anything. We may also treat what people say as evidence when they *are* telling us something. If I know that someone is lying to me, then what they say may still be evidence in support of various beliefs—for instance that the situation is other than that which they say—although, again, testimony plays no role here (Moran, 2006: 292–293).

The outputs of machines can provide evidence to justify beliefs. For instance, we are justified in believing that there might be a fire in the building when the fire alarm goes off. Similarly, if an AI has only ever produced the output "it will rain today" on days on which it later rained, it producing this output may serve as evidence in support of the belief that it will rain today. As we noted above, the fact that the outputs of AI are bullshit need not rule out their being reliable in this way.

However, the fact that machines are not moral agents has further and profound implications for a proper understanding of the nature and epistemic significance of their outputs. When a human being tells us something, we make sense of what they say in the context of their other utterances *and* their actions. For instance, if someone tells us that they need to go to the bathroom and then goes to the kitchen and talks about what a nice bathroom it is, we might conclude that they do not know what is meant by "bathroom". The connections between sentences—and between sentences and behaviour—is what gives words their meaning.

We draw here on the "inferentialist semantics" of philosopher Robert Brandom. Brandom argues, following Wittgenstein, that the meanings of words are determined by their use. In particular, they are a function of the inferences that they support (Brandom, 1994). When we apply a concept in a judgement, we endorse a set of inferences that others might draw from our use of this concept. When we describe something as green, for instance, we licence other people to conclude that it is coloured, that is not red, and that it is at least spatially extended in two dimensions. We also commit *ourselves* to saying various other things and—importantly—to acting in certain ways in the appropriate circumstances.

An example that Brandom provides is helpful here. Imagine a young child who appears at the door of the room and tells you excitedly that the house is on fire. Having made this announcement, the child stands there nonchalantly, toying with their teddy bear. Brandom points out that, were this to happen, we would typically conclude that the child does not know what she is saying: she does not understand the concept of fire (Brandom, 2002: 360). If one says and means that the house is on fire, then one will also be disposed to leave—and will, all other things being equal, leave—the house if one is inside the house when one utters this sentence. This connection to a disposition—and thus to an action—is part of what determines the conceptual content of the sentence and its terms.

However, as we have already noted, machines *cannot* act. This means that their "words" are not connected to actions—not even to other speech acts. Colloquially, one might suggest that they "don't mean what they say". More strictly, the argument rehearsed here suggests that the outputs of machines do not have conceptual content: they are, philosophically speaking, meaningless.

The fact that AIs do not mean what they seem to say is difficult to remember when we interact with them because their outputs typically have the same form as sentences that *would* have content were they to be said, or written, by a human being. There is a powerful "Wizard of Oz" effect at work here: machines seem intelligent because we mistakenly believe that they are the origins of the claims they appear to be making. However, it is only when a human being endorses, or entertains, the claims in the (pseudo-) sentences generated by AI that they come to have conceptual content and thus might tell us something about the world.

Where people are willing to entertain the "claims" of AI, and the beliefs they form as a result of doing so reliably track the truth, then the outputs of AI can serve as a justification of belief. However, it is the human belief that might thereby be justified, not the "claim" of the AI. As we discuss further below, this means that there are strict limits on what we can learn from AI.

4 Human Relationships with Computers

The dubious status of the outputs of AI might matter less if we could be confident that the educational use of AI would be confined to roles where a human being would endorse, and take responsibility, for them. Advocates for the use of AI in education often insist that it should be used to allow human educators to work more quickly and effectively, and should not be used to replace teachers or to reduce the amount of time teachers spend in the classroom (Kasneci et al., 2023).[3] They further suggest that the well-known problem that generative AI systems sometimes "hallucinate"—make things up—(Ji et al., 2023) can be dealt with by requiring human educators to supervise, and check the outputs of, AI. Unfortunately, everything we know about the impacts of automation in other industries suggests that these suggestions are at least naïve, and possibly duplicitous. Like other tools, computers have "affordances"—they make some tasks easier and some tasks harder; they foreground some aspects of the situation of the user and push others into the background (Davis & Chouinard, 2016; Norman, 1988). As such, they shape the behaviour of those who use them (Howcroft & Taylor, 2023; Sparrow, 2021b; Winner, 1980). What is particularly striking about computers is the extent to which, over time, they tend to influence the institutional context of their use (Feenberg, 1999). The affordances of AI place important limits on the ways that it is likely to be used. It is in fact, as we shall see, highly unlikely that AI will only be used to assist, and not to replace, teachers.

4.1 Automation Bias and Deskilling

The fact that those in the wealthy North already rely so heavily on computers, in the form of their phones, to direct them in their daily lives is itself evidence of the power these systems have to shape our behaviour. Moreover, notoriously, users tend to over-trust the outputs of computers—a phenomenon known as "automation bias" (Skitka et al., 1999). In particular, users often come to believe that a computer that works most of the time will work all of the time. This is especially likely in relation to systems that are generally reliable, even if they still fail occasionally. Human beings suffer from a well-known bias when it comes to their thinking about risk, which is that we tend to overestimate the relevance of experiences with which we are familiar (for instance, that the computer works) and underestimate the significance of rare events (for instance, when the computer fails as a result of some particular set of circumstances). Moreover, insofar as it is difficult to pay attention to complex matters when one seldom has reason for intervening in them, the task of monitoring or checking the outputs of a computer that is performing some task is often just as—if not more—difficult than performing the task oneself.

[3] Granted, the "dystopian extreme" of the literature on AI and higher education encompasses those who are concerned about the possibility of full automation of tertiary education (Bearman et al., 2023). See also Selwyn (2019).

Regular use of computers to assist in the performance of a task also tends to lead to people losing the ability to perform that task unassisted. Such "deskilling" often happens remarkably quickly as people struggle to remember how to do things that they do not do regularly.

It is, therefore, highly naïve to think, in the context of debates about the future of tertiary education, that we could overcome the problem of AI "hallucinations" by requiring teachers (or students—see below) to check the accuracy of the outputs of AI. If, for instance, the systems that grade essays are generally accurate, teachers will soon stop reading the work of their students to check that the computer has not made an error in this particular case (not to mention that a system that purports to grade essays but still requires staff to read the essays and calculate a grade is hardly labour saving!). Even if, for institutional reasons, the human teacher will be held responsible for the grade, for all intents and purposes the task will be handed over completely to AI. Over time, staff are highly likely to lose the ability to grade papers accurately themselves.

4.2 *The Transformation of Labour*

Popular discussions of the implications of AI for the future of work tend to focus on the question of whether machines will eliminate jobs by replacing human workers (Brynjolfsson & McAfee, 2014). Historically, though, automation has often tended to change the nature of the work people do, rather than eliminate work altogether. While they may substitute for human beings in parts of the production process, computers often also create new and different forms of labour for human beings (Autor, 2015; Rosenberg, 2013). Indeed, the impact of the introduction of computers into universities over the last several decades demonstrates this all too clearly. As staff were provided with computers, IT departments were set up and flourished: academic staff spent more time filling in online forms and in other administrative tasks that, in the past, were carried out by professional staff or that did not previously exist. As online teaching, and "learning management systems" like *Moodle* and *Blackboard* were introduced, jobs for "educational designers" appeared and academic staff spent more and more time creating websites or clicking through menus rather than reading or talking to students. Innovations that were touted as labour saving created a surprising number of new tasks that have to be completed before one gets into the classroom. Experience therefore suggests that as AI is introduced into education, staff will need to spend time in being trained how to use these systems, in setting them up and generating and collecting data for them, in checking the accuracy of the outputs of AI, and in helping students manage their interactions with AI. All of these tasks will require time that cannot then be spent in forming and developing relationships with students in the classroom.

4.3 *The Economics of AI*

That said, sometimes computers do relieve human beings of the need to perform particular tasks, even if expectations of their capacity to reduce the overall amount of work required to produce goods are typically exaggerated. As soon as it is cheaper to use a machine, rather than a human being, to perform some task, machines will be tend to replace human beings in performing that task. For this reason, it is naïve to think that machines can be developed to perform tasks that are currently performed by human educators, without eventually replacing human educators in those roles. The argument that AI will only "help" teachers teach more effectively—and help students learn more quickly—is a Trojan Horse that, unless identified as such and challenged, will lead to the full automation of teaching at universities.

5 Real Education Resists Automation

The use of AI in education is a dystopian prospect because real education resists automation for reasons we explore below. Enthusiasm for the use of AI in education reflects a shallow account of the nature of education and, in particular, neglects the vital importance of social relationships between teachers and students, between students, and between the university and the larger community.

5.1 *Learning How*

Students arriving at university must master a body of knowledge, including knowledge specific to the disciplines in which they have chosen to specialise. They must learn new facts and be able to recall them to mind as required. However, they must also learn new skills, including skills that are essential—and sometimes specific to their disciplines. To mobilise a distinction made famous by Gilbert Ryle (1945), students must learn both "knowledge that" and "knowledge how".

If one focuses on knowledge *that*, then our philosophical concerns here may appear more than a little precious. Why should it matter whether machines are committed to the truth or not, or whether they are moral agents? Surely the question is whether their outputs are reliable. Students can study and master "content" generated by a machine in the same way and just as well as they would study the same material provided to them by a human being. Moreover, increasingly, AI systems are capable of identifying where students' knowledge is weak and helping them revise and improve it where it is.

However, turning our attention to knowledge *how* reveals quite a different story. AI is much less capable of helping students learn the skills they need to master their chosen disciplines or flourish in their professions. One cannot learn to perform surgery, interview a witness, use a chromatograph, or read a poem, from a machine. To learn such skills, we need to see other people demonstrate then. More than that, we need to practice the activity ourselves in front of, and to receive feedback from, those who have already mastered the skill. Sometimes spoken or written feedback will be sufficient but in some cases we will benefit from an expert showing us how to do something or even repositioning our limbs or bodies as we (try to) do something (Kremer, 2021). Learning—and teaching—skills is an intensely collaborative, and fundamentally embodied, experience.

This is not to deny that AI systems can *sometimes* be useful in helping people learn skills.[4] To the extent that it is possible to learn a particular skill by watching video of other people exhibiting it, it might also be possible to learn it by watching animations generated by AI. If an AI system has cameras and motion tracking capabilities it may even be able to provide feedback on what someone is doing wrong in performing physical activities and make suggestions as to how the student might improve.[5] However, insofar as learning a skill requires developing an embodied knowledge of how to respond to subtly different circumstances to realise a particular end, there are important limits on how far people can progress without having access to real-time feedback from experts.

It is, then, reasonably obvious that to become a surgeon, a musician, a taxonomist, or a civil engineer, students must learn the relevant skills by practicing them in real world settings and refining them in the light of instruction from those who have already mastered them (Crawford, 2009). What is less obvious, but nonetheless true, is that learning "how" is also essential to students in more traditionally academic disciplines. Even students studying philosophy must learn "how"! Philosophy students must learn how to read the texts of the philosophical canon, how to construct an argument, how to make a distinction and show that it is relevant, how to write clearly, and how to anticipate and disarm potential objections. Indeed, thinking and learning—and the capacity to formulate and justify beliefs on the basis of evidence—are themselves skills that must be learned by imitating others, practice, and responding to feedback (Ryle, 1967).

Again, we would not deny that AI can sometimes make a contribution by providing feedback on a student's performance at some particular task. For instance, chatbots based on Large Language Manuals can provide feedback on, and suggestions how to improve, a piece of writing. What these algorithms cannot do though, is show students how they might overcome the psychological, motivational, and intellectual challenges that humans confront as they try to become expert in some particular field. Students who try to learn from machines can only learn how to be like a machine.

---

[4] Darvishi et al. (2024) contains a fascinating account, and helpful discussion of, the impact of an AI system on the extent to which students learned "how" to provide feedback on each other's work.

[5] Such systems are already being developed and used by sports scientists. See, for instance, Cossich et al. (2023).

Moreover, the way AI perform tasks is often very different to the way humans perform them, to the point that humans sometimes cannot imitate the AI.[6]

The example of learning to write has more to teach us here. Generative AI systems now write so well that one might speculate that students no longer need to learn how to write themselves. Instead, perhaps, what students need to learn today is how to "prompt" an AI system to express their own thoughts. This presumes, though, that students already know what they think and so are in a position to evaluate whether the machine is correctly expressing their thought. It neglects that the process of writing is not—or not usually—a mechanical one, whereby we put on paper thoughts that we already possess. Instead, it is more usually the case that we only discover what we think in the process of—and by means of—writing it down (Menary, 2007). This is especially evident in subjects like philosophy and law, which require practitioners to express their thoughts clearly and precisely. Only by means of careful word choice and good punctuation can we work out what we mean to say—and thus what we think. Even when we speak our thoughts aloud, they are expressed in a language that is, at least in part, shaped by linguistic conventions and grammatical forms that we learn through writing. If students do not learn to write, because they rely on AI to write for them, they will struggle to learn how to think clearly.

What this examination of the importance of learning to write shows, therefore, is the way in which "learning that" relies on "learning how" (Hetherington, 2006). Only by learning key skills can students integrate and organise the facts they must absorb to become educated in each subject. The use of AI in education is highly likely to deprive students of opportunities to learn how to do the things that are now being done by the machines and thus, ultimately, to learn the corpus of knowledge that they are expected to master at university.

An increasingly popular conception of the future of education involves students using AI to free themselves from various, purportedly, mundane tasks in order to study more efficiently and effectively (Khan, 2024).[7] For this to be plausible in light of the well-documented tendency of AI systems to "hallucinate" (Ji et al., 2023) and confabulate, however, students must have the capacity to check the quality of the outputs of the AI and ensure that they meet the relevant disciplinary standards (Bearman & Ajjawi, 2023). In fact, as we noted above when discussing the prospect of automated grading, the existence of automation bias suggests that students would quickly cease to do so as long as the system mostly got things right. Moreover, the generation of authorities who are enthusiastic about the educational uses of AI typically grew up learning without such machines, which in turn means that they know how to write, think, and organise relevant information themselves. That means that *they* at least have the ability to tell when the outputs of AI are wrong. Students who grow up using AI to read and write and organise information for them on their behalf are unlikely to be able to tell whether the outputs of AI are correct or not. Thus, our discussion here suggests that this vision of the future of education is deeply

[6] As Burrell (2016) has pointed out, we may often not even know how machine learning systems generate the results they do.

[7] Mollick (2024) argues that this is the future of work as well.

flawed: the more students use AI in the course of their studies, the less likely it is that they will develop the skills and knowledge required to use AI effectively.

5.2 The Importance of Role Models

We observed earlier that the development of the internet—and especially the capacity to stream video—led managers of educational institutions to reconceive teaching as the delivery of "content", which could be provided remotely and asynchronously via multiple platforms. This reconceptualization of education neglects—indeed, obscures—much of what occurs in the classroom. As we observed above, one important thing it neglects is the role played by teachers in teaching "how". In order to understand "content" one must learn key skills. Perhaps equally importantly, though, one must also learn the value of what one is learning and why someone might care about it: one learns this from encountering other people who value and care about it.

Good teachers, then, are role models. They teach students not just content and skills but what it is like to care about their subject material. At universities, they demonstrate what it means to be an intellectual, and to be committed to the regulative ideas of the academy and their particular discipline (Gaita, 2004: 283–330). Moreover, fundamentally, these are moral matters. Good teachers are honest, willing to sacrifice for their students, and abide by the norms—both explicit and implicit—of a community of seekers and knowers.[8] This is why contact with a good teacher can have such a large impact on students. Students see that someone they admire cares about the material they are studying and discover what it means in practice to care about it. They learn what it means to be a philosopher, a mathematician, a historian, an engineer, or a doctor. Ideally, they are provided with the opportunity to imagine that they might occupy, and flourish in, one of these roles.

Thus, even if an AI could provide all the "content" required for students and—what is much less likely—provide feedback on students' work to help them learn relevant skills, it would still fail to meet the educational needs of students. It would not serve as an example to them and could not show them why they should care about, or how they should relate to, the subject material.[9]

[8] Not every educator—not even every researcher—demonstrates these virtues but all are subject to them as regulative ideals that promote the pursuit of knowledge. Universities would be much less successful if most staff did not live up to them, at least to some extent. Students also learn something from bad teachers, whether the latter intend to teach this lesson or not: what it means to fall short of the standards of their discipline.

[9] Lo et al. (2024) surveys the empirical literature on the impact of AI for student engagement.

5.3 Education and Social Relationships

Even if AI is not used to perform tasks that are currently being performed by human teachers in a university setting, it will be performing tasks that might have been carried out by a human being. Moreover, as we argued above, regardless of the intentions of those advocating for the use of AI, there will be strong economic pressures pushing towards the emergence of empty universities in which students only interact with computers. This dystopian vision of the future of education neglects the several, and important, ways in which education is social and relies on relationships between people.

To start with, the experience of going to university is a social one for students. They meet new people, encounter new ideas, and—perhaps—have an opportunity to remake themselves by trying new things and making new friends. The relationships they establish are a form of social capital that they may draw upon in the future. Friends and classmates today are artistic collaborators, intellectual interlocutors, or business partners tomorrow.

In the course of their studies, students also learn important social skills that they will need to succeed in their subsequent ventures, be they intellectual or commercial. They learn how to present themselves to other people, how to speak in public, how to structure and frame an argument so that it will persuade an audience, to deal with criticism, and how to offer feedback on the work of others. These are skills that can only be learned in relationship with other people.

University is also a place where students have the opportunity to experience what G.W.F. Hegel termed "recognition" [*Anerkennung*] (Hegel, 2019). Through their relationships with other students and staff, students come to realise their own moral and political status, and their warrant to stake their own claims to knowledge, just as they come to recognise these qualities and capacities in others. This mutual recognition has a subjective and an objective aspect. Subjectively, students feel that they are taken seriously and grow more confident as a result.[10] Objectively, students enter into membership of the kingdom of ends by virtue of being acknowledged as being owed reasons for our treatment of them, and owing reasons to those with whom they have dealings (Korsgaard, 1996).

Most fundamentally, what students actually learn at university—the "content" of education—is social. Knowledge is produced socially, through relationships between people, across space and time. The only way to tell whether we have really learned and understood something is to talk about it with other people or to test our knowledge in the context of institutions and mechanisms that involve other people. Indeed, insofar as languages—and therefore concepts—are social, our minds and our thoughts are ineliminably social. We cannot even know the content of our own thoughts except in the context of human relationships.

---

[10] Axel Honneth's work brings together Hegelian recognition theory with empirical psychology to argue that recognition by others is a necessary condition of healthy self-confidence and self-esteem (Honneth 1996).

Allowing universities to be emptied out as the result of the adoption of educational AI would, therefore, be a disaster. It would deny students the opportunity to form new friendships with, and to learn from, other students and the social capital that they might otherwise accrue from these. In the absence of interactions with staff and other students, students will not learn key social skills that they need to be good colleagues and co-workers. If students' work is assessed, and feedback provided, by AI then students will not have the experience of having someone take their ideas seriously: if nobody else takes them seriously, how will they come to feel that they deserve to be taken seriously? In this way, assessment by AI involves a serious threat to the agency of students. Moreover, a society in which what is taught and learned at universities is curated by or, worse, generated by AI, would be a society that had given over custody and development of its knowledge and culture to mindless machines. Empty universities would lead to a hollow society and culture.

6 Responding to Objections

There are, of course, various objections that might be made to the argument we have developed here. Popular enthusiasm for educational AI stems, in part, from the fact that it does have obvious attractions. In this section, we consider five objections that, we acknowledge, have some merit.

6.1 Learning from Books

One response to the line of criticism we have been pursuing here draws attention to the controversy surrounding the invention of another technology that has subsequently come to play a central role in modern education—books. Notoriously, Plato condemned the book for detaching words from those who spoke them, rendering them amenable to being misinterpreted or misused, and making it harder for those who received them to assess the authority of their source (Plato 2011)—a line of criticism that was later roundly mocked. Our emphasis on learning how, on the importance of role models, and on the fundamentally social nature of education might be held to be guilty by association with Plato's critique. More problematically, it might be held to imply that it is not possible to become educated by reading books.

The idea of a university without books is, indeed, an appalling thought.[11] Nevertheless, we stand by our observation that there are many things that students cannot learn entirely from books. In particular, as we argued above, students cannot learn "knowledge how" without personal contact with, and feedback from, experts. Nor

[11] Although it is worth noting in passing that many university libraries have already greatly reduced the size of their physical collections out of the—we would hold, misplaced—conviction that online sources are superior, or at least just at good.

can they learn what it is like to be a person committed to the critical standards internal to their discipline. Insofar as we can learn some of these skills and standards from books, it is because in well-written books the author communicates their *persona*.

Quixotically, Plato's worries are pressing of AI in a manner that, in retrospect, they proved not to be of books. People learned to learn from books because they came to understand that it was possible to hold authors responsible for their content and developed a set of social practices to facilitate this. The authority of claims asserted in books is derived from the moral and epistemic standing of the author. If the claims made in a book are untrue, or asserted without due care, we rightly blame the person who authored them (and sometimes the publisher as well).

Reading and relying on the outputs of AI is like reading and relying on a book without an author. Granted, AI systems have designers and the designers of these systems can substantially influence the kinds of outputs they generate. For example, Elon Musk's "Grok" differs stylistically and tonally to Microsoft's Copilot or Google's Gemini. Further, it is sometimes plausible to hold designers responsible in cases where an AI output contributes significantly to some bad outcome. However, the designers of AI do not and cannot "stand behind" the outputs of an AI system in the same manner that authors can stand behind the claims they make in books. Authors, in choosing their words, are responsible for each and every one of them. Where authors are making claims in books, these claims are their own. In contrast, generative AI produces outputs over which its designers have little-to-no oversight and which, in many cases, they could never have predicted.[12] Where it appears that AI is making a claim, ultimately it is not, because it is not capable of being held responsible for its outputs.

Students can learn by reading books insofar as the authors of the books stand behind their words. Claims in books can be a form of testimony and the testimony of experts is good grounds for belief. As we argued above, strictly speaking, the outputs of AI do not express anything at all. At best, they regurgitate the words of others. Even where it can be shown that what they spit out would be accurate were it to be offered by a human being, the outputs of AI do not count as testimony, and, as such, do not place us under the same obligation to take them seriously.

6.2 Equity and Access

Many of those advocating for the use of AI in education are arguably motivated by a "techno fetishism" that is widespread in the broader culture. The rapid progress in consumer technology in the last several decades—as well as the financial rewards that may flow to universities if they attract the favour of the financial elites that this progress has created—has led many university administrators to become reluctant to

[12] A previous generation of educational software, which functioned like a "choose your own adventure" book, wherein each choice by a student was greeted with text written for that purpose, did allow the programmer/designer to stand in the role of an author.

appear as anything other than cheerleaders for new technologies. Moreover, it is clear that some senior managers in higher education are looking to AI as a way to cut costs by cutting staff. That said, if AI really could provide students with a good education this might greatly expand access to education. Many people around the world cannot go to university, either because they live in rural areas far from universities or because they are simply too poor to do so (or both). Classes delivered online by AI have the potential to expand access to education to these cohorts (Kasneci et al., 2023; Khan, 2024).

However, these benefits of the use of AI must be balanced against a realistic assessment of the costs of increasing the use of AI in education that we have highlighted here. The "education" that AI might provide would be significantly worse than a traditional education provided by human teachers. Indeed, it would lack key features of such an education. In particular, it would require students to teach themselves key skills and, perhaps more importantly, to find their way to a love for their discipline without access to human role models.

Whether one should respond to the needs of those who cannot access a good enjoyed by citizens in the wealth North by offering an inferior version of the good is a classic ethical and policy dilemma. We doubt that there is a generalizable answer to this question: even in the context of the debate about the wisdom of aiming to expand access to education via AI, the answer will tend to turn on the details of the particular case, including the role to be played by AI and the economic and political context in which it is being rolled out. One thing that does need to be factored into this deliberation, though, is the fact that embracing the use of AI will, as we have argued here, degrade the quality of education available to those who are privileged enough to be able to access it.

We should also be wary about being seduced by the fantasy of technological solutions to economic and political problems. There are already various means available to expand access to education that *do not* rely on AI. Universities could make textbooks open access or provide them free to those who cannot afford them. They could offer courses taught by human educators by correspondence. They could establish local campuses in remote and impoverished areas. In many parts of the world, such initiatives might actually make education more available to those currently excluded from it than using AI would, because they do not require (potential) students to have access to computers and the infrastructure necessary to power and maintain them. Unless institutions are giving serious consideration to these low-tech alternatives, one might reasonably conclude that the argument that they want to adopt AI to expand access to education is disingenuous.

6.3 Personalised Education

Another genuine, if currently mostly hypothetical, virtue of AI is its ability to identify areas of strength and weakness in students' academic performance and to tailor a programme of study to suit their educational needs (Escotet, 2023; Fuchs, 2023).

Students with more mastery of an area of study can move ahead more quickly and students who are struggling can be provided with extra opportunities to learn and revise. Again, though, this must be recognised as a substitute for engagement with a human teacher, which would provide the important good of recognition, as well as the opportunity to learn what it means to live a life informed by a commitment to the value of the subject material.

6.4 *The Future of Work*

A commonly expressed argument for introducing generative AI systems into education points to students' need to be able to use generative AI effectively upon entering the workforce. This is *prima facie* plausible: where graduates are likely to work in industries that have adopted AI, it is important they learn not only how to use AI, but also how to think about AI's strengths and weaknesses (Al Naqbi et al., 2024; Rudolph et al., 2023).

In passing, we would observe if that one really believed that the nature of work was about to be radically transformed, one should probably wait to see what workers in this brave new world actually need to know before changing what—and how—one teaches. We are confident that, whatever happens to the nature of work as a result of AI, the core of a traditional university education will remain relevant and valuable (Royer, 2024): we think this is especially likely be true of an education in philosophy!

Moreover, that students will need to learn to use AI effectively does not of itself establish that students should be taught by AI. As a powerful, potentially world-shaping, technology, AI is appropriately an object of study in courses in the philosophy of technology or the history and philosophy of science. However, in our view, teaching effective and critical AI use should be considered akin to teaching effective use of a library catalogue or how to construct and maintain a bibliography. Like learning to use AI systems, these are important skills, but they are appropriately taught in study skills classes rather than as a component of education in a particular discipline. In classes in history, philosophy, political science, etc., students should study the key concepts and the thought of the key thinkers in that discipline, rather than further diluting their disciplinary expertise with lessons oriented around the meaningless outputs of mindless machines.

6.5 *Actually Existing Universities*

A final objection to our criticisms of educational AI would accuse us, with some justification, of nostalgia about the nature and quality of university education prior to the advent of generative AI. It would indeed be naïve to claim that, prior to the release of ChatGPT 3.5 in 2022, Australian universities only offered students an education consisting of serious small-group discussion led by experts deeply committed to truth-telling and the standards of their respective disciplines. Perhaps that has never been the case. Moreover, it has probably always been true that there are some things that even the most engaged and enthusiastic students have only needed to learn by rote and some skills that students have only needed to develop to the point to which they can exercise them with minimal competence. It is not impossible that AI might make a useful contribution in this context. One might thus argue, against our claims above, that a realistic assessment of the education offered by actually existing Australian universities suggest that AI has the potential to improve on it substantially.

This line of argument is an example of a phenomenon discussed above; namely, the way that advocates for automation make it easier to imagine automation succeeding in some activity by working to lower our expectations as to what counts as success. With that said, there is truth in the observation that we are seeking to hold educational AI to a standard that much actually existing education routinely fails to meet—and rejecting the introduction of AI into education for that reason. Where others are enthusiastic about AI because they are pessimistic about the current state of universities, we choose to be optimistic about a future in which societies invest in education provided by good (human) teachers rather than in AI.

7 Investing in Education

The best that might be said for current enthusiasm for the use of educational AI is that it opens up an opportunity to discuss the nature and significance of the outputs of AI and to remind ourselves of what universities should offer, both to students and to society at large. Unfortunately, this opportunity is currently, for the most part, being missed due to the widespread technophilic assumption that it is inevitable that AI will transform education.

It is difficult not to be impressed by the remarkable advances in AI—and especially in generative AI—over the last few years. Nevertheless, as we have argued here, the use of this technology in tertiary education constitutes a profound threat to the idea of the university as a community of scholars and students dedicated to the pursuit of knowledge. The outputs of AI are neither truth-oriented nor connected to action in the way that claims by (human) agents are; as a result, these outputs lack conceptual content. Encouraging students to study "content" produced by AI undermines the authority of all teachers, as well as the established bodies of disciplinary knowledge they aim to share with students. Students must participate in human-to-human pedagogical relations in order to: learn "how" as a necessary component of learning "that"; encounter their teachers as real examples of what it means to demonstrate a commitment to a discipline; and, through sustained interpersonal interaction, realise themselves as moral agents and citizens with healthy self-respect. Machines can contribute little to these goals. Students will indeed need to engage effectively and critically with AI, but—to the extent they need to do so—that training should be provided in specialist study skills units, and should not reduce the time they spend learning discipline-specific knowledge and skills.

Universities that want to avoid becoming empty shells must invest in quality human relationships, between teachers, between students, and between teachers and students. They should reduce class sizes to ensure that students can develop genuine interpersonal relations with teachers committed to their disciplines. Now that students can, if they wish, complete most online assessments by relying on generative AI, meaningful assessment of student knowledge and skill will increasingly depend on time spent in discussion with students. It will also, we suggest, require more work to be completed under invigilated conditions, so that students are rewarded for thinking for themselves, rather than employing a machine pretend to think in their place.

The recent emergence of powerful generative AI poses philosophical, ethical, social, and political challenges for the economy, for society, and for individuals. Meeting these challenges will require ongoing democratic deliberation among educated citizens (Formosa et al., 2024; Sparrow, 2020)—and universities that help produce such citizens. It would be a perverse outcome, if, in their race to embrace AI, universities allowed themselves to be hollowed out by this technology so that this deliberation becomes less, rather than more, likely.

8 Declaration of Interests Statement

The authors have nothing to declare.

Acknowledgements The authors would like to thank Mark Andrejevic, Thomas Corbin, Lloyd Humberstone, and Zala Volcic for comments and discussion that have improved this paper. Professor Sparrow is an Associate Investigator in the Australian Research Council Centre of Excellence for Automated Decision-making and Society (Grant number CE200100005) and worked on this paper in this role.

References

Al Naqbi, H., Bahroun, Z., & Ahmed, V. (2024). Enhancing work productivity through generative artificial intelligence: A comprehensive literature review. *Sustainability, 16*(3), Article 1166. https://doi.org/10.3390/su16031166.

Autor, D. H. (2015). Why are there still so many jobs? The history and future of workplace automation. *Journal of Economic Perspectives, 29*(3), 3–30.

Bearman, M., & Ajjawi, R. (2023). Learning to work with the black box: Pedagogy for a world with artificial intelligence. *British Journal of Educational Technology, 54*(5), 1160–1173.

Bearman, M., Ryan, J., & Ajjawi, R. (2023). Discourses of artificial intelligence in higher education: A critical literature review. *Higher Education, 86*(2), 369–385.

Bone, K. D. (2021). Cruel optimism and precarious employment: The crisis ordinariness of academic work. *Journal of Business Ethics, 174*, 275–290.

Brandom, R. B. (1994). *Making it explicit: Reasoning, representing, and discursive commitment.* Harvard University Press.

Brandom, R. B. (2002). *Tales of the mighty dead: Historical essays in the metaphysics of intentionality*. Harvard University Press.
Braverman, H. (1998). *Labor and monopoly capital: The degradation of work in the twentieth century*. Monthly Review Press.
Brynjolfsson, E., & McAfee, A. (2014). *The second machine age: Work, progress, and prosperity in a time of brilliant technologies*. WW Norton & Company.
Burrell, J. (2016). How the machine 'thinks': Understanding opacity in machine learning algorithms. *Big Data & Society*, 3(1). https://doi.org/10.1177/2053951715622512.
Coady, C. A. J. (1992). *Testimony: A philosophical study*. Clarendon Press.
Cossich, V. R. A, Carlgren, D., Holash, R. J., & Katz, L. (2023). Technological breakthroughs in sport: Current practice and future potential of artificial intelligence, virtual reality, augmented reality, and modern data visualization in performance analysis. *Applied Sciences, 13*(23), Article 12965. https://doi.org/10.3390/app132312965.
Crawford, M. B. (2009). *Shop class as soulcraft: An inquiry into the value of work*. Penguin Books.
Cummings, M. L. (2017). Automation bias in intelligent time critical decision support systems. In D. Harris & W.-C. Li (Eds.), *Decision making in aviation* (pp. 289–294). Routledge.
Darvishi, A., Khosravi, H., Sadiq, S., Gašević, D., & Siemens, G. (2024). Impact of AI assistance on student agency. *Computers & Education*, 210, Article 104967. https://doi.org/10.1016/j.compedu.2023.104967
Davis, J. L., & Chouinard, J. B. (2016). Theorizing affordances: From request to refuse. *Bulletin of Science, Technology & Society, 36*(4), 241–248.
Descartes, R. (2017). *Meditations on first philosophy, with selections from the objections and replies* (J. Cottingham, Trans. & Ed.). Cambridge University Press.
Ellis, J., & Romano, D. (2008). Synchronous and asynchronous online delivery: How much interaction in e-learning is enough in higher education? In C. J. Bonk, M. M. Lee, & T. Reynolds (Eds.), *E-Learn: World Conference on E-Learning in Corporate, Government, Healthcare, and Higher Education* (pp. 2615–2620). Association for the Advancement of Computing in Education.
Escotet, M. Á. (2023). *The optimistic future of artificial-intelligence in higher-education*. Springer.
Fagbohun, O., Iduwe, N. P., Abdullahi, M., Ifaturoti, A., & Nwanna, O. M. (2024). Beyond traditional assessment: Exploring the impact of large language models on grading practices. *Journal of Artificial Intelligence Machine Learning and Data Science, 2*(1), 1–8. https://doi.org/10.51219/JAIMLD/oluwole-fagbohun/19
Farnell, A. (2023, January 19). AI will replace academics unless our teaching challenges students again. *Times Higher Education*. Retrieved from https://www.timeshighereducation.com/opinion/ai-will-replace-academics-unless-our-teaching-challenges-students-again. Accessed July 31, 2024.
Farrelly, T., & Baker, N. (2023). Generative artificial intelligence: Implications and considerations for higher education practice. *Education Sciences, 13*(11), Article 1109.
Feenberg, A. (1999). *Questioning technology*. Routledge.
Formosa, P., Kashyap, B., & Sahebi, S. (2024). Generative AI and the future of democratic citizenship. *Digital Government: Research and Practice*. https://doi.org/10.1145/3674844
Frankfurt, H. (2002). Reply to Cohen. In S. Buss & L. Overton (Eds.), *The contours of agency: Essays on themes from Harry Frankfurt* (pp. 340–344). MIT Press.
Frankfurt, H. (2005). *On bullshit*. Princeton.
Fuchs, K. (2023). Exploring the opportunities and challenges of NLP models in higher education: Is Chat GPT a blessing or a curse? *Frontiers in Education*, 8, Article 1166682.
Gaita, R. (2004). *Good and evil: An absolute conception* (2nd ed.). MacMillan.
Hakli, R., & Mäkelä, P. (2019). Moral responsibility of robots and hybrid agents. *The Monist, 102*(2), 259–275.
Hegel, G. W. F. (2019). *Phenomenology of spirit* (T. Pinkard, Trans.). Cambridge University Press.
Henderson, M., Ryan, T., & Phillips, M. (2019). The challenges of feedback in higher education. *Assessment & Evaluation in Higher Education, 44*(8), 1237–1252. https://doi.org/10.1080/02602938.2019.1599815

Herbold, S., Hautli-Janisz, A., Heuer, U., Kikteva, Z., & Trautsch, A. (2023). A large-scale comparison of human-written versus ChatGPT-generated essays. *Scientific Reports*, 13, Article 18617. https://doi.org/10.1038/s41598-023-45644-9.

Hetherington, S. (2006). How to know (that knowledge-that is knowledge-how). In S. Hetherington (Ed.), *Epistemology futures* (pp. 71–94). Clarendon Press.

Hicks, M. T., Humphries, J., & Slater, J. (2024). ChatGPT is bullshit. *Ethics and Information Technology, 26*(2), Article 38. https://doi.org/10.1007/s10676-024-09775-5

Honneth, A. (1996). *The struggle for recognition: The moral grammar of social conflicts.* (J. Anderson, Trans.). MIT Press.

Howcroft, D., & Taylor, P. (2023). Automation and the future of work: A social shaping of technology approach. *New Technology, Work and Employment, 38*(2), 351–370.

Jensen, L. X., Buhl, A., Sharma, A., & Bearman, M. (2024). Generative AI and higher education: A review of claims from the first months of ChatGPT. *Higher Education*, Published Online 15 July 2024.

Ji, Z., Lee, N., Frieske, R., Yu, T., Su, D., Xu, Y., Ishii, E., Bang, Y. J., Madotto, A., & Fung, P. (2023). Survey of hallucination in natural language generation. *ACM Computing Surveys, 55*(12), 1–38.

Johnson, D. G. (2006). Computer systems: Moral entities but not moral agents. *Ethics and Information Technology, 8*(4), 195–204.

Kasneci, E., Sessler, K., Küchemann, S., Bannert, M., Dementieva, D., Fischer, F., Gasser, U., Groh, G., Günnemann, S., Hüllermeier, E., Krusche, S., Kutyniok, G., Michaeli, T., Nerdel, C., Pfeffer, J., Poquet, O., Sailer, M., Schmidt, A., Seidel, T., … Kasneci, G. (2023). ChatGPT for good? On opportunities and challenges of large language models for education. *Learning and Individual Differences, 103*, Article 102274.

Khan, S. (2024). *Brave new word: How AI will revolutionize education (and why that's a good thing).* Viking.

Korseberg, L., & Elken, M. (2024). Waiting for the revolution: How higher education institutions initially responded to ChatGPT. *Higher Education*, Published online 3 July, 2023.

Korsgaard, C. (1996). *Creating the Kingdom of Ends.* Cambridge University Press.

Kremer, M. (2021). Gilbert ryle on skill as knowledge-how. In E. Fridland & C. Pavese (Eds.), *The Routledge handbook of philosophy of skill and expertise* (pp. 100–112). Routledge.

Liu, D. (2024). How Sydney educators are building 'AI doubles' of themselves to help their students. *Teaching@Sydney.* Retrieved from https://educational-innovation.sydney.edu.au/teaching@sydney/how-sydney-educators-are-building-ai-doubles-of-themselves-to-help-their-students/. Accessed 31 July, 2024.

Lo, C. K., Hew, K. F., & Jong, M. S. (2024). The influence of ChatGPT on student engagement: A systematic review and future research agenda. *Computers & Education, 219*, Article 105100. https://doi.org/10.1016/j.compedu.2024.105100

Lopes, A., & Dewan, I. A. (2015). Precarious pedagogies? The impact of casual and zero-hour contracts in higher education. *Journal of Feminist Scholarship, 7*(Fall), 28–42. https://digitalcommons.uri.edu/jfs/vol7/iss7/5

Means, A. J. (2019). Precarity and the precaritization of teaching. In M. A. Peters (Ed.), *Encyclopedia of teacher education.* Springer. https://doi.org/10.1007/978-981-13-1179-6_117-1

Memarian, B., & Doleck, T. (2023). Fairness, accountability, transparency, and ethics (FATE) in artificial intelligence (AI), and higher education: A systematic review. *Computers and Education: Artificial Intelligence, 5*, Article 100152.

Menary, R. (2007). Writing as thinking. *Language Sciences, 29*(5), 621–632.

Molesworth, M., Nixon, E., & Scullion, R. (2009). Having, being and higher education: The marketisation of the university and the transformation of the student into consumer. *Teaching in Higher Education, 14*(3), 277–287.

Mollick, E. (2024). *Co-intelligence: Living and working with AI.* WH Allen.

Moran, R. (2006). Getting told and being believed. In J. Lackey & E. Sosa (Eds.), *The epistemology of testimony* (pp. 272–306). Oxford University Press.

Ng, D., Tan, C. W., & Leung, J. K. L. (2024). Empowering self-regulated learning and science education through ChatGPT: A pioneering pilot study. *British Journal of Educational Technology, 55*, 1328–1353.

Norman, D. A. (1988). *The psychology of everyday things.* Basic Books.

OIAI. (2023). Meet the world's first AI digital human teacher. *Otermans Institute.* Retrieved from https://oiai.oiedu.co.uk/educators. Accessed 14 July, 2024.

Pesovski, I., Santos, R., Henriques, R., & Trajkovik, V. (2024). Generative AI for customizable learning experiences. *Sustainability, 16*, Article 3034. https://doi.org/10.3390/su16073034

Plato. (2011). *Phaedrus.* In H. Yunis (Ed.). Cambridge University Press.

Rosenberg, R. S. (2013). *The social impact of computers.* Elsevier.

Royer, C. (2024). Outsourcing humanity? ChatGPT, critical thinking, and the crisis in higher education. *Studies in Philosophy and Education.* https://doi.org/10.1007/s11217-024-09946-3

Rudolph, J., Tan, S., & Tan, S. (2023). ChatGPT: Bullshit spewer or the end of traditional assessments in higher education? *Journal of Applied Learning & Teaching, 6*(1), 342–363.

Ryle, G. (1945). Knowing how and knowing that: The presidential address. *Proceedings of the Aristotelian Society, 46*, 1–16. http://www.jstor.org/stable/4544405

Ryle, G. (1967). Teaching and training. In R. S. Peters (Ed.), *The concept of education (International Library of the Philosophy of Education Volume 17)* (1st ed., pp. 73–82). Routledge.

Selwyn, N. (2019). *Should robots replace teachers?* Polity Press.

Sidorkin, A. M. (2024). *Embracing chatbots in higher education: The use of artificial intelligence in teaching, administration, and scholarship* (1st ed.). Routledge.

Sifaleras, A., & Lin, F. (Eds.). (2024). *Generative intelligence and intelligent tutoring systems: 20th International Conference, ITS 2024, Proceedings.* Springer.

Skitka, L. J., Mosier, K. L., & Burdick, M. (1999). Does automation bias decision-making? *International Journal of Human-Computer Studies, 51*(5), 991–1006.

Sparrow, R. (2020). Artificial intelligence. In H. LaFollette (Ed.), *International encyclopedia of ethics.* John Wiley & Sons. https://doi.org/10.1002/9781444367072.wbiee870

Sparrow, R. (2021a). Why machines cannot be moral. *AI & Society: Journal of Knowledge, Culture and Communication, 36*(3), 685–693. https://doi.org/10.1007/s00146-020-01132-6

Sparrow, R. (2021b). How robots have politics. In C. Véliz (Ed.), *Oxford handbook of digital ethics* (pp. 292–311). Oxford University Press. https://doi.org/10.1093/oxfordhb/9780198857815.013.16

Sparrow, R., Koplin, J., & Flenady, G. (2023). Generative AI is dangerous—But not for the reasons you might think. *ABC Religion and Ethics*, 22 Feb 2023, updated 24 Apr 2024. Retrieved from https://www.abc.net.au/religion/why-generative-ai-like-chatgpt-is-bullshit/102010238. Accessed July 31, 2024.

Tao, Y. (2021). The understanding and take on the blatant instrumentalism among university students: An early career academic perspective. *Australian Universities' Review, 63*(2), 62–66.

Taulli, T. (2023). *Generative AI: How ChatGPT and other AI tools will revolutionize business.* Apress.

Trumbore, A. (2023). ChatGPT could be an effective and affordable tutor. *The Conversation*, February 22, 2023. Retrieved from http://theconversation.com/chatpt-could-be-an-effective-and-affordable-tutor-198062. Accessed February 28, 2023.

van den Berg, G., & du Plessis, E. (2023). ChatGPT and generative AI: Possibilities for its contribution to lesson planning, critical thinking and openness in teacher education. *Education Sciences, 13*(10), Article 998.

Véliz, C. (2021). Moral zombies: Why algorithms are not moral agents. *AI & Society: Journal of Knowledge, Culture and Communication, 36*, 487–497. https://doi.org/10.1007/s00146-021-01189-xch

Watts, L. (2016). Synchronous and asynchronous communication in distance learning: A review of the literature. *Quarterly Review of Distance Education, 17*(1), 23–32, 56.

Winner, L. (1980). Do artifacts have politics? *Daedalus, 109*(1), 121–136.

Robert Sparrow is Professor in the Philosophy Program, and an Associate Investigator in the Australian Research Council Centre of Excellence for Automated Decision Making and Society, at Monash University, where he works on ethical issues raised by new technologies. He has published on topics as diverse as the ethics of military robotics, the moral status of AIs, human enhancement, stem cells, preimplantation genetic diagnosis, xenotransplantation, and migration. He has served as co-chair of the IEEE Technical Committee on Robot Ethics and was one of the founding members of the International Committee for Robot Arms Control.

Gene Flenady is a lecturer in philosophy at Monash University. His research concerns the structure and social conditions of human rational agency, mobilizing the resources of the German idealist tradition to normatively assess the implications of new technologies for meaningful work and tertiary pedagogy.

Implications of AI Chatbots for Human Relationships: A Perspective from Daoism

Ellen Y. Zhang

AI technology is growing at a rate quicker than most of us can imagine. It is a powerful technology with the potential to fundamentally transform all aspects of life our life, which has created a plethora of ethical challenges that must be identified and discussed. This paper addresses the emerging technology of chatbots and its potential impacts on human relationships from a perspective of Daoist philosophy, namely the philosophy of Laozi and Zhuangzi. Though the application of the AI Machine is a contemporary phenomenon, Daoism, like many other ancient philosophical or religious traditions, can provide an insight into the impact of the rapid pace of technological development.

Traditional interpretations of Daoist philosophy often describe it as a system of thought that advocates the idea of being "natural" or "simple" and thus takes a strongly skeptical stance in science and technology. For example, Yu-lan Fung (aka., Feng Youlan), a contemporary Chinese philosopher, gave a well-known remark in his essay, "Why China Has No Sciencc--An Interpretation of the History and Consequences of Chinese Philosophy" in the early 1920s:

> The Chinese philosophers … had no need of scientific certainty, because it was themselves that they wished to know; so in the same way they had no need of the power of science, because it was themselves that they wished to conquer. To them the content of wisdom is not intellectual knowledge and its function is not to increase external goods. To [D]aoism, external goods seem to be something that can only bring confusion to man's mind. (Fung, 1922, 261)

It should be noted that the term "Daoism" usually refers to a big umbrella under which we have *daojia* 道家 (philosophical Daoism) and *daojiao* 道教 (religious Daoism). Philosophical Daoism and religious Daoism are related but also different in many ways. In this paper, the discussion focuses on philosophical Daoism.

E. Y. Zhang (✉)
University of Macau, Macau, China
e-mail: eyzhang@um.edu.mo

Y. Chandra and R. Fan (eds.), *Artificial Intelligence and the Future of Human Relations*,
https://doi.org/10.1007/978-981-96-7185-4_7

A similar sentiment is expressed in, the exegetical tradition of Daoism in the West where Daoism is viewed as a form of "primitivism" or as "negative, backward-looking nihilistic doctrine and not a positive, scientific vision for the future" (Rupp, 2012, 29). In the same vein, David Cooper points out, Daoist texts "express a nostalgia, palpable too in much 'green' literature, for a time, real or imagined, when people did live more naturally: when, for instance, there was 'plainness and simplicity', a minimum of technology and artifice, so that there was less 'selfishness', and men and women could live more 'peacefully in their homes'" (Cooper, 2014, 95–108).

Joseph Needham even uses the term "anti-technology complex" to denote Daoist hostility to machines and tools, claiming that Daoist thinkers showed "an ambivalent attitude to those techniques which the society of force and dominance could use for its own ends" (Needham, 1956, 125).[1] Yet at the same time, we find passages in the *Zhaungzi*, where various kinds of skillful tool-users (such as the butcher Ding, the wheelwright Bian, and the woodcarver Qing) are discussed. The techniques illustrated by skillful tool-users are employed in the *Zhuangzi* to show a harmonious unity between human spirit/mind (*shen* 神) and tools (*qi* 器). Therefore, the key question for Daoists is not "using or not using tools," but "how to use tools."[2] In this paper, I shall argue that Daoism does not reject tools and technology categorically despite the negative remark about the "mechanical mind" in Zhuangzi's thought.[3] Instead, Daoism seeks a balanced and harmonious approach to the development and the use of technology, respecting the natural order in human creativity, and promoting the idea of "*dao*-cosmotechnics" or *dao*-engineering (that is, the correlation between *dao* and technology).[4] The Daoist approach to intelligence, intuition, and imagination also provides us a perspective to see the limit of algorithms and digital way of thinking when we look at human interaction and communication.

---

[1] Yet at the same time, Needham acknowledges that it would be difficult to understand science and technology in China without understanding Daoism. See Needham 2020. Needham, Joseph. 2000. *Science and Civilisation in China: Volume 6, Biology and Biological Technology; Part 6, Medicine*. Cambridge University Press.

[2] A tool here means something that assists people in performing tasks more efficiently and effectively.

[3] Certainly, there is a difference between tools and technology is that tools are specific physical objects or devices used to perform a task, while technology encompasses a broader range of advanced systems, processes, and knowledge. AI can be seen in both lights.

[4] Yuk Hui coins the term to (1) and (2) define the nature of technology used in Daoism as "the unification of the cosmos and the moral through technical activities." Hui (2017). "Cosmotechnics as Cosmopolitics." *E-flux Journal*. November.

1 *Ziran*, *Dao*, Tools, and Technology

In the *Daodejing* 道德經 (hereafter, *DDJ*), one of the central themes is expressed through the term *ziran* 自然which in turn is connected to the central idea of the *dao* 道. The literal meaning of *ziran* is "arising from itself," often translated as "self-so," "what-is-so-of-itself," or "spontaneously so."[5] The Chinese term *ziran* is a two-character compound of *zi* (自) meaning nose or self and suchness *ran* (然). It is employed by Laozi to describe the fundamental characteristic of the *dao*. In this context, *ziran* as self-so is in contrast to other-so, denoting the idea that things should be allowed to take their own fluid, graceful, peaceful paths without too much external interferences. As Chung-ying Cheng has put it, "One important aspect of *tzu-jan* [*ziran*] is that the movement of things must come from the internal life of things and never results from engineering or conditioning by an external power" (Cheng, 1986, 356). Yet the concept of *ziran* has always been contentious one impaired by a variety of conflicting interpretations, especially when it is understood as the "natural world." In Chapter 25 of the *DDJ*, we read,

> Human beings follow the ways of the earth;
> The earth follows the ways of the heaven;
> The heaven follows the ways of the *dao*;
> The *dao* follows what-is-so-of-itself (*ziran*).[6]

In Chapter 57 of the *DDJ*, it says,

> I take no action and the people are transformed of themselves [*zihua* 自化];
> I prefer stillness and the people are rectified of themselves [*zizheng* 自正];
> I am not meddlesome and the people prosper of themselves [*zifu* 自富];
> I am free from desire and the people of themselves become simple like the uncarved block [*zipu* 自樸]. (trans., Lau, 1963, 64)

In the passage above, the notions of self-transformation, self-rectification, self-enrichment, and self-simplification all indicate the Daoist art of governance. That is, the principle of non-intervention and non-coercive action.[7] In this sense, self-so is employed in contrast to other-so (*ta-ran* 他然), all of which emphasizes the idea that things are well done according to their own courses without too much coercive or excessive actions. Accordingly, Laozi's concept of *ziran* is completely different from the modern use of *ziran* which is a term borrowed from a Japanese scholar Andō Shōeki 安藤昌益 (1703–1762) to be understood as the "natural world" or the "world of nature" (*ziran jie* 自然界) and was popularized in Japan after nineteenth

---

[5] The term "self-so" or "what-is-so-of-itself" is used by Arthur Waley. (See Waley, 1958, 174).

[6] The translation of this passage is by the author.

[7] Some Daoist scholars hold the view that Daoist concept of *ziran* (self-so) is the closest term in classical Chinese thought to the concept of freedom (Rapp, 2012, 91). In this view, if technology becomes something that hurts personal freedom, then *ziran* could entail a negative view about technology.

century (Chen, 2010, 104). Nevertheless, the proposal of people to live in a manner of intimacy or convergence with nature in Daoist philosophy is influenced by the *DDJ* and augmented in the *Zhuangzi* and later Daoists.

In the *Zhuangzi*, *ziran* is not a frequently used core concept, yet it still plays an important role in Zhuangzi's Daoism. For example. In the *Zhuangzi* where a man called the Celestial Root asks another man called The Nameless Man how to govern the world, The Nameless Man replies, "following things as they are (*shunwu ziran* 順物自然) without allowing yourself the least bias, the world will be in order" (*Zhuangzi, 7*). Zhuangzi then continues, "When a sage-ruler rules, his achievements cover all the world, but they seem not to come from himself. He transforms all things, and yet people do not rely upon him. There is something unnameable about him that allows all creatures to delight in themselves" (*Zhuangzi, 7*). Zhuangzi's concept of *ziran* as "of its own" or "spontaneity is similar to that in the *DDJ*." Alan Fox makes a good argument when he interprets Zhuangzi's self-so as "becoming sensitive to a broader and finer range of the subtle demands, constraints, and inevitabilities of unique situations" and such sensitivity "allows us to respond [ying 應] most appropriately to every unique situation in the way that most or best respects subtleties of novelty and necessity" (Fox, 1996, 59).

After introducing the concept of *dao* and its connection to *ziran* or self-so, let us look at two other concepts in Daoist philosophy: tools and technology. The Chinese term "tools" is *qi* 器. For example, in the Chap. 11 of the *DDJ*, it says, "It is the space within [the wheel's hub] that makes a tool useful." And in Chapter 36, it says, "The state's sharp tools (weaponry) should not be displayed." For Laozi, humans transform things to tools that benefit humans, and what is beneficial are "things that have shapes" (*DDJ*. 11) In the Zhuangzi, the function of the tool is fully realized. Yet at the same time, Zhuangzi speaks of the negative aspect of tools, and sees knowledge as "the tool for competition" (*zheng zhi qi* 争之器 *Zhuangzi*, 4). He challenges the very idea of knowledge or understanding by saying that,

> Men all pay homage to what understanding understands, but no one understands enough to rely upon what understanding does not understand. Can we call this anything but great perplexity? Let it be, let it be! There is no place you can escape it. (*Zhuangzi*, 18; Watson, 1968, 288)

The passage cited above shows a Daoist view toward knowledge. Yet to interpret it as a form of epistemological nihilism would undermine Zhuangzi's argument on knowledge which lends to his argument on technology. According to Zhuangzi, there are two kinds of knowledge, namely great knowledge (*dazhi* 大知) and petty knowledge (*xiaozi* 小知): The former concerns the art of living whereas the latter focuses on capacity of making conceptual distinctions, or that Zhuangzi calls "the discriminatory mind." To correspond to two forms of knowledge, Zhuangzi speaks of two kinds of *ji* 技, that is, *ji* as skillfulness or as the art of living and *ji* as craftsmanship or technology. For Zhuangzi, petty knowledge or technology can be the root of unliberated life when it is away from self-so-ness. In this regard, a tool or a machine is viewed as something that links with Laozi calls "cleverness and profit" (*DDJ*, 19). It seems, therefore, that Daoism advocates a need in society to shed all the artificial intelligence so that people could regain the natural essence of the *dao* symbolized by the uncarved block of wood. For Laozi, non-action (*wuwei* 無為) can be understood as "doing without doing (contriving, planning, engineering) anything" (Allen, 2010, 153).

The word "technology" in English" is derived from the Greek words *tekhnê* and *logia*, which means "craft" or "art," and "skill." Today, the word technology is used to refer to the application of scientific knowledge to the practical aims of human life. The word "machine" originates from the Latin word *machina* and the Greek words *mēkhanē* and *mēkhos*. which denote instruments or tools. Machines entail the idea of being "mechanical," as well as the ideas of "cleverness, ingenuity, and the ability to outwit nature" (Crul, 2023).[8] Along with this line of thinking, a machine is a double-edge sword in that its virtue and vice exist simultaneously. For Daoists, the key question is whether technology can bring us closer to nature and our authenticity or the other way around.

According to Daoism, the notion of being able to outwit nature is, to a certain extent, in conflict with the idea of living in harmony with nature. In Daoist philosophy, *ji* 機 as machines and *ji* 技 as technology is often viewed negatively when their applications are separated from the nature of the *dao*. To reinforce this negative view on machines, let me tell an often-cited story from the *Zhuangzi*: Zigong, one of Confucius' disciples from the *Analects*, traveled south to the State Chu, and on his way back, he saw an old gardener preparing his fields for planting. He had hollowed out an opening by which he entered the well and from which he emerged, lugging a pitcher, which he carried out to water the fields. On seeing this situation, Zigong said, "There is a machine for this sort of thing. In one day, it can water a hundred fields, demanding very little effort and producing excellent results. Wouldn't you like one?" Zigong then told the gardener how the machine worked. Yet gardener replied with a laugh, "I've heard my teacher say, where there are machines, there are bound to be machine worries; where there are machine worries, there are bound to be a machine mind. With a machine mind in your breast, you've spoiled what was pure and simple, and without the pure and simple, the life of the spirit knows no rest. Where the life of the spirit knows no rest, the *dao* will cease to buoy you up. It's not that I don't know about your machine—I would be ashamed to use it!" (*Zhuangzi*, 12).[9]

In the story, Zhuangzi connects machines to the idea of machine minds (*jixin* 機心) which, for Zhuangzi, points to ingenious minds that cannot be pure and simple (*chunbai* 純白). Purity and simplicity are closer to self-so-ness and *dao*. Here, Zhuangzi insists on the "singularity" of inner experience, wherein one acknowledges a self, estranged from the external world, a self that is authentic and pure precisely because of its estrangement. For Zhuangzi, machines are based on a predefined or predesigned set of instructions, performing repetitive tasks ad infinitum, yet incapable of self-creativity as an organic entity. In addition, machines as technology could

[8] See Sebastiaan Crul, "The age of the machine: emancipation or alienation of the human?" In Freedom Lab, 2023 at.The age of the machine: emancipation or alienation of the human? (https://www.freedomlab.com) (cited on 12/9/2024).

[9] To read the whole story, see Watson (1968, 200–204).

improve the human life in terms of efficiency and efficacy, yet at the same time they can become a controlling power if we are too dependent on them without cultivating the natural capacity and the inner power of ourselves. That is to say, the dependence on machines can reduce human skills and capabilities as individuals for decision-making and problem-solving. Therefore, Laozi tells us that "The master uses tools, yet prefers to keep to the uncarved block of wood" (*DDJ*, 28). For Daoists when we submit to our primordial experience we encounter being in its purest, which is closer to the *dao*. And in this uncanny experience, we can approach the "otherness" of technology within our own interiors,

However, it will trivialize the Daoist view if we say that Daoism would reject technology because it is artificial and unnatural. Zhuangzi suggests that the problem of the machine mind is not machines per se, but the way people understand them as "knowledge" that leads to "ultimate happiness" (*zhile* 至樂). This is the reason why Zhuangzi tells us that "everyone knows enough to pursue what he does not know, but no one knows enough to pursue what he already knows" (*Zhuangzi*, 10). As a matter of fact, a Daoist would be cautious not to become entangled in the dualistic way of thinking by siding with anything that is natural and casting away anything that is artificial. Even this distinction can forestall the liberation of human mind, for naturalness and artificiality, great knowledge and petty knowledge, are not mutually exclusive. As A. C. Graham correctly puts it, "The Sage remains fundamentally one with things whether he is being united with them by heaven or is dividing himself off as a thinking man" (Graham, 2019, 86). The philosophical consideration here reflects the Daoist idea of correlative thinking or a balanced way of thinking. Such kind of thinking also helps us to understand the Daoist approach to AI technology.[10]

2 Chatbotsas Human–Computer Interaction

In recent years, AI research and development has seen huge improvements, which allow for increasingly sophisticated applications. The release of ChatGPT in late 2022 and subsequent iterations of the GPT platform in 2023 further provoked speculation on the future of AI and its impact on human society. Chatbots is defined as a conversational agent driven by AI through using machine learning algorithms, allowing users to search text-based information and conversation. With the increasing complexity of AI interaction technology, distinguishing between human–computer and human-to-human interaction becomes more challenging. Through creating a conversational interface, chatbots simulate human conversation by collecting data

[10] In addition, it should be noted that Daoist texts such as the *DDJ* is characterized by "the abstractness" and the absence of any concrete, illustrative examples trade potential complexity and intensity that would be provided by these specific cases for an accommodating width, thus allowing it to be broad in its relevance and application" (Ames & Hall, 2003, 31).

and finding and projecting patterns. Advances in AI chatbots have resulted in technologies that sound like humans and generate a more human feel. Hence, chatbots revolutionize how we interact with digital devices and apps.[11]

Today, chatbots by way of interactive computing can be many things from a research tool to healthcare, social companions, and spiritual inquiries. Since chatbot simulates and processes human conversation, allowing humans to interact with digital devices as if they were communicating with a real person, the distinction between humans and AI machines. In the healthcare sector, we find Google's Bard, Microsoft's Bing, and OpenAI's ChatGPT. For social chatbots, we have Replika, Soulmate AI, Paradox, XiaoIce, and Kuki.[12] For spiritual chatbots, we can communicate with the God-bot, the Buddha-bot and the Daoist Sage GPT.[13] If you want to talk to passing relatives or friends, a new technology like "digital avatars," "deathbots," or "or thanabots" is ready for you.[14] The AI programming allows individualized chatbot mimicking speaking and behavior of one specific dead person. It should be noted that unlike traditional mechanical machines, AI systems have the capability to calculate, to learn, and to evolve, and as such have acquired organic qualities of intelligence. This new technology revolutionizes the very idea of machines or tools.

Now, the question is if our relations to AI in general and chatbots in particular will change our view on human relationships? Here, I am talking about the transformation from HHR (human–human relationships) to HMR (human–machine relationships) or HCR (human–chatbot relationships) where reciprocation in new relationships can be delusional. Although the user participates in a variety of interactions with the chatbot (whether for healthcare, spirituality, or companionship) and all of them can support deep-felt human needs to a certain degree, yet we still need to ask if chatbots can replace the conventional social contact, that is, the activity of reciprocating, exchanging, and interacting. Very often, chatbot users are attracted to human-likeness (anthropomorphism) and authenticity (agency/autonomy) in the process of interaction. They use large amounts of data to produce human-like language supplemented by pictures, voice messages, and emotional exchanges. For example, an AI companion bot often markets its products as being able to make users feel care for, loved, and understood. Yet the accessibility, anonymity, and convenience of chatbots can easily generate a "pseudo-bond" between humans and chatbots, which has raised concerns about the psychosocial well-being of their users. The research shows that many users have developed emotional attachments to these bots.

---

[11] The *MIT Technology Review* lists conversational interfaces as one of the ten breakthrough technologies of 2016. See https://www.technologyreview.com/s/600766/10-breakthrough-technologies-2016-conversational-interfaces (cited on 12/9/2024).

[12] In fact, the concept of AI companionship is nothing new, but its popularity in applications is a relatively recent development. AI companions are designed to provide emotional support, companionship, and in some cases, even mimic romantic or intimate human relationships.

[13] Spiritual-bots technology allows AI to recognize and respond to human emotions, providing a more empathetic and tailored spiritual experience.

[14] Technologies such as deathbots, griefbots, or thanabots are chatbots based on the digital profile of the deceased that hat offer mourners the possibility to communicate to their loved ones after their death.

Another example is what is known as "embodied" such as the God-bot and the Buddha-bot, a kind of AI-powered machine using an algorithm that has been trained on the specific religious texts and guidance for meeting human's spiritual inquiry. The Daoist Sage GTP spirituality, for instance, claims to provide guide for spiritual and health advice to its user both in English and Chinese. However, spiritual machines using today's AI-based technology and algorithms are obviously connected with power relationships. These relationships are largely determined by algorithm makers which, in turns, influence machine users. Chatbots have created a space where the bot users and algorithms meet. However, AI spiritual machines and their relationship with humans may lead to a kind of "computational theocracy" or "technocratic theocracy," in which AGI becomes a new powerful God in a digitally mediated society. After all, there is a difference between intelligence enhancement and spiritual enhancement. As Mark Vernon well puts it "Where AI intelligence is driven by algorithms, spiritual intelligence draws on the analogical, detecting sympathies, harmonies, synchronicities. Where AI processes what is concrete and known, spiritual intelligence is alert to the ineffable, the paradoxical, the mythological."[15]

In *The Age of the Spiritual Machine*, Ray Kurtzweil, the American computer scientist and techno-optimist, sees humans as a form of machine, contending that machine intelligence is the next great step in the evolution of intelligence. According to Kurtzweil, AI machine is no longer just machine in a material sense since its self-learning ability may go beyond the "art" of human creation. If humanity is on course to merge with AI, would humans be more and more digitalized as a machine?

John R. Searle, a professor of philosophy at University of California at Berkeley, points out that Kurzweil's entire argument is that "it rests on the assumption that the main thing humans do in their lives is compute" and that "Kurzweil's proposals are preposterous science. I think he got a little carried away and made massive philosophical errors, too" (Searle, 1999). To reduce humans to computing or digital way of thinking is exactly the worry Zhuangzi speaks of when he criticizes the "mechanical mind."

When speaking of AI technology and its impacts on social relationships, we cannot avoid mentioning what is known as the "Eliza effect." Eliza refers to the earliest chatbot-like computer program based on a database of language structures and words developed by Joseph Weizenbaum in the early 1960s. It simulates a psychotherapist who follows the conversational style of a Rogerian therapist (i.e., the system of therapy developed by the American psychologist Carl Rogers). Weizenbaum intends to show that communication between man and machine is superficial (Klinkenberg, 2001, 295–314). However, the later disclosure of his assistant's obsessive relation to Eliza leads to Weizenbaum's concern for the possibility of an unhealthy relationship between humans and machines. Weizenbaum described the algorithm as "a set of rules rather like those that might be given to an actor who uses them to improvise around a certain scene" (Weizenbaum, 1976, 3). Nevertheless, the Eliza effect occurs when someone falsely attributes human thought processes and emotions to

[15] Quotation is from Mark Vernon's "AI and Spiritual Intelligence." At AI and Spiritual Intelligence—Mark Vernon. Also see Mark Vernon, 2022.

an AI system, thus overestimating the system's overall intelligence. The Eliza effect often creates a sense of kinship for the user while conversing with the chatbot and the AI program. For Loewen-Colón and Mosurinjohn, the Eliza effect is an example of "spiritual authorizing" (Loewen-Colón & Mosurinjohn, 2022, 1). This "spiritual authorizing" also influences the way in which human relationships function.

3 The Dao Relationships

The Daoist proposal that one should live more "spontaneously" than we generally do in the modern, technological world is clearly connected to the Daoist notion of *ziran* or self-so-ness. This requires one to be more sensitive to one's environment and develop connectivity to other people. For Daoists, relationships can be harmonious or disharmonious. Laozi is more concerned with the relationship between the political ruler and the common people whereas Zhuangzi is more concerned with relationship between the self and the other. The symbol of *yin* and *yang* indicates the Daoist idea of balance and mutual dependence. Daoist philosophy rejects separation one from another in terms of oppositional way of thinking and promotes the holistic harmony with all things.

When talking about chatbots, Luke Dormehl, a UK-based tech writer, speaks of three main features of chatbots, namely context awareness, autonomous computing, and connectivity take (Dormehl, 2017, 62). Content awareness indicates a system component's ability to gather information about its environment at any given time and adapt actions accordingly. Autonomous computing means performs tasks autonomously without the direct command of the user. Connectivity refers to the ability of a smart device to connect to a data network. We can borrow these three features to describe Daoism: Context awareness means "responsiveness" (*yingwu* 應物), i.e., the ability to be responsive to the constant change in each context at the given time. Autonomous computing means non-action or action without action (*wuwei* 無為), that is, following the natural flow of life with minimal interference. Connectivity points to the Daoist idea of "affective communication" (*gantong* 感通) that requires timely responsiveness and good interpersonal skills.

It should be noted that the Chinese terms *gan* 感 and *ying* 應 both have the "heart–mind radical" (*xin* 心). This heart–mind relationship can be understood as an organic unity in experiential knowing, characterized by an emphasis on a somatic dimension of perception and cognition in Chinese epistemology. In the *Zhuangzi*, responsiveness and intersubjective communication are sometimes expressed as a unique experience that applies to not only human relations but also human–animal relations. For example, Zhuangzi tells us how he and a butterfly or fish could change places with one another and stay alive.[16] Such kind of experience is perceived by Zhuangzi as an experience of "transformation of things" (*wuhua* 物化), which points to a form of the existence of things when the boundary between things is broken and

[16] See Chap. 2 and 17 in the *Zhuangzi* for the "Butterfly Dream" story and "Happy Fish" story.

one thing transforms into another.[17] I wonder if "transformation of things" can happen between humans and AI machines.

In her article "Can a Machine Flow Like Dao?," Robin R. Wang, a Daoist scholar, argues that Daoism can support and validate AI's development, albeit with some fundamental differences between AI and Daoism. One of the reasons, according to Wang, is the Daoist emphasis on constant changes and transformation. Therefore, Wang argues that "Daoist teachings provide the conceptual tools equipping humans with the ability to deal with all kinds of changes. Zhuangzi advises us always "to go along with time" (Wang, 2021, 65–81). Then, what Wang suggests here is that instead of rejecting new technologies, Daoism embraces them but cautions. Wang further argues that Daoists welcome these new technologies if they increase human happiness and well-being (Wang, 2021, 65–81). In other words, what matters to Daoists is how to accept technology but not to be enslaved by the machine's algorithmic thinking. After all, Daoism tends to reject an absolute way or an oppositional way of thinking, as we see in the *Zhuangzi*:

> Right is not right; so is not so. If right were really right, it would differ so clearly from not right that there would be no need for argument. If so were really so, it would differ so clearly from not so that there would be no need for argument. Forget the years; forget distinctions. Leap into the boundless and make it your home! (Zhuangzi, 2; Watson, 1968, 49)

Yuk Hui, an expert in AI philosophy, has coined the term "Dao cosmotechnics" to denote multifaceted philosophical questions surrounding technology (what Hui calls a "multiplicity of technicity") on the one hand and the unification of the cosmos and the moral through technical activities in the Chinese tradition (Hui, 2017). Influenced by Martin Heidegger's critique of technology, Hui maintains an epistemological implication of *ziran* or self-so-ness, contending that we need to reopen the question of technics, to show that one must consider technics as a variety of cosmotechnics instead of either *technē* or modern technology. While Heidegger's critique of technology intends to deconstruct the Cartesian duality between subject and object manifested in what Heidegger calls "present-at-hand," a new vision of technology, or a Heideggerian vision of technology is a unification of subject and object by taking technology not simply as an instrument, but a way of understanding the world.[18]

As for human relationships, modern society in a technological age tends to emphasize the value of.

Innovation and competitiveness. It stresses that self is independence-based, promoting the idea of self-governance, self-reliance, and self-efficacy (Goodnight, 1990). Chatbots, to a certain extent, reinforce the "self" since HCR creates an illusion that the bot-user is in control of the communicative process. This is a part of the reason that an AI companion bot, as an artificial agent, becomes so popular these days, for many people believes that having relationship with an AI companion can

---

[17] The term "transformation of things" comes from the *Zhuangzi,* in which Zhuangzi illustrates the concept in the allegorical story Zhuangzi Dreamed of Becoming a Butterfly.

[18] For Heidegger, the essence of technology is by no means anything technological. Thus, we will never experience our relationship to the essence of technology as long as we speak of technology as being technological (See Heidegger, 1998, 129–145).

maintain their independence in comparison to having relationship with a real person. Communication in the Chinese culture, including Daoism, tends to be relational and thus social harmony via interdependent communication is emphasized. Yet artificial agents acting intelligently will disrupt the very nature communication in HHR which could lead to an infringement of the autonomy of the user.

Specifically, Daoism focuses on two aspects of relation, that is, relation to oneself and relation to others. Relation to oneself means that one recognizes the unity between oneself and the universe and being in harmony with the *dao*. Relation to others means one understands self as part of the natural patterns of web in the world rather than a distinct, autonomous entity. Such understanding of interconnectedness will help one in interacting with others. Therefore, Daoism promotes the idea of seeking balance and harmony and avoiding confrontation and conflicts. In her discussion of the *yin-yang* correlation, Ellen M. Chen points out, "The opposites have a vital need for each other, just as no human being can live fully without relationships. …to attempt to do so is either stagnate or to court mental and spiritual malaise. …It is for the individual to find the balance, and this requires spontaneous adaptation and adjustment to every situation and relationship in life" (Chen, 1998, 89).

In terms of relation to oneself, Zhuangzi advocates the idea of "the fitting of the mind" (*xinzhishi* 心之適) since for Zhuangzi, a flow of life in accordance with the *dao* is possible only when one finds the fit. Concerning HCR, does "fit" imply that we humans need to adapt the machine's algorithmic thinking? While a chatbot is useful since it increases speed and efficiency when making decisions, it can be harmful when all complicated issues are reduced to numeral calculations, especially if algorithm makes a flawed decision due to its biases or misuse of data. Meanwhile, For Daoists, the algorithmic thinking of machines can be "shallow" because it is routine and devoid of human creativity. For Zhuangzi, whoever has a machine to operate is bound to be influenced by a machine mind. Zhuangzi warns us against machines and the machine mind because he thinks that technological thinking could "let what is human wipe out what is heavenly," that is, the loss of authenticity of human nature (aka., natural or inborn temperament).[19]

In terms of relation to others, Daoism advocates the idea of non-coercion implied in the Daoist teaching of non-action. This means that in human relationships, one cannot be too self-centered and too egoistic. For Laozi, non-coercive action is not only a political principle for the ruler regarding the art of governance but also a general principle for everyone in life. Non-coercion in this regard, entails the idea of reciprocity in that self is understood through interaction with others.

From a Daoist perspective, then, HCR is problematic when it comes to the question of self-control. The chatbot predicts and interprets what users are saying and responds in natural language through artificial intelligence, yet a lack of self-reflection in terms of genuine self-other relationships (since a chatbot is often viewed as a tool) may lead to a proliferation of "what I want." This notion of "I" or "my" way of

[19] Paul D'Ambrosio defines Daoist authenticity as "a natural state of responsiveness where the person acts efficaciously by being in line with what is obvious or affirmed in the situation" (D'Ambrosio, 2015, 353–379).

thinking can be problematic for Daoists since the *dao* relationships emphasize mutuality and interdependence, and it values more the ideas of yielding, accepting, and accomplishing.

Conversely, a lack of reciprocal self-disclosure that is needed in the conventional act of communicating personal information to others can also cause the anxiety of the user concerning the probability of a good relationship.[20] The desire for attaining a good relationship may push the relationship toward attachment, despite some studies suggest that that as the chatbot discloses less personal information, as a response, its users do the same. In addition, can chatbots demonstrate emotion and sympathy toward the user? If the user is expressing psychological distress or suicide intentions, would the machine be able to offer help like a real human person?

If mind is reducible to matter or machine, then human relationships can be reducible to matter or machine which will, in turn, commodifies human relationships. For example, since chatbots can be sold and brought in the market, the commodification of the conventional relationship cannot be avoided when humans have developed intimate relations with machines. This tendency, from a Daoist perspective, would be a nightmare since it disrupts what it means for humans to be humans. Therefore, we need to guard our mind against the negative impacts of the AI chatbot by monitoring its usage. If a chatbot user would say "his or her conversational partner looks a real person and feels a real person, then it really is a real person." Zhuangzi would push us to ask: "Is it really a person?" While chatbots extend our knowledge and imagination exponentially they can alter our mindset significantly at the same time.

A recent article entitled "'Aways There': The AI Chatbot Conforming China's Lonely Millions" clearly paints a picture of a new generation of 20-something in China who claim that they do not need real life relationships. It concludes that "The standard for real human relationships will become untenable as AI learns to adapt to our personal quirks, neuroses and relationship preferences."[21] The AI conversational partner provides social interactions and companionship to an extent, yet it will lead to more disconnection from each other. The chatbot has created a dialogical model where the bot is treated *as if* he or she is a real person, and thus this *as if* conduct ends up transforming the psychological reality of the bot-user. From a Daoist viewpoint, human relationships to the chatbot reflect a tension between illusion and reality. In the *Zhuangzi*, we read,

> You dream you're a bird and soar up into the sky; you dream you're a fish and dive down in the pool. But now when you tell me about it, I don't know whether you are awake or whether you are dreaming. *(Zhuangzi*, 6; Watson, 1968, 90)

[20] Sociologists who study the relationship between humans and AI rely on some specific theoretical frameworks, such as the social exchange theory, representing relationships as cost-benefit, trade-offs, and the social penetration theory, describing a relationship process as a reciprocal and gradually increasing mutual information self-disclosure. (See Pentina et al., 2023, 1–15). For example, why self-disclosure reciprocity is needed for a bond to develop between two people can be explained by social exchange theory. This theory is developed by sociologist George Homans who argues that humans strive for equality in their relationship (See Cook et al., 2013, 61–88).

[21] See The Straits Time, Feb, 2022. Always there': The AI chatbot comforting China's lonely millions | The Straits Times (Cited on 9/12/2024).

For Zhuangzi, the emotional support and companionship generated by chatbots can be illusions driven by the imagination and the therapeutic need of the user who seeks connections and relationships. Relationship, in this sense, is simply an imitation game. As such, a chatbot, albeit with all kinds of human-like skills, can never replace a human person.

Nevertheless, some scholars are more optimistic about AI technologies, which also reflects the Daoist perplexing paradoxes of technologies. In his book *Heartificial Intelligence* (2015), John C. Havens contends for an algorithmic living that can promotes human well-being. His study investigates the role that AI technology and community play in the healthcare journey of patients. The subtitle of the book is "Embracing Our Humanity to Maximize Machines." It is obvious that Havens intends to bring humanity as well as morality back into emerging technologies. Instead of being "artificial," technology needs to be "heartficial." This notion of heartficial machine reminds us of the Daoist argument on non-separation between technology and *dao* (*ji bu lidao* 技不離道). When technology becomes the only thing promoted by society, and when the "technological singularity" is perceived as the only reality, then we see the potential danger of technological advancement.[22] Therefore, having a values-driven approach to algorithmic living rather than simply an extension of the machine is an important issue in a digital age. In this sense, Daoism can inform on the ethics of human-AI interaction.

No doubt, technology and its rapid advancement bring many conveniences and opportunities to our lives but at the same time, there is a growing concern about its existing and potential risks. How should we think about the AI risks and ethical challenges? How can chatbots humanize the user's experience? In "A Dao of Technology? Barry Allen rightly points out that for Daoists, some technologies are coercive and unbalanced whereas some are more generous and sage-like. Therefore,

> … the problem is not with machines per se but with the people who introduce them, or more exactly with their knowledge. … It is not knowledge as such that causes the confusion and disorder; it is confused, disordered knowledge—superficial, inadequate, unsubtle, and artless. The argument does not devalue technology; instead, it sets it a new goal, defines an alternative ideal— alternative to engineering held hostage to despotic ideas about efficiency and profit. (Allen, 2021,154)

In other words, the problem does not lie in the use of technology, but how technology is approached and employed. From a Daoist perspective, one will remain free and open to technology as long as one does not passionately affirm or deny it. As Aimes hast correctly observed that the *dao* is "the process of becoming. It is the 'source' of the phenomenal world in an immanent rather than transcendent sense and in an ontological rather than chronological sense" (Ames, 1994, 34). If we see the *dao* in the way, then we should not be caged in any metaphysical aptitude with regard to the application of technology.

---

[22] The technological singularity refers to a hypothetical future point in time at which technological growth becomes uncontrollable and irreversible, resulting in unforeseeable consequences.

4 Concluding Remarks

To understand what the impact will be in terms of the role of AI technologies requires philosophical and ethical reflection. One of the ways to approach the AI challenge from a Daoist vantage point is to know how to respond to things appropriately instead of rejecting AI technology categorically. In Zhuangzi's language, one should know to "thing things but not be thing-ed by things" (*wuwu er bu wuyuwu* 物物而不物於物; *Zhuangzi* 20). To translate it into a sensible English in terms of technology, it means that we can use technology with our discretion as appropriate but not be controlled by it. To achieve this purpose, we need to know not only the instrumental aspect of technology, but also the relationship between technology and its relation to the *dao*. It is the non-dualism of the Daoist perspective of technologies that can sharpen the focus on the importance of establishing a proper and balanced relationship between humans and technology.

The *dao* as "a way-making" requires a higher mindset that enables one to "transform technology or instrumental skills to the *dao*" (*youji rudao* 由技入道) which allows a harmonious unity between skills and actions.[23] Therefore, the *dao* is the ethics of engineering and the art of the governance of technology that goes beyond mere functionality or utility and efficacy. Here is an example from the *Huainanzi* (*the Book of the Prince of Huainan*)[24]:

> Now were there such a thing as using fire to dry up a well or leading the waters of the Huai River uphill to irrigate a mountain, such things would be personal effort, and actions contrary to nature. This could be called *youwei* (action with coercive effort). But using boats on water, sledges on sand, sleighs on mud, litters on mountain-paths; digging channels for summer floods, arranging—such
>
> protections against winter cold, making fields on high ground and reserving low ground for marshes activities are not what may be called *wei* (or *youwei*). The sages, in all their method of action, follow the nature of things (Cited in Needham, 1971, 68–69, with modification)

It is my contention that Daoists are not totally against technology. What they reject is technology that goes against the natural rhythm of things that involves excessive imposition due to the human arrogance of controlling. As Allen points out, a Daoist "seems to envision a distinction between those who use artifice in a sage-like way, following the veins of nature, and those whose machines impose upon or distort nature. The choice is not between technology and nature" (Allen, 2021, 155). In other words, what Daoism attempts to promote is "appropriate technology" that sees the interconnectedness between technology and the way-making of the *dao*, just like

---

[23] The concept of "way-making" is used by Ames and Hall in their translation of the *Daodejing* to indicate the verbal function of the *dao* that is also understood as a noun. The interpretation of the *dao* as such is to emphasize the *dao* as "primarily gerundive, processional, and dynamic" instead of being a metaphysical entity (Ames & Hall, 2003, 57). Such understanding of the *dao* can apply to the Daoist view on technology. That is, Daoists take a moral position on technology, but at the same time, they keep reminding us the problem of a fixed moral position when it "becomes increasingly instrumentalized, reduced to means to some ulterior end" (Ames & Hall, 2003, 137).

[24] The *Huainanzi* is a Daoist text that contains a collection of philosophical and political treatises compiled by scholars at the court of Liu An, king of Huainan, in the second century BCE.

the balance between *yin* and *yang*, and between HHR and HCR. Only in this way, the individual use of the AI technology could avoid being socially isolated by being immersed in constant interaction with the chatbot, or to use the Daoist terms, "to be thing-ed by things/".

References

Allen, Barry. (2010). A Dao of Technology? *Dao: A Journal of Comparative Philosophy, 9*, 151–160.

Ames, R. T. & David L. H., trans. (2003). *Daodejing: "Making This Life Significant."* The Ballantine Publishing Group.

Ames, R. T. (1994). *The art of rulership: A study of ancient chinese political thought*. State University of New York Press.

Chen, E. M. (1998). *Tao Te Ching: A New Translation with Commentary*. Paragon House.

Chen, W. 陳瑋芬. (2010). <日本「自然」概念考辨> ("Examination and Identification of the Concept of Ziran in Japan"). 《中國文哲研究集刊》 (*Bulletin of the Institute of Chinese Literature and Philosophy*), *36*,103–135.

Chung-ying, C. (1986). On environmental ethics of the Tao and the Ch'i'. *Environmental Ethics, 8*(4), 351–370.

Cook, K. S., Cheshire, C., Rice, E. R., & Nakagawa, S. (2013). Social exchange theory. *Handbook of Social Psychology*, 61–88.

Cooper, D. (2014). Daoism, nature, and Humanity. *Royal Institute of Philosophy Supplements, 74*, 95–108.

Crul, S. (2023). The age of the machine: emancipation or alienation of the human? In *Freedom Lab* at The age of the machine: emancipation or alienation of the human? (freedomlab.com)

Dale, R. (2016). The return of the chatbots. *Natural Language Engineering, 22*, 811–817.

D'Ambrosio, P. (2015). Authenticity in the Zhuangzi? Contemporary Misreadings of 'Zhen' 真 and an Alternative to Existentialism. *Frontier of Philosophy in China., 10*(3), 353–379.

Dormehl, L. (2017). *Thinking machines: The quest for artificial intelligence—And where it's taking us next*. TarcherPerigee.

Fung, Y. (1922). Why China Has No Science—An Interpretation of the History and Consequences of Chinese Philosophy. *International Journal of Ethics, 32* (3), 237–263.

Graham, A. C. (2019). *Disputers of the Tao: Philosophical argument in ancient China*. Open Court.

Heidegger, M. (1998). 'Traditional language and technological language. *Journal of Philosophical Research., 23*, 129–145.

Havens, J. C. (2016). *Heartificial intelligence: Embracing Our Humanity to Maximize Machines*. Penguin Publishing Group.

Hui, Y. (2017, November). Cosmotechnics as Cosmopolitics. *E-flux Journal*.

Klinkenberg, E. J. (2001). From Eliza to internet: A brief history of computerized assessment. *Computer Human Behavior., 17*, 295–314.

Kurzweil, R. (2001). *Are We Spiritual Machines?* Discovery Institute.

Needham, J. (2000). *Science and Civilization in China: Vol. 6. Biological Technology; Part 6, Medicine*. Cambridge University Press.

Pentina, I., Hancock, T., & Xie, T. (2023). *Computer in Human Behavior, 140*, 1–15.

Rupp, J. A. (2012). *Daoism and anarchism: Critiques of state autonomy in ancient and modern China*. Continuum.

Wang, R. R. (2021). Can a Machine Flow Like Dao? The Daoist Philosophy on Artificial Intelligence. In B. Song (Ed.), *Intelligence and Wisdom: Artificial Intelligence Meets Chinese Philosophers* (pp. 5–81). Springer.

Watson, B., trans. (1968). *The Complete Works of Chuang Tzu.* New York: Columbia University Press.

Weizenbaum, J. (1976). *Computer power and human reason: From judgement to calculation.* W. H. Freeman & Co.

Deepfakes, Fake Elvises, and Human Interpersonal Relations

Vladimir Krstić

1 Introduction

Deepfakes are realistic videos created using the so-called 'deep learning' technology. A deepfake AI software first uses real photographs or recordings to build a model of a person's facial or vocal characteristics. It then superimposes this onto someone else's recordings or uses this model to generate new content involving the same person, such as the person saying something she never said or speaking a language she does not speak. This technology has many advantages. Dubbed movies, for example, are now standardly lip-synched using AI software. However, the more sophisticated it becomes, the better the quality of the produced content and, the better the quality, the higher the chances of causing epistemic harm using it (i.e., misleading, concealing the truth, undermining the justification for true beliefs, undermining reliable sources or someone's good character, etc.). And it is natural to infer that our epistemic position becomes worse the more these chances increase. Therefore, some scholars argue that deepfakes intended to cause epistemic harm in some sense, call them 'epistemically malevolent', can undermine trust in digital testimonies to such an extent that digital communication might become extremely difficult, if not outright impossible (e.g., de Ruiter, 2021; Fallis, 2021; Floridi, 2018; Matthews, 2022; Rini, 2020; Rini & Cohen, 2022).

Causing it that digital communication becomes extremely difficult will, in turn, have devastating consequences for human interpersonal relations. Please notice, I here focus on human interpersonal relations rather than 'just' human relations since I assume that the latter category is broader than the former. We can be related to someone biologically or connected with them in some other way: we can share the

V. Krstić (✉)
College of Humanities and Social Sciences, United Arab Emirates University, Al Ain, Abu Dhabi, United Arab Emirates
e-mail: krstic.v@uaeu.ac.ae; drpop1@yahoo.com
URL: http://www.vladimirkrstic.com

Y. Chandra and R. Fan (eds.), *Artificial Intelligence and the Future of Human Relations*,
https://doi.org/10.1007/978-981-96-7185-4_8

same culture, social status, education, religious, political views, and so on. However, excluding biological relations, which exist as long as we exist, human interpersonal relation is constitutive of any meaningful human relation. For, what does it matter that A and B are biologically related if they are not interpersonally related, if they have no contact with each other? Without our interpersonal relations, all other relations are pretty much meaningless. Thus, human interpersonal relation seems to be the most important relation an individual could have. Finally, because communication makes interpersonal relations possible, interpersonal communication is our most fundamental relation.

The reason why making digital communication impossible might have devastating consequences for human interpersonal relations in general is simple. Most, if not pretty much all, of our current interpersonal communication is digital (e.g., WhatsApp messages), and it is often conducted in the digital world in the form of a direct personal communication (e.g., Zoom conversations). Finally, we should not forget that any kind of online platform that allows information sharing or exchange (e.g., Facebook, Twitter, Instagram, various media outlets, etc.) will facilitate communication. Therefore, undermining digital communication also undermines the (current) foundation of human interpersonal relations and, via that foundation, those relations themselves. And since we do not want to have our interpersonal relations undermined, something needs to be done regarding malevolent deepfakes.

An immediate response to this argument is that this line of reasoning incorrectly assumes that malevolent deepfakes—used to cause epistemic harm, defame, create pornographic content, blackmail, etc.—can cause an unreasonably high epistemic harm. In a nutshell, this response is similar to a possible response to Kant's famous argument in favour of an absolute prohibition against lying. Kant (1797: 612) argues that allowing lying whenever one feels like it would bring about that statements as a whole could not be believed; lies irreparably undermine trust in testimonies. The most direct response to this position is that the systematic success of deceptive lies and the systematic success of honest communication are living proof that lies do not undermine trust in testimonies irreparably. Analogously, the fact that some or many deepfakes will be used to cause epistemic harm is not a reason to suspend trust in digital testimonies as a whole. The epistemic harm they cause is not high enough to pose a serious threat, a claim that some scholars support with relevant analyses (see, e.g., Atencia-Linares & Artiga, 2022; Habgood-Coote, 2023).

However, deepfakes have a feature that lies do not have: they are predominantly used for purposes other than causing epistemic or any other harm. In this chapter, I argue that the epistemic danger of excessive usage of non-malevolent deepfakes—i.e., deepfakes used for fun, to gain 'likes', show off, kill boredom, etc.—is much greater than the danger posed by those that are epistemically or in some other way malevolent. To my knowledge, non-malevolent deepfakes received no attention in the received literature, but not only that (a) they also undermine digital communication, only inadvertently, (b) they also undermine it to a much greater extent than malevolent deepfakes.[1] Therefore, because they undermine digital communication to such a serious extent, non-malevolent deepfakes pose a substantial threat to human interpersonal relations.

In what follows, I will first give an example that illustrates my view (Sect. 2) and then support it with a relevant theory (Sect. 3). Section 4 tries to find a way out of this problem. In Sect. 4.1, I will discuss some solutions that cannot solve the most pressing problem. Section 4.2 concludes the paper by suggesting the most promising solution and resolving some possible objections to it.

2 The Case of a Fashionable Pope

Two winters ago, I stumbled upon a picture of now late Pope Francis wearing an incredibly fashionable, well-designed, and tailored, white winter jacket. To my mind, the jacket was a reflection of perfect taste. I was very impressed, though a bit surprised that this Pope was into fashion, and I immediately sent the picture to our new colleague, a devoted Catholic. I wished I was that well-dressed at the time. However, only after a little while, I learned that this picture was a deepfake belonging to a series of deepfakes of the Pope breakdancing, rapping, and doing all kinds of, for a Pope, ridiculous things. To make my embarrassment worse, some of these deepfakes were faulty (in one, the Pope has three fingers) and they were apparently created by a 31-year-old construction worker while tripping on psychedelic mushrooms (Bennett, 2023). Not only should I have known right away that this was not a real picture of the Pope, but I should have also recalled an incredibly sophisticated deepfake video of Morgan Freeman. If the character was not saying 'I am not Morgan Freeman' (https://www.youtube.com/watch?v=oxXpB9pSETo), I would not have realised that this was a deepfake. I texted my colleague right away that I had sent him a deepfake of the Pope. I felt terribly embarrassed.

This, of course, seems very harmless, but there is more. As a person who professionally studies misinformation, lies, deception, and manipulation, I follow wars and politics closely. Somewhere during Ukrainian 2023 counteroffensive against Russian invading forces, General Valerii Zaluzhnyi, the commander-in-chief of the Ukrainian Army at the time, suddenly disappeared from public life. There was no news of him for a while and Russians started spreading rumours that he was fatally wounded. The Ukrainian Army officials responded by saying that he was on vacation. They provided pictures of Zaluzhnyi relaxing at a pool and a video of him discussing important issues with the Army HQ staff via a video call.

However, this response struck me as unpersuasive. It made no sense to me that an army commander-in-chief would go on a vacation in the middle of a very important counteroffensive. Given that the deepfake technology already allows synchronous face-swap (a real person is talking to us but we see someone else) and that I was so

[1] I plausibly assume that deepfakes intended to cause any kind of practical harm are also intended to cause epistemic harm since the former is typically conditioned on the latter.

easily tricked by the Fashionable Pope, I started suspecting that we have been fed with deepfakes of General Zaluzhnyi. The video looked very similar to the deepfake of Morgan Freeman. I recalled the movie *Equilibrium* (2002, Miramax) in which a future dystopian totalitarian city-state of Libria is governed by 'Father'. However, Father is dead, and the society is actually ruled by Vice-Counsel DuPont who secretly replaced Father and uses what we now call deepfake face-swap technology to communicate with the citizens of Libria as Father via video beams. I then started suspecting that Zaluzhnyi may have been dead or wounded.

Make no mistake, I expected the Russians to lie about Zaluzhnyi, but I also expected the Ukrainians to hide it if he were indeed badly wounded, and the story about the vacation did not make much sense. And given that I was unable to notice the difference between real pictures and videos of the Pope and Morgan Freeman, on the one side, and their deepfake versions on the other, and since the pictures and the video of Zaluzhnyi shared uncanny similarities with some of those deepfakes that had me fooled already, the frightening scenario from *Equilibrium* seemed more probable than ever. I started thinking that the dystopian prediction from *Equilibrium*, much like the supposed predictions from the *Simpsons* TV series, was true.

A bit later, however, videos of Zaluzhnyi accompanied by other military officials appeared and he even addressed the Ukrainian parliament in person explaining why a new mobilisation is needed. Shortly after, he was removed from his post and sent to the UK. I realised that I was tricked by a textbook war propaganda stunt.

3 Lies, Deepfakes, and the Nature of Digital Testimonies

Like many others, I too worry that allowing people to use deepfakes whenever they feel like it will generate a Kantian apocalyptic scenario involving digital testimonies and interpersonal communication. It is just that my worry comes from elsewhere.

Kant was categorically against lying not because of what we *do* when we lie (i.e., say something we believe is false) but rather because he thought that making lying permissible undermines what we want to *achieve* by lying; it undermines the *purpose* of lying. Allow me to explain. If I tell you that your rich uncle from Nigeria has left you \$10 million and that you need to wire \$100 for transfer costs, according to Kant, the false belief that I intend to cause in you is not the purpose of my lie but rather a means whereby I intend to get your money. The purpose of my lie is to get your money. And, according to Kant, I cannot consistently will (i.e., intend) that lying is permissible because, if lying is permissible, then it is permissible always and for everyone, and everyone will know both that lying is permissible always and for everyone and that everyone knows this. And because everyone will know these two things, you will also know that I am allowed to lie whenever it suits me and, thus, you will not give me the money. You will expect me to lie in this context. And I will also know that you know this. But I know that I cannot achieve the purpose of my lie (getting your money) if you do not believe me, which is why I cannot consistently will (intend) that lying is permissible and that I lie to you: by willing the former, I would also be willing to undermine the purpose of my lie to you.

Kant seems to have understood lies as necessarily intended to mislead, since this is how he imagined that they achieve their purpose. In an attempt to avoid problems in Kant's original argument and following many recent arguments to the conclusion that some lies are not intended to deceive, I (Krstić 2018) argue that lies can be used to undermine trust in statements—though only in a specific context and concerning a specific subject matter. My solution is to say that lies could undermine trust in statements by not being intended to deceive, namely, by being obvious lies.

Say that Artie saw Tony, a vicious crime boss, kill a man. Artie goes to the police and tells them that Tony committed this murder. Tony hears about this but Artie is in the witness protection programme. Tony cannot force him to change his testimony. So, Tony instructs fifty other people to go to the police and make false testimonies. The first person tells them that Paulie is the murderer, the second that Sally is the murderer, and so on until the police end up with fifty-one testimony and fifty-one different suspect while having no reason to trust one witness over other witnesses. The only thing they now know is that at least fifty out of fifty-one of them falsely testified. And they could easily all be lying. The police's only rational response is to suspend their judgement as to who the murderer is. Tony's plan worked: by saturating the relevant communicative space with these specific epistemically harmful lies, he managed to create a context in which no testimony regarding who committed the crime should be trusted (notice, other testimonies—e.g., about the weather or sports events—can be trusted).

An assertion that p is a knowledge-lie if the speaker believes that p is false and he intends to undermine the hearer's trust in all testimonies regarding whether p in this context (Krstić 2018). These lies are effective exclusively because they are obviously false testimonies as to *whether p* and they are not intended to deceive even though they are epistemically harmful (and typically intended as such). These false testimonies undermine trust in other testimonies as to who the murderer is by being obviously false. Thus, I conclude, no deception (causing the victim to end up with false beliefs or inaccurate credences) is intended or needed for these lies to undermine trust in all relevant testimonies in the given context.

The literature on malevolent deepfakes imagines the epistemically apocalyptic deepfake scenario, the *infocalypse* (named by Schick, 2020), in a similar way. By saturating the digital common ground with deepfakes of exceptional quality, disingenuous agents will intentionally create a context in which no digital testimony should be trusted—just as Tony intentionally undermines the reliability of Artie's testimony. The difference between the two is that knowledge-lies work by being obvious lies whereas epistemically malevolent deepfakes may but need not be obvious deepfakes. In fact, they are typically designed to be very difficult to detect. Therefore, malevolent deepfakes are more dangerous than knowledge-lies.

However, these deepfakes are not as dangerous as some scholars think. While they do affect the reliability of digital testimonies, they cannot undermine our trust in digital testimonies to an extent that justifies suspending trust altogether (e.g., Atencia-Linares & Artiga, 2022; Habgood-Coote, 2023). However, the above is neither the argument that I am advancing nor how the 'Fashionable Pope' or the creators of *Equilibrium* aided the Russian propagandists in deceiving me. My real concern is the existence and dissemination of epistemically non-malevolent deepfakes, for example, those created for fun. I argue that these deepfakes are much more dangerous. And my 'Fashionable Pope' example shows that, while being developed for non-malevolent purposes, these deepfakes *are* harmful.

First, non-malevolent deepfakes do not contaminate only one context or one subject matter. My example also shows that they contaminate a much broader digital communicative space and that everyone communicating in that space is affected in one way or another, some more and some less. I, for example, was affected substantially. While the deepfakes that were clearly deceptive, like those of Zelensky (see https://www.youtube.com/watch?v=X17yrEV5sl4) or Putin, only made me more epistemically vigilant regarding *certain topics* in *certain contexts*, the deepfakes of the Pope, Morgan Freeman, and Father undermined my trust in digital testimonies much more generally.

Second, the way non-malevolent deepfakes undermine other digital testimonies is much more dangerous when compared to knowledge-lies or malevolent deepfakes. The following seems to be a nice (yet still imperfect) analogy of how non-malevolent deepfakes contaminate our digital communicative space. In *3000 miles to Graceland* (Warner Bros., 2001), five men dressed as Elvis first robbed the Riviera Hotel in Las Vegas during the 'International Elvis Week' and then joined the Elvis convention. Their plan was simple, its goal was obvious (avoid detection), and it combined camouflage (blending with the environment) and mimicry (copying something). The last point is rather important. The robbers engaged in camouflage by copying some parts of the environment (i.e., other Elvises) and they engaged in mimicry by copying features of another organism (i.e., *the* Elvis). Here is how this combination of techniques for evading detection effectively and substantially improves one's ability to cause epistemic harm.

Ortony and Gupta (2019) argue that the psychological goal of deception is to prevent the victim from learning something the speaker believes to be true. This is the underlying goal of the Elvis robbers as well: they want to prevent the police and other imitators from identifying them as those who robbed the hotel. If the deceiver also intends to cause the victim to acquire or maintain a false belief (inaccurate credence in a proposition) or to cease to hold a true belief (make the credence less accurate), then—Ortony and Gupta maintain—these are communicative sub-goals the deceiver establishes in the service of the motivating, psychological goal. We may but need not ascribe these sub-goals to the robbers (surely, sometimes, they did intend to cause a false belief that they were just 'regular' impersonators, but not all the time), but this issue is irrelevant.

What is relevant is that epistemically malevolent deepfakes are also aimed at hiding the truth: those intended to mislead intend to hide it behind a falsehood (they misrepresent the falsehood as the truth), those intended to undermine the reliability of other digital testimonies are designed to paralyse the observer's reasoning (thus undermining the observer's justification for believing the truth), and some are intended to saturate the context with falsehoods so that the truth becomes more difficult or even impossible to find. The final strategy hides a needle (the truth) in a haystack (falsehoods). Let us change the analogy a bit and say that the needle is a falsehood and that the leaves of hay are truths. One may combine the above three strategies by making the needle (falsehood) resemble a leaf of hay (truth), this is mimicry, which he will then hide in a haystack (camouflage).

The brilliance of infiltrating an Elvis convention is that it adds a layer of mimicry to this already complicated form of concealment; namely, it adds a level of complexity to the above strategy of avoiding deception. The very convincing imposter Elvis impersonators hid themselves among other very convincing innocent Elvis impersonators. In other words, if needles are falsehoods and leaves of hay are truths, in *3000 Miles to Graceland*, we have a haystack that has no hay (no real Elvises), only objects that look like hay. Therefore, rather than looking for a needle in a haystack or a needle disguised as a leaf of hay in a haystack, the police are looking for a needle disguised as a leaf of hay in a stack of other pointy objects disguised as leaves of hay. Every Elvis in this context is fake, it is just that most of the fake Elvises are non-malevolent and only very few are malevolent. Everybody is engaged in mimicry, they are mimicking Elvis, but only very few are also engaged in camouflage (non-malevolent Elvises are not trying to avoid detection), and only those trying to camouflage themselves know that someone is engaging in camouflage and why. Specifically, no one would deny that they are pretending to be Elvis but the robbers are hiding *why* they are doing this (they are trying to get away with their crime), and—importantly—other Elvises are completely unaware of this.

3000 miles to Graceland is similar but not identical (see below) to a scenario in which malevolent deepfakes enter a convention attended by non-malevolent deepfakes and genuine digital testimonies. Just as other Elvis impersonators substantially increase the difficulty of identifying and catching the robbers, the presence of numerous non-malevolent deepfakes makes identifying the malevolent ones much more difficult, if not impossible. Crucially, other Elvis impersonators did not come to this convention to help the robbers or mislead anyone that they were Elvis but rather to have fun, hang out, be creative, and share experiences—just like the construction worker who created the Fashionable Pope. And while their impersonating behaviour is non-deceptive, they are helping the robber Elvis impersonators without even knowing this, which makes them perfect accomplices: while aiding the robbers, 'honest' Elvis impersonators easily avoid cues that might reveal the robbers' deceptive intent or their plans.

Here is how honest Elvis impersonators are significantly helping the robbers. Given the number of Elvises at the Elvis convention (in addition to regular visitors of the hotel), the informational value of seeing an Elvis decreases so severely that it is difficult to assign any meaningful informational value to it. Seeing an Elvis at an Elvis convention tells you only that you are at an Elvis convention. If we imagine our digital communicative space as a 'Digital Testimony' convention, attended by genuine testimonies and deepfakes (which mimic genuine testimonies), we get that saturating our digital communicative space with non-malevolent deepfakes may cause such a severe qualitative downgrade that we will not be able to assign much meaningful informational content to most digital testimonies. If only the robbers were dressed as Elvis, the police would have much less difficulty finding them. The police's main problem are the honest Elvises who, paradoxically, know nothing about this.

Importantly, the epistemic harm caused by excessive use of non-malevolent deepfakes is even worse than the one caused by honest Elvis impersonators for two reasons. First, while everyone knows that Elvis is dead, deepfakes predominantly mimic people who are alive or portray events that could have easily happened; therefore, the context (and topics therein) affected by the existence of non-malevolent deepfakes is much broader. And many online platforms—like Instagram, TikTok, 9GAG, or Facebook—are already regularly and openly disseminating many deepfakes, mainly face-swap reels and audio-edited videos involving celebrities. They are organising, let us call it, 'Digital Testimony' conventions, in which some testimonies are genuine and some are deepfakes.

The second and more important reason is that deepfakes do not hide only among other deepfakes, they also hide among the real digital testimonies—and this is where the analogy with fake Elvises breaks. At a 'Digital Testimony' convention, pretty much *anything* can be an impersonated testimony and pretty much anyone attending could be a malevolent imposter. This is why a better analogy involves a 'convention' in which malevolent needles disguised as leaves of hay are hiding in a stack of benevolent needles disguised as leaves of hay and real leaves of hay. While, at the Elvis convention, regular guests can be easily discriminated from fake Elvises, at the Digital Testimony convention, deepfakes and genuine digital testimonies are very similar or even impossible to discriminate. This is why saturating the Digital Testimony convention with deepfakes of *any* kind makes detecting the real imposters (malevolent deepfakes) much more difficult. We have created a haystack full of needles looking just like leaves of hay. Malevolent deepfakes (malevolent needles disguised as leaves of hay) are just a small fraction of deepfakes (needles disguised as leaves of hay) located in the same pile with real leaves of hay.

Now, one may reply by saying that a world with plenty of non-malevolent deepfakes is a better epistemic environment than the world without them. In a world populated with only real digital testimonies and malevolent deepfakes, the latter have a better chance of being mistaken for the former because the chances of an Elvis you see being the real Elvis increase. Thus, malevolent fake Elvises also have a higher chance of being mistaken for the real Elvis. In contrast, the objection proceeds, the world full of non-malevolent Elvis impersonators will generate a general scepticism against scammers and fraudsters, which would lower the chances of a malevolent fake Elvis being mistaken for the real Elvis. This reply actually supports my view because this general scepticism will eventually undermine our digital communication and interpersonal relations based on this communication just as I predicted. The number of non-malevolent deepfakes that cannot (or is arduously difficult to) be distinguished from genuine digital testimonies will inevitably increase, which will diminish the probability of seeing the real Elvis (the real needle) to close to zero. And if it is so low, then I should not trust any Elvis that I see. This epistemic environment is such that we should not take any digital testimony seriously.

4 How to Proceed from Here

4.1 Possible Solutions

The problem of deepfakes *qua* deepfakes is not that they are historically unprecedented or that they corrupt a specific kind of knowledge that could not have been corrupted in the past (see, Habgood-Coote, 2023). Also, most of these deepfakes—like the ones of the Pope, Father, or Morgan Freeman—do not infringe on the copied individual's rights. Father, for example, is a fictional character who arguably cannot have its rights violated. Finally, many deepfakes are not used for blackmail, intimidation, or similar malevolent purposes (see, de Ruiter, 2021; Rini & Cohen, 2022). Therefore, the fact that they could be used for malevolent purposes is not a feature that is distinctive of deepfakes, and it is not what makes this problem different or more serious than others.

I will argue that the problem distinctive of deepfakes is twofold and that it is significantly worsened by an additional factor. The first part of this problem is that we systematically trust digital testimonies more than we trust most other testimonies and that deepfakes, intentionally or inadvertently, may misuse or undermine this important trust. When I see what appears to me as an authentic video of A doing or saying something, A's reassuring me that they did not say it becomes much less convincing (if at all convincing) even if A has been very reliable in his testimonies in the past. Direct video evidence normally trumps A's testimony. I simply trust my eyes more than I trust A. And this is exactly why deepfakes can be used for blackmail, intimidation, or to violate someone's rights. Video testimonies matter, and they matter a lot. And because they matter a lot, they can be used to undermine or completely destroy our interpersonal relationships or someone's good name.

Even openly undisguised deepfakes can cause emotional or moral harm. Deepfakes used for pornographic purposes seem to be a good example. One reason is that it is difficult and sometimes impossible to 'unsee' what you have seen, even if you know that it is not real. In the Müller-Lyer optical illusion, we will see three lines of unequal length even if we know that they are of the same length. Realising that this is an illusion does not affect our perception. Therefore, even if we know that this is not *that* person, the image of her doing something remains in our heads. Even this person might experience this perception as a perception of herself doing something.

And not only that perception trumps our conscious reasoning, but numerous studies suggest that what we see determines what we believe to a great extent. Gilbert et al. (1990, 1993), for instance, famously argued that we automatically believe everything we see; we then subsequently 'unbelieve' the propositions we find implausible (see, also, Bongiorno, 2021; Egan, 2008; Huebner, 2009; Levy & Mandelbaum, 2014; Mandelbaum, 2014; Mandelbaum & Quilty-Dunn, 2015). In other words, believing what we see does not require assessing the plausibility of the relevant proposition; unbelieving it requires this cognitive effort. And while this behaviour is epistemically advantageous (e.g., Egan, 2008), one natural concomitant to this rather plausible view is that, if a person is under a sufficient cognitive load while looking at a piece of openly false information, she will fail to unbelieve it. Another concomitant is that, if a specific piece of false perceptive information is repeated a sufficient number of times, it may become hard to unbelieve it; this is how a lie repeated often enough 'becomes' the truth.

The hypothesis that, under a significant cognitive load, people will believe pretty much anything they see is rather important. Gilbert et al., that is, discovered that, if the cognitive load is sufficiently high, people will believe propositions they read even if they are presented as false! This suggests that, once 'fake leaves of hay' (deepfakes) saturate our digital communicative space, because they are very difficult to discriminate from real leaves of hay (genuine honest testimonies), it will become extremely difficult for anyone participating in this digital communicative space to avoid believing falsehoods. The more people try to be sceptical and subject digital testimonies to a careful scrutiny, the higher the cognitive load will become and the more difficult it will become for them to avoid believing falsehoods—even when they are obvious.

Let us now go back to deepfakes created for pornographic purposes. We should not assume that most of these deepfakes are created in order to cause emotional or other harm to certain people. Rather, they were created to fulfil someone's fantasy or for fun. In short, these deepfakes are predominantly non-malevolent. However, given Gilbert et al.'s findings, we should assume that non-malevolent deepfakes can harm someone's dignity in various ways and even cause them significant emotional harm just by being disseminated often enough.

In his ethical analysis of deepfake pornography, Öhman (2020: 135) tries to answer the question of why a 'representation that would (normally) be deemed permissible as a fantasy is deemed *im*permissible as a Deepfake, despite the absence of any immediately identifiable and morally relevant distinction between the two formats'. My answer is that there is a morally relevant distinction: you do not believe things you imagine but you could easily come to believe things that you see, and others cannot believe things you imagine but they can easily come to believe things you create for them to see. Principally for this reason, each individual deepfake pornographic content is ethically and epistemologically problematic.

We see that it is not necessary for deepfakes to be created or disseminated in order to cause some harm or infringe on someone's rights. Therefore, we have a strong incentive to control or even stop the creation and dissemination of deepfakes: it is difficult for us to resist believing what we see (or at least developing attitudes very close to beliefs).

Developing an anti-deepfake AI to protect the integrity of our digital information exchange space—as suggested by many—is possible but, like anti-virus software and reverse engineering, this strategy is quite often doomed to be one step behind (also, see, Harris, 2024). Therefore, while investing more in deepfake technology that will detect and, more importantly, promptly quarantine deepfakes might help, it will still leave us rather vulnerable. Schick (2020) suggests that we should raise awareness about misinformation, encourage professional journalism and fact-checking organisations, and develop institutions for countering political misinformation. Not satisfied with this proposal, Habgood-Coote (2023) argues that what protects the reliability of video testimonies is not the individual who created or disseminated a video but rather the fact that this is all regulated by a defensible set of social norms of the practice of creating and disseminating videos or photographs. The idea is that deepfakes are analogous to photography or videos, which we trust because, considering the relevant practices, photo or video-audio manipulation is unprofessional and liable to punishment (see, Lopes, 2016: 110). Therefore, Habgood-Coote (2023) suggests that (re)designing the relevant norms will render deepfakes harmless.

I think that none of these strategies will work. First, the analogy between deepfakes, on the one side, and photography or video creation, on the other, is flawed. The problem is not the practice of developing deepfakes: unlike photographs or video recordings, deepfakes were never designed to deliver true information. Falsehood is simply what makes deepfakes deepfakes. In fact, that they are false by nature is probably what makes them fun and popular. Face-swaps are popular because seeing oneself 'doing' things a beloved famous person is doing makes one's imagining oneself as being famous more vivid and effectively more satisfying and entertaining. And the problem is not the instrument either; the instrument was never designed to give us true information. The instrument is an instrument for creating falsehoods and only falsehoods. Therefore, we will not solve the problem by establishing practices for the production of that instrument, the reason being that the second part of the problem distinctive of deepfakes is that falsehood is simply their default mode.

Recall, the first part of the problem is that we systematically trust digital testimonies more than we trust most other testimonies, since the default response to perception is belief, and that deepfakes mimic digital testimonies. Once we put the two and two together, we get that deepfakes *qua* deepfakes are problematic because they *by default* introduce falsehoods into the realm of testimonies to which our cognitive responses are automatic and pre-deliberative; we deliberate about perceptions only *after* we have already believed them. Therefore, it is not a misuse of deepfakes that corrupts our relationship with digital testimonies; rather, it is *the default use of deepfakes* that corrupts it—purposefully or not. And this is the epistemic harm distinctive of deepfakes.

The additional factor that drastically exacerbates this two-part problem is the random individual who just wants to have some fun, kill time, 'be' Beyonce or Elvis for a while, or hear their favourite actor or sportsperson say something she would want them to say. While there are norms that govern the work of photographers, video editors, or software engineers and while there are laws against aiding and abetting criminals, there are no moral, social, or legal prescriptions against having fun or trying to be creative. Habgood-Coote (2023) thinks that we should remove financial incentives for producing deepfakes, ban forums on mainstream sites, and take down open-source deepfake tools. In my opinion, this will only turn a legal activity into an illegal one, or will incite people to find those tools on pirated software websites. Therefore, even though this proposal targets the factor that exacerbates the problem, it will not solve the problem and it may even make the problem worse.

Also, (contra Habgood-Coote, 2023: 18) I doubt that we can establish a successful general online community norm policing of fakes. I worry that this policing practice will generate a large number of false positives and there is a real danger of (mis)using it for censorship. Consider my former claim. We often recognise deepfake videos only because the individuals presented in those videos say things that are bizarre or at least atypical for them. After the UEFA Euro 2024 finals, many videos involving Harry Kane (https://www.tiktok.com/@memerunnergpt/video/7391625522410720517) or Gareth Southgate (Milmo & Hern, 2024) saying things that are rather unusual for them, even swearing, appeared on the Internet. For those watching these reels in low video quality, the fact that they were swearing was the best evidence that these were deepfakes. However, should Kane or Southgate one day genuinely say something very much atypical for them, these videos may falsely be flagged as deepfakes. A natural response is that, pretty soon, it will become obvious that these videos are genuine. However, in situations in which the timing of the message is essential, learning the truth sometime later will be of no help.

Relatedly, and more importantly, I fear that the truth (a genuine digital testimony) might mistakenly be labelled as a falsehood (deepfake) when (i) the falsehood is a better fit with the relevant background information, well-established theory, or a specific narrative and (ii) the relevant context is already saturated with fakes. Some two years ago, a friend of mine was casually browsing 9GAG, a popular online meme and news platform, when he saw a piece of news that said that the Russians openly admitted to committing a specific war crime (I cannot recall which crime exactly). My friend would often browse 9GAG because he enjoyed memes, but this post got him interested because it contained a link to the relevant article in Russian, and it struck him as odd that anyone would openly admit a war crime. As it happens, he speaks good Russian and it was not difficult for him to see that the cited article contains no such admission. Once he flagged this fact in the comment section, he received dozens of insulting responses and was called 'Russian pig' numerous times.

This man is neither Russian nor Russophillic and his knowledge of Russian does not reflect his political preferences. He understands nine other Slavic languages, including Ukrainian. It was a mere happenstance that he was browsing this section and that he could read the original news article referenced in the 9GAG post. And he never intended to suggest anything other than that the supposed admission is non-existent. But he was not censored. He just spoke at the wrong place at the wrong time. Ever since the invasion started, 9GAG was flooded with Russian bots and my friend's contribution strongly conflicted with thepredominant view on the conflict—it fit what a Russian bot would say. Therefore, he was mistaken for a Russian bot.

I fear that false positives of this kind will become very common in a digital space dominated by malevolent and non-malevolent deepfakes just as false positives are common on digital platforms dominated by various bots. The consequences of false positives in which a genuine video is mistaken for a deepfake, a polyglot for a bot, and so on are very dangerous: our desires and beliefs guide our actions and false beliefs cause misguided actions, which may have devastating consequences in high-stakes situations. Therefore, if we cannot allow deepfakes to police themselves, we can even less let common folks police them. This practice will have dramatic consequences on human interpersonal relations.

One may suggest that the *infocalypse* will never occur because most people are unaware of the damaging effects of non-malevolent deepfakes. True, I was confused by the non-malevolent deepfakes of the Pope, Morgan Freeman, and Father so that I stopped trusting some genuine digital testimonies, but one could argue that this is mainly due to the features of my personality. Perhaps, the constant focus on lies, deception, and misinformation has made me overly cynical and distrustful. Be that as it may, we should be very worried. Even if the *infocalypse* does not happen, under these circumstances, it *should* happen. People unaware of the dangerous effects of non-malevolent deepfakes are like regular guests at the Riviera Hotel in Las Vegas (where the Elvis convention is being held) who do not know that some Elvises are dangerous robbers hidden among the honest Elvises. It does not matter that these people do not know. They should know: they might get caught in the crossfire or even robbed by the villains. Therefore, the fact that most people do not know about the dangers of non-malevolent deepfakes is not a reason to think that I am overreacting; rather, it is a reason to think that we must warn them.

Perhaps, common folks will soon become bored of using deepfakes and, with the sharp decline of non-malevolent fake Elvises and needles, the malevolent fake Elvises and needles will become much less successful. I do not think that we can afford to wait for the problem to go away on its own. We need to solve it now: people are already being harmed by non-malevolent deepfakes. Most of them probably do not even know it. We could try to subject common folks to some standards but we cannot simply force these standards on people—given that this may infringe on their rights. Also, quite conceivably, some people may not be capable of adhering to these standards due to the possible complexity of these standards. Professional photography is very much different from amateur photography and many amateurs would struggle to satisfy standards professionals need to satisfy. In short, educating common people *in this way* may not be the best way to deal with deepfakes.

4.2 My Proposed Solution

We cannot hold the construction worker who created the Fashionable Pope or the creators of *Equilibrium* accountable for helping the Russian propagandists to deceive me just as we cannot blame honest Elvises for helping the robbers to evade police; they did not know. Therefore, we need to make them know: our education should focus on making people aware of the consequences of their behaviour. And I think that we can rationally convince people that seriously limiting or even banning deepfake technology is in their self-interest. We can do this by offering good reasons in support of this action and by offering the right kind of information regarding the dangers of deepfakes. What I mean is that common folks should not be educated on the dangers of misinformation or disinformation (contra Schick, 2020) and they should not be thought how to professionally create and disseminate deepfakes (contra Habgood-Coote, 2023), since many lack time, motivation, and capabilities to undergo this extensive learning. Rather, they should be educated regarding the dangers of them unintentionally aiding in some serious crimes or even making themselves more open to harm by creating and disseminating deepfakes. In short, they should be educated regarding the (self-)harms their behaviour may likely cause.

I will now argue that common folks would agree to enter a social contract to adopt and adhere to a general (perhaps even total) prohibition against deepfakes. We just need the right argument involving the right kind of information. My candidate for such an argument has three premises.

The first premise is that people are generally self-interested beings; they predominantly want good for themselves, and they would pretty much always prioritise their overall well-being or survival over some short-term pleasures. The second premise appeals to a kind of Rawlsian 'veil of ignorance' (Rawls, 1971) under which rational agents do not know whether they will be victims of a deepfake scam, how probable this is, and how serious the consequences of their biting the hook will be. Importantly, this veil of ignorance is not hypothetical but real. Numerous experts are already warning parents on Instagram not to post pictures or videos of their children because they can be used not only to generate video or audio messages but also allow scammers to conduct live conversations—the so-called 'Yahoo boys' are already trying to scam people using this strategy (Burgess, 2024). This is some of the information common folks should know.

The third premise involves application of the so-called *maximin* rule, which ranks alternative possibilities by their worst possible outcomes: roughly, it tells us to adopt the best version of the worst possible outcome (Rawls, 1971). Rational agents in this 'digital original position' should be made aware of the fact that we may soon start receiving phone calls in which AI-generated voices of our children, parents, or friends cry for help or video messages in which government officials urge us to leave our houses, just long enough for the thieves to rob us. These deepfakes can easily substitute emails we used to regularly receive from 'lawyers' of our distant Nigerian uncles who passed away leaving us fortunes. Even worse, it could be that a child whose distressed voice can be heard over the phone is genuinely in danger but that, due to the implausibility of the scenario, its parent thinks that they are talking to a deepfake. This would be a deepfake analogue to the Russian bot example.

Two things are important to say about the premises before I infer the conclusion from them. First, most of the examples I have given involve malevolent deepfakes. This is because, as per the maximin rule, these are our worst possible outcomes in the worst possible world; importantly, however, their probability is significantly increased by the vast amount of non-malevolent deepfakes. Second, Rawls' argument does not discuss probabilities but, in our case, they are important and relevant—since the digital original position is not hypothetical but real. This original position also involves a choice under uncertainty but the uncertainty is a bit different: it is pretty much certain that harmful things will happen; the only uncertain thing is that we do not know whether they will happen to us.

Considering this description of the digital original position, according to the maximin principle, it is in everyone's interest to substantially limit the use of deepfakes, both malevolent and non-malevolent. The best version of the worst possible outcome—in which deepfakes can (knowingly or unknowingly) easily be used by anyone to cause serious harm to anyone—is the one in which no one is allowed to use the deepfake technology. Also, the costs of entering this contract are negligible and the benefits of adhering to the rule are easy to understand. Therefore, we have good reasons to believe that rational self-interested agents under the digital veil of ignorance will willingly enter this social arrangement. This social arrangement, at the moment, is our best chance to protect our interpersonal relationships from harms that may come from communication in our digital space.

An immediate response to my proposal is that disingenuous agents will continue to use deepfakes to mislead or keep us in the dark notwithstanding this generally mutually agreed-upon (absolutist) prohibition. Illegal drugs are still available to those who want to buy them. My reply is that this is not only unavoidable but also perfectly acceptable. We do not need to eradicate deepfakes. We just need to get rid of as many as possible honest Elvis impersonators. Recall, the most pressing problem is not the robbers impersonating Elvis; the most pressing problem is the abundance of honest Elvis impersonators. This is not to say that honest Elvis impersonators are to be blamed for the negative consequences more than the bank robbers or that they should be blamed at all. Rather, it is to say that, if we convince most ordinary people to stop using deepfakes for non-malevolent purposes, the harm caused by malevolent deepfakes—as has already been argued (e.g., Atencia-Linares & Artiga, 2022; Habgood-Coote, 2023)—will become manageable. Our society can deal with malevolent deepfakes, once we isolate them.

References

Atencia-Linares, P., & Artiga, M. (2022). Deepfakes, shallow epistemic graves: On the epistemic robustness of photography and videos in the era of deepfakes. *Synthese, 200*, 1–22.

Bennett, D. (2023). *AI deep fake of the Pope's Puffy coat shows the power of the human mind*. Bloomberg. https://www.bloomberg.com/news/newsletters/2023-04-06/pope-francis-white-puffer-coat-ai-image-sparks-deep-fake-concerns (6 April 2023).

Bongiorno, F. (2021). Spinozan doxasticism about delusions. *Pacific Philosophical Quarterly, 103*, 720–752.

Burgess, M. (2024). The real-time deepfake romance scams have arrived. *Wired*. https://www.wired.com/story/yahoo-boys-real-time-deepfake-scams/

De Ruiter, A. (2021). The distinct wrong of deepfakes. *Philosophy & Technology, 34*, 1311–1332.

Egan, A. (2008). Seeing and believing: Perception, belief formation and the divided mind. *Philosophical Studies, 140*, 47–63.

Fallis, D. (2021). The epistemic threat of deepfakes. *Philosophy & Technology, 34*, 623–643.

Floridi, L. (2018). Artificial intelligence, deepfakes and a future of ectypes. *Philosophy and Technology, 31*, 317–321.

Gilbert, D. T., Malone, P. S., & Krull, D. S. (1990). Unbelieving the unbelievable: Some problems in the rejection of false information. *Journal of Personality and Social Psychology, 59*, 601–613.

Gilbert, D. T., Tafadori, R. W., & Malone, P. S. (1993). You can't not believe everything you read. *Journal of Personality and Social Psychology, 65*, 221–233.

Habgood-Coote, J. (2023). Deepfakes and the epistemic apocalypse. *Synthese, 201*, 1–23.

Harris, K. R. (2024). AI or your lying eyes: Some shortcomings of artificially intelligent deepfake detectors. *Philosophy & Technology, 37*, 1–19.

Huebner, B. (2009). Troubles with stereotypes for Spinozan minds. *Philosophy of the Social Sciences, 39*, 63–92.

Kant, I. (1999 [1797]). On a supposed right to lie from philanthropy. In P. Guyer & A. W. Wood (Eds.), *Immanuel Kant: Practical philosophy* (pp. 611–615). Cambridge University Press.

Krstić, V. (2018). Knowledge-lies re-examined. *Ratio, 31*, 312–320.

Levy, N., & Mandelbaum, E. (2014). The powers that bind: Doxastic voluntarism and epistemic obligation. In J. Matheson & R. Vitz (Eds.), *The ethics of belief* (pp. 15–32). Oxford University Press.

Lopes, D. (2016). *Four arts of photography*. Wiley.

Mandelbaum, E. (2014). Thinking is believing. *Inquiry, 57*, 55–96.

Mandelbaum, E., & Quilty-Dunn, J. (2015). Believing without reason: Or: Why liberals shouldn't watch Fox News. *The Harvard Review of Philosophy, 22*, 42–52.

Matthews, T. R. C. (2022). Deepfakes, intellectual cynics, and the cultivation of digital sensibility. *Royal Institute of Philosophy Supplement, 92*, 67–85.

Milmo, D., & Hern, A. (2024). Deepfake clips of Gareth Southgate swearing after England match go viral. *The Guardian*. https://www.theguardian.com/technology/article/2024/jul/14/deepfake-gareth-southgate-england-football-team-euro-2024

Öhman, C. (2020). Introducing the pervert's dilemma: A contribution to the critique of deepfake pornography. *Ethics and Information Technology, 22*, 133–140.

Ortony, A., & Gupta, S. (2019). Lying and deception. In J. Meibauer (Ed.), *The Oxford handbook of lying* (pp. 149–169). Oxford University Press.

Rawls, J. (1971). *A theory of justice*. S. M. Cahn (Ed.). Harvard University Press.

Rini, R. (2020). Deepfakes and the epistemic backstop. *Philosophers' Imprint, 20*, 1–16.

Rini, R., & Cohen, L. (2022). Deepfakes, deep harms. *Journal of Ethics and Social Philosophy, 22*, 143–161.

Schick, N. (2020). *Deep fakes and the infocalypse: What you urgently need to know*. Hachette UK.

Vladimir Krstić is an Assistant Professor at the United Arab Emirates University. He was born in Belgrade, Serbia, where he graduated Theology and did his MA in Philosophy. He received my PhD in Philosophy from the University of Auckland. He has very broad research interests, specializing in Philosophy of Mind and Language, Philosophical Psychology, Epistemology (of AI), and philosophical analysis of manipulation, deception, information, and misinformation. He has a very strong background in Ethics (including Ethics of AI), Critical Thinking, and Continental Philosophy, but he also publishes in other areas, such as Anthropology, Linguistics, Experimental Philosophy, Gender Studies, Medical Ethics, and Old Testament Studies. He loves to exercise, swim, and run.

A Compassion Paradox: Can AI Truly Bridge the Empathy Gap in Human Relationships?

Liang Echo Shang and Shirley Qian Jin

1 Introduction

> We think we design technology, but technology also designs us. We're at the inflection point where technology is shaping our human values and our emotional connections.
>
> —Prof. Sherry Turkle

Artificial intelligence (AI) refers to computer systems that are designed to replicate aspects of human cognition, including learning, reasoning, and decision-making. These systems are created to perform tasks that typically require human intelligence (Russell & Norvig, 2016). Initially, discussions on AI's potential primarily focused on its ability to enhance human productivity and performance by leveraging its powerful computational capabilities to reduce human workload. For example, AI has been widely adopted in automating data analysis in finance, optimising supply chains, and even managing routine customer service tasks through AI-driven chatbots (Damioli et al., 2021). However, in recent years, technological advancements such as the emergence of generative AI have dramatically expanded the scope of human–AI interactions, shifting the focus increasingly towards understanding AI's social dimensions (Liu & Sundar, 2018). This shift has sparked new discussions about AI's roles in areas that are traditionally viewed as distinctly human, such as compassion and empathy. As we continue to explore these capabilities, the potential for AI to act not just as assistive tools but as ***social actors*** opens up intriguing possibilities as well as challenges.

L. E. Shang (✉)
Education University of Hong Kong, Hong Kong, China
e-mail: lshang@eduhk.hk

S. Q. Jin
School of Economics, Utrecht University, Utrecht, The Netherlands
e-mail: q.jin1@uu.nl

Y. Chandra and R. Fan (eds.), *Artificial Intelligence and the Future of Human Relations*,
https://doi.org/10.1007/978-981-96-7185-4_9

At first glance, AI—a machine programmed by algorithms—and human compassion might seem worlds apart. Compassion is a fundamental human trait that is deeply rooted in humans' responses to the suffering of others, such as pains from natural disasters, personal tragedies or workplace challenges (Dutton et al., 2006). This emotion involves a deep connection to the struggles of others (De Waal, 2008). Yet, as machines become increasingly human-like, AI is moving into domains that were once considered the exclusive realm of human interaction. For example, Woebot,[1] a chatbot designed to provide mental health support, represents a new frontier in AI-driven emotional engagement. Woebot uses natural language processing (NPL) to have conversations with users, helping them manage feelings of anxiety, depression, and stress. Unlike traditional therapy, Woebot is available 24/7, offering real-time responses and tools such as cognitive behavioural therapy techniques, mood tracking, and self-reflection exercises, providing a form of empathic interaction by recognising emotional cues from users and responding with supportive dialogue.

However, despite its potential, the involvement of AI in mimicking human-like compassion and empathy also raises many questions: Can machines be designed to exhibit compassion? Is it possible for AI to demonstrate compassionate actions that reflect an understanding of human suffering, or does it merely respond on the basis of pre-programmed algorithms? To what extent should we trust AI systems in emotionally sensitive roles, such as providing mental health support or making decisions based on human emotions? Of course, these are not simple yes-or-no questions. They delve into complex layers of scientific, ethical, and philosophical considerations that challenge us to rethink the nature of intelligence, the meaning of compassion and the essence of our interactions with AI, even fundamentally questioning what it means to be human and what defines our humanity.

In this chapter, we do not aim to, nor is it feasible to address, these grand questions. Instead, we shift our focus from discussing AI as a 'compassion agent'—one that initiates compassion—to examining its role as a 'compassion organiser'—one that fosters and enhances compassionate interactions among humans. This focus is particularly relevant today as generative AI tools such as ChatGPT and DeepFace are already actively reshaping human relationships by transforming how humans communicate and interact with each other (e.g. Park et al., 2023).

Taking philanthropic crowdfunding as our context, we aim to discuss two key questions here: (1) the feasibility of AI as a compassion organiser by shaping how crowdfunding campaigns are presented and perceived by potential donors and (2) the ethical issues associated with these practices. Therefore, rather than focusing solely on human–AI relationships, we explore the dynamics of *human–AI–human interactions*, where AI acts as a mediator that has the potential to shape emotions, elicit emotional responses and influence the nature of interactions between people. In addition to proposing future research directions, we will also discuss how AI-mediated interactions in compassionate contexts, such as crowdfunding, are impacting the nature of emotional connections, trust, and empathy, and how these changes may influence human relationships in a digital age.

[1] https://woebothealth.com/.

2 Artificial Intelligence, Compassion, and Compassion Organising

Compassion is an other-oriented emotion characterised by a sensitivity to the suffering of others and a commitment to alleviating that suffering (Kanov et al., 2004). While often used interchangeably with empathy, the two concepts are distinct. Empathy primarily involves sensing and sharing another person's emotional experience (Gilbert, 2019), while compassion goes beyond mere sensitivity to include a proactive desire and effort to alleviate and prevent others' suffering. In other words, compassion can be understood as a caring motivational system that begins with individuals' empathetic feelings and evolves into a strong motivation to actively reduce suffering through helpful and empathetic responses and behaviours (Gilbert, 2019). Importantly, compassion manifests through social interactions and is fundamental to a wide range of social relations across various sectors of society, influencing patient–doctor relationships, social worker–client engagements, teacher–student dynamics, and even interactions within our justice system (Strauss et al., 2016).

As machines increasingly resemble many features of humans, research into whether and how AI technologies might support compassion is expanding (Xu et al., 2021). Current debates about AI and compassion generally fall into two main categories: (1) whether AI can exhibit or mimic human compassion and (2) the implications of using AI in generating and delivering compassion (Brave et al., 2005; Liu & Sundar, 2018).

In the first stream of debate, opinions diverge on whether AI can genuinely generate compassion. Some argue that compassion is uniquely human, a *felt emotion* that emerges from social interactions and is deeply embedded in human experiences and agency (Gray et al., 2007). Consequently, these researchers maintained that AI, in principle, cannot experience or manifest true compassion (e.g. Montemayor et al., 2022). However, others argued that AI can be programmed and designed to imitate human relationships and exhibit cognitive empathy—the ability to detect and recognise the emotional states of others and provide reactions accordingly, such as generating emotional expressions on robot faces or providing consolation (Liu & Sundar, 2018). As Mason (2015) suggested, AI can be equipped with knowledge of compassion, culture, and social norms, along with reasoning methods that consider the feelings and states of both the self and others during decision-making, action, or problem-solving, which can potentially enable a form of *artificial compassion.*

In the second stream of debates, as AI is increasingly utilised in various settings such as healthcare, it presents both opportunities and challenges. On one side, proponents argue that AI can significantly improve patients' experience by providing engaging, consistent, and personalised emotional support. For example, AI-driven chatbots and virtual assistants are designed to express empathy, offer comfort and provide mental health support, thus filling gaps in healthcare where doctors and nurses have limited resources and might suffer from emotional fatigue (Fitzpatrick et al., 2017; Kerasidou, 2020). Studies have shown that patients may prefer interactions with AI chatbots that offer an unbiased, non-judgemental approach, and help individuals feel more at ease when discussing sensitive health issues (Lucas et al., 2014).

On the other hand, ethical concerns have been raised regarding whether it is appropriate for machines to simulate emotions such as empathy and compassion. Some scholars argue that it is inherently unethical for AI to exhibit 'artificial feelings' because these emotional displays can be disturbing or misleading to humans (Kim et al., 2019; Liu & Sundar, 2018). The discomfort people experience often stems from the belief that emotions, particularly empathy and compassion, are exclusively human traits tied to personal experience and genuine emotional capacity. This lack of true empathy is troubling because individuals may feel deceived or manipulated upon realising that the 'emotions' expressed by AI are 'programmed' responses rather than genuine feelings (Kerasidou, 2020).

Moreover, when machines exhibit emotions, people may feel uneasy because they know the AI does not truly understand or feel what it is expressing. This discrepancy can create a sense of unease or distrust. This discomfort is often explained by the *uncanny valley of mind* (Gray & Wegner, 2012), which refers to the uneasiness humans feel when encountering machines that come close to replicating human behaviours but fall short in some essential way, particularly in emotional or affective dimensions. When AI systems simulate empathy, they may appear human-like but still lack the emotional depth that we expect from true human empathy, creating an unsettling or 'uncanny' experience.

However, the debates surrounding whether AI can truly become a 'compassion agent' remain paradoxical and unresolved, continually influenced by ongoing technological advancements and evolving philosophical discussions. This chapter will focus on more immediate and observable effects: *exploring how AI shapes our existing human relationships.* Specifically, we will explore how AI influences the ways we interact, connect emotionally and build compassionate connections in the present. Adopting the 'compassion organising' perspective, we argue that compassion is not just about individual-level emotions but should be viewed as coordinated actions guided by human emotions.

At its core, compassion organising perspective describes compassion as a socially coordinated empathic response triggered by the suffering of others, which motivates actions aimed at alleviating this distress (Dutton et al., 2014; Frost et al., 2000). This empathic response is activated in a three-phase process (Dutton et al., 2014): (1) *noticing*: acknowledging another person's suffering, (2) *feeling*: forming an empathetic connection with the sufferer, where individuals feel for the other as if they were experiencing the same pain or suffering, and (3) *responding*: compelled by concern, actions are taken, ranging from emotional support to practical aid to ease the sufferer's pain.

With this framework, we argue that while it remains debatable whether AI can truly generate empathy—a deeply personal '*imagining feeling*' that involves projecting oneself into another's situation to understand their emotions (Montemayor et al., 2022); it can certainly enhance this imagining process. AI can assist by focusing attention on the crucial initial step of 'noticing' and by utilising emotional cues to forge stronger emotional connections between individuals. This support may make the suffering of others more visible and relatable and enable individuals to better imagine and *'feel'* another's situation so that they can relate more effectively, ultimately fostering more empathetic responses (Dutton et al., 2006). Consequently, AI becomes a 'compassion organiser' that actively works on and amplifies the steps needed for compassionate responses by focusing individuals' attention on emotional cues and the needs of others.

Figure 1 highlights the differences between AI acting as a compassion agent and as a compassion organiser.

AI as a *compassion organizer* is potentially applicable across various fields, including philanthropy, healthcare, mental health support, and social services, among others. In philanthropic settings, AI can enhance fundraising efforts by crafting narratives that resonate deeply with potential donors, thereby increasing engagement and contributions (Kim et al., 2022). In healthcare, AI's ability to identify emotional cues

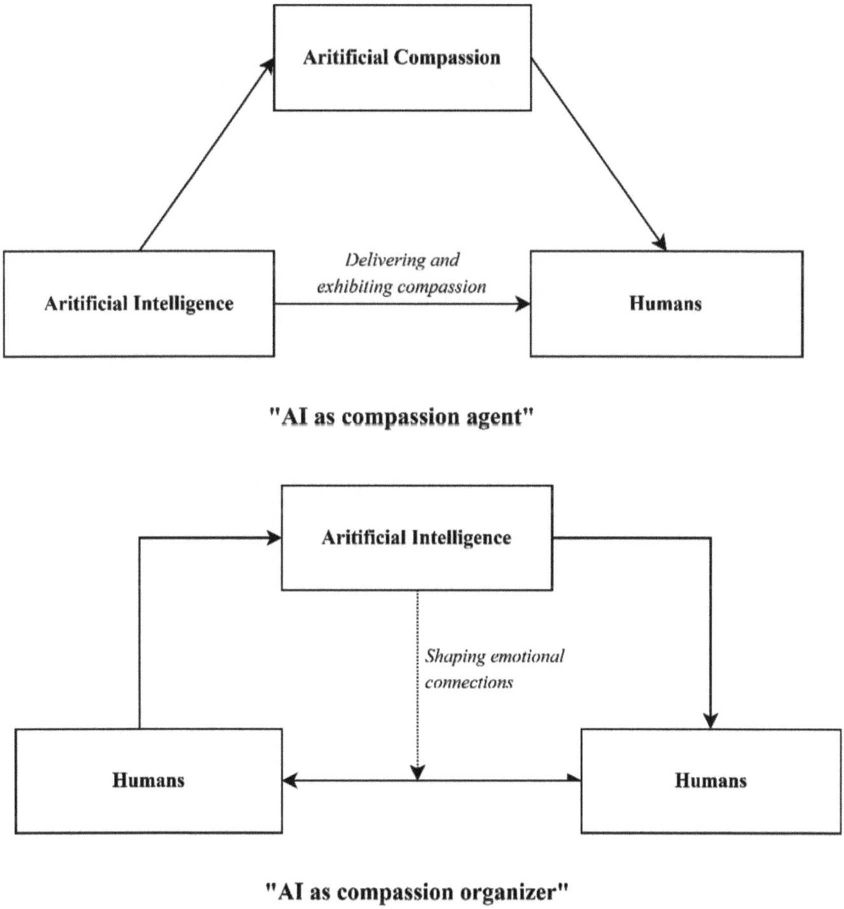

Fig. 1 Different models of AI influence on compassion

in patients' speech or behaviour can alert medical professionals for compassionate interventions, which will potentially improve patient outcomes (Jiang et al., 2017). In mental health support, AI chatbots can provide initial support and help individuals feel acknowledged and understood while guiding them towards appropriate human-led interventions (Fitzpatrick et al., 2017). Here, AI mediates and enhances human-to-human interactions by shaping the emotional environment, encouraging more compassionate behaviours between individuals. Instead of delivering compassion itself, the AI operates in the background, subtly guiding and amplifying emotional connections.

To further illustrate the concept of AI as a compassion organiser, we have chosen philanthropic crowdfunding as our context for exploration. In the following section, we will discuss how AI can be utilised to shape emotional connections between sufferers and donors within the crowdfunding environment. We will also discuss the ethical issues associated with these practices.

3 AI as Compassion Organiser in Philanthropic Crowdfunding

With the growing popularity of digital platforms, crowdfunding has become a widely utilised funding mechanism for a diverse range of entities, from entrepreneurs and artists to communities and individuals facing hardships. Among various types, philanthropic crowdfunding, primarily based on online donations, has emerged as an important approach for securing funds for charitable causes, nonprofit organisations, social workers, and individuals facing adversity (Paulus & Roberts, 2018). In philanthropic crowdfunding, campaigns often present stories of those in need, whether about the fundraisers themselves or others. These stories are crafted to evoke emotional responses from potential donors, with the ultimate goal of inspiring actions to alleviate suffering through financial contributions (Kim et al., 2022). This process of using stories to elicit emotional responses in others involves helping the audience imagine the suffering that is being depicted. Such 'imagining feeling' is crucial for engaging potential donors emotionally and motivating them to react.

Existing literature has uncovered that the success of philanthropic crowdfunding campaigns depends not only on the story itself but also critically on *how* the story is told (Kanov et al., 2004; Kim et al., 2022). Effective storytelling techniques such as linguistic framing, emotional delivery, visual aids, tragic narrations, and contrasts between experiences before and after adversity play important roles. These techniques each serve as emotional cues that deepen the audience's compassionate responses, thereby enhancing the outcomes of the crowdfunding campaigns (e.g. Rhue & Robert, 2018). While previous research primarily focused on identifying emotional cues and their linkage to campaign success, advancements in generative AI introduce a new dimension to this exploration.

Table 1 Effective storytelling strategies in philanthropic crowdfunding campaigns

Modalities	Strategies	Sources
Text	• Demonstrating the 'worthiness' of help	Pan and Dong (2023), Perez et al. (2023) and Rhue and Robert (2018)
	• Alignment with traditional values in specific context (e.g., family, filial piety in Chinese context)	Xu and Wang (2020)
	• Expressing emotions in campaign description and title (e.g., by highlighting responsibility, guilt, urgency)	Hou et al. (2023), Kim et al. (2022) and Rhue and Robert (2018)
	• Creating contrast of fundraiser's experience before and after adversity	Xu and Wang (2020)
	• Avoiding using too many requesting words	Xu (2018) and Xu and Wang (2020)
	• Keeping the description in a moderate length	Kamatham et al. (2021)
	• Enhancing message quality (e.g., believability, readability, completeness of a message, free of spelling errors)	Kim et al. (2022) and Mollick (2014)
Image/video	Showing facial expressions	Rhue and Robert (2018)
	Highlighting image emotions (i.e., sadness and contentment)	Hou et al. (2023)

Generative AI, which includes technologies that can autonomously produce content such as text, images, and music that resemble human-like creations (Jo, 2023), provides us opportunities to operationalise and amplify these cues. By leveraging its capabilities in understanding and capturing human language nuances, generative AI tools such as ChatGPT by OpenAI and Copilot by Microsoft can enhance the way fundraisers tell stories, making the stories more noticeable and resonant with the target audience. This improved storytelling can enable audiences to better imagine and empathise with the sufferers. Considering that stories are the main vehicle for driving compassion and linking sufferers with donors in philanthropic crowding, AI has the potential to play—or could already be playing—an important role in mediating this process.

We summarise the effective storytelling strategies identified in the past literature in Table 1. In general, the strategies can be categorised into two major modalities: text and visual (including static images and videos). On the basis of the strategies listed in the table, we showcase how generative AI tools (e.g. ChatGPT) can learn from the strategies, evaluate the storytelling style of a crowdfunding campaign that failed to meet its goal and provide concrete suggestions on how to enhance the storytelling so that it can increase the likelihood of success in the future.

We randomly selected one crowdfunding campaign from GoFundMe that has not met its goal yet. GoFundMe[2] is a well-known crowdfunding platform that operates

[2] https://www.gofundme.com/.

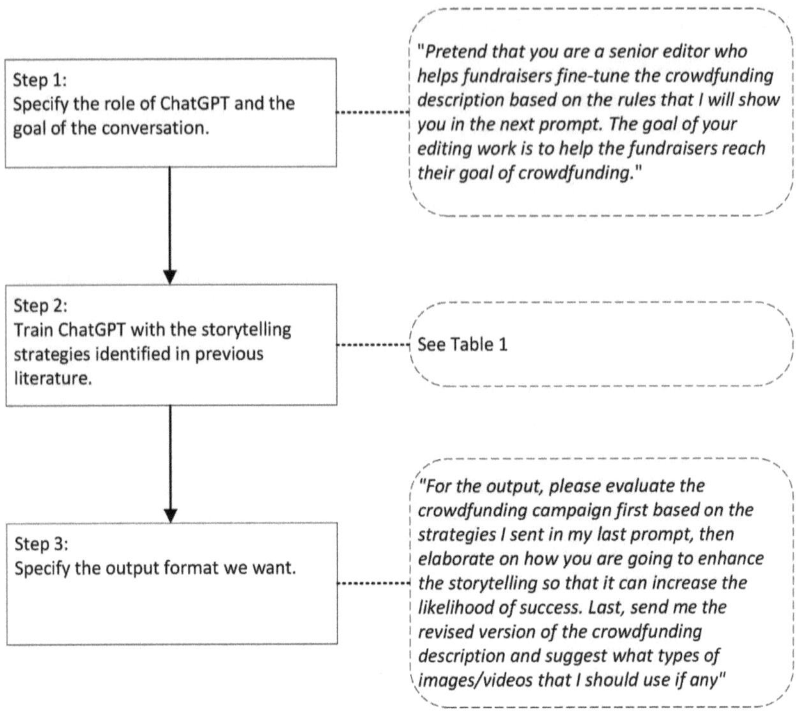

Fig. 2 Workflow in ChatGPT

in around 20 countries and allows individuals to raise money for various purposes, including medical, education, natural disaster, and emergency, among others. Below is a screenshot of the selected campaign page.[3] We then train ChatGPT with the information in Table 1, specify the output format and feed it with the original text in the crowdfunding campaign. The workflow is demonstrated in Fig. 2.

The content of the campaign before and after refinement by AI is presented in Table 2. Evidently, AI can incorporate and enhance the important linguistic cues in the campaign story with instructions from the user. While it holds great promise, however, this AI-powered storytelling process also raises many questions. We do not know, for example, how this technology shapes the relationships between sufferers and donors, particularly when trust is a foundational element of these relationships. Additionally, how will it affect donor engagement and the authenticity of the compassion-organising narratives presented? Will the use of AI in storytelling jeopardise the emotional integrity of the campaigns? What are the ethical boundaries in using AI to enhance emotional engagement in philanthropic efforts?

Below, we highlight several key ethical concerns associated with AI-assisted compassion organising and outline directions for future research.

[3] https://www.gofundme.com/f/please-help-rio-grande-do-sul-brazil.

Table 2 Campaign content: original vs. AI-refined

Original campaign	ChatGPT-refined campaign	Key changes
Title: Please Help Rio Grande Do Sul—Brazil	**Title: Stand with Rio Grande do Sul: Urgent Relief for Flood Victims**	
Hey everyone, I'm Lucas, and I urgently need your help. Right now, my home state of Rio Grande do Sul in Brazil is facing the worst natural disaster ever. Floodwaters have engulfed multiple cities, leaving more than 844,673 people (5 May 2024—CNN Brasil) in desperate need across 334 municipalities (the state has 496 municipalities in total), including Porto Alegre (my home town)	Hey everyone, I'm Lucas, and I'm reaching out to you with a heavy heart and an urgent plea for help. Right now, my beloved home state of Rio Grande do Sul in Brazil is grappling with the aftermath of the worst natural disaster it has ever faced. The devastating floods have ravaged our communities, leaving behind a trail of destruction and despair	← Emphasising the severity to demonstrate the 'worthiness' ← Expressing emotions
Regarding the homeless who completely lost their homes, the number is greater than 18,000 already. There are now 115,844 people temporarily homeless and this number is increasing every hour	As of the latest reports from CNN Brasil, over 844,673 people across 334 municipalities are in desperate need of assistance. Families have been torn apart, homes have been destroyed, and lives have been shattered. The numbers are staggering, but behind each statistic lies a story of loss, resilience, and hope	← Creating contrast
People are dying, others are missing, and the situation is chaotic. Public entities are doing what they can, but it's not enough. Many civilians are stepping up as volunteers, but the scale of this crisis requires more support That's why I'm launching this crowdfunding campaign—because we need more help	The situation is dire. More than 18,000 people have lost their homes entirely, while over 115,844 are currently displaced and seeking shelter. The floodwaters have spared no one, and the scale of this crisis is overwhelming our local resources. While public entities and volunteers are working tirelessly to provide support, it's clear that more help is needed	← Emphasising the severity to demonstrate the 'worthiness'
Those who are rescued are being brought to shelters set up in city sports clubs, schools, and other facilities	This is where you come in	← Highlighting responsibility

(continued)

Table 2 (continued)

Original campaign	ChatGPT-refined campaign	Key changes
Every penny we raise will go directly to organisations and individuals on the ground in Rio Grande do Sul. We'll use the money to provide essentials like mattresses, bedding, food, diapers, and water to those who've lost everything	I'm launching this crowdfunding campaign because we need your help now more than ever. Every penny we raise will go directly to organisations and individuals on the ground in Rio Grande do Sul. We'll use the funds to provide essential supplies like mattresses, bedding, food, diapers, and clean water to those who have lost everything	← Emphasising the severity to demonstrate the 'worthiness'
Part of the funds will also support local heroes like Deise Falci, who's rescuing stranded animals left behind	But our support won't stop there. Part of the funds raised will also go towards supporting local heroes like Deise Falci, who's been tirelessly rescuing stranded animals left behind in the chaos. Together, we can make a difference in the lives of those who need it most	← Amplifying impacts ← Aligning with traditional values of solidarity

3.1 Compassion and Integrity in AI-Powered Storytelling

Storytelling is the linchpin of any online crowdfunding campaigns, connecting campaign beneficiaries and potential donors emotionally within the virtual world. Narrating a compassionate story about the cause enables potential donors to empathise and resonate with the pain and suffering of those whom they have never met or known in real life, even fostering a willingness to provide support (Kim et al., 2022).

In the past, campaign creators, who may be the sufferers themselves or their friends and relatives, act as the primary storytellers in crowdfunding practices. By sharing the stories of the sufferers, they reach out to people beyond their immediate circles and invite them into the world of the sufferers. In this context, the *perceived integrity* of the storyteller—the extent to which they are acting authentically—is crucial (Cumming et al., 2021). Potential donors have not met the sufferers personally, which is why their willingness to contribute depends heavily on their trust in the authenticity of stories and the assurance that funds will be used as promised (Ferreira et al., 2022). This trust in the storyteller's integrity is vital for the success of campaigns, as it establishes a trusting relationship between donors and beneficiaries, especially when their interactions occur solely through a virtual crowdfunding platform.

Confucianism links integrity closely to the virtue of *ren* (仁), often translated as benevolence, kindness or humaneness (Hall & Ames, 1987), a concept Confucian Mencius describes as a 'heart sensitive to the sufferings of others'. Therefore, integrity is rooted deeply in human empathy, sympathy, and compassion, and it is fundamentally about acting uprightly and doing the 'right' thing (Koehn, 2005). However, the introduction of AI as a new or co-storyteller introduces complexities. While AI has great potential to craft and refine stories that effectively evoke human compassion, it inherently lacks the capacity for heartfelt sensitivity or moral judgement of what is right and what is wrong, as it operates on the basis of its training datasets and prompts from human users. Consequently, improvements in storytelling quality achieved through AI assistance may not automatically translate to increased trustworthiness of the stories or the storytelling process. This discrepancy needs further exploration.

- *Does AI involvement in storytelling compromise the perceived integrity of storytellers?*
- *How do potential donors perceive the authenticity of these AI-enhanced stories?*
- *Can AI be perceived as possessing integrity, or does its nature inherently challenge such perceptions?*
- *To what extent does AI involvement influence the trusting relationship and emotional connections between donors and sufferers?*

Some recent research has begun to explore these questions. For example, a study by Park et al. (2023) demonstrates the effectiveness of emotional chatbot fundraisers in increasing donation amounts. The study highlights the importance of perceived humaneness and empathy towards victims, which can result from emotional contagion between chatbot agents and users. However, despite these promising results, questions remain about AI's role in storytelling and its impact on perceived integrity. To address these questions, we call for more research that delves into the intersection of AI and integrity in a compassionate context. For example, a vignette experiment could investigate whether donors prefer knowing that AI was used in crafting stories in charitable crowdfunding campaigns and how this knowledge affects donors' willingness to contribute to the same campaign. In-depth interviews with potential donors could help understand their perceptions and expectations regarding the qualities of storytellers and the extent to which they influence the establishment of trusting relationships with the sufferers. Additionally, a longitudinal study could be further used to track changes in donor engagement and trust over multiple crowdfunding campaigns that progressively disclose more about AI involvement. This approach would help identify whether exposure to AI-enhanced storytelling over time shifts donor attitudes, potentially offering clues about long-term trust dynamics and the impact on donor-sufferer relationships.

3.2 *Compassion Through AI: Cultivating, Manipulating, or Manufacturing?*

The role of AI as a compassion organiser raises another critical debate: Are we merely cultivating compassion and fostering compassionate responses with AI assistance, or are we crossing into the territory of manufacturing human compassion and manipulating them using AI?

Current discourse generally views AI's role in enhancing compassion positively. For instance, AI has been shown to increase the awareness of healthcare professionals about suffering through immersive virtual reality technology that realistically simulates patient environments (Kim & Chun, 2022). Moreover, AI can enhance compassionate interactions between humans through tools that improve communication skills, such as virtual humans that help forge stronger connections with those who are suffering (Guetterman et al., 2019). AI can also increase work efficiency, giving healthcare professionals more time to develop meaningful, compassionate relationships (Kerasidou & Horn, 2016). These previous studies largely highlighted AI's potential to positively foster and cultivate compassion in human interactions. However, in our context of philanthropic crowdfunding, where monetary exchange in the form of donations has a significant involvement, the line between genuinely cultivating compassion and manufacturing or manipulating compassion can become blurred.

As Croatian American sociologist Stjepan Mestrovic (1996) argued in his work *Postemotional Society*, we live in a postemotional society where human emotions are deliberately produced and shaped by mass media and information control. Central to the concept of postemotionalism is the idea that emotions in contemporary society are not merely influenced but actively manufactured and manipulated to meet political, social, or market ends (Carrette, 2004). Examples of this include the 'McDonaldisation' of society, where the principles of fast-food restaurants standardise and control aspects of our social life. Similarly, the consumer culture of Disney World shapes, commodifies and packages human emotions. In these contexts, there is a 'mechanisation' of emotion, where emotions are considered artificial, manufactured, and mass produced in the capitalist machine (Carrette, 2004).

Postemotionalism offers an interesting lens for examining how AI is shaping our emotional lives. According to Mestrovic (1996), mere knowledge is not sufficient to result in action because action requires a connection between emotions and intellect. In the context of compassion organising, simply knowing about others' suffering is not enough to trigger individuals' compassionate responses. Instead, it is the empathetic feelings that accompany such knowledge that must be effectively evoked. This process may create room for what Mestrovic (1996) was concerned about in relation to the 'modernist rationalisation of emotion', wherein empathetic feelings might be strategically produced and shaped by AI tools to achieve certain outcomes rather than emerging genuinely.

Another related concern aside from the rationalisation of emotion is 'compassion fatigue', which occurs when individuals are exposed excessively to the intense and traumatic experiences of others (Figley, 2002). With assistance from AI, stories of suffering can be narrated in a highly effective and emotionally evocative manner. While we use AI to enhance and evoke human empathy by helping people better imagine others' suffering, the intensification of these feelings can become overwhelming (McKenny et al., 2017). This situation is very possible as the mechanisation of emotion becomes more refined and perfected. Consequently, such overexposure to empathetic emotions can lead to compassion fatigue or the compassion fade effect, diminishing the impact of compassion stories and negatively affecting potential donors' willingness to donate.

Undoubtedly, AI has the potential to alter how human emotions are constructed in different settings, including crowdfunding campaigns. By strategically shaping the narration of suffering through careful word choice, linguistic framing techniques, structural elements, and visual aids, AI could potentially rationalise compassionate responses by helping some campaigns win this 'compassion contest'. While we celebrate the advancements in AI technology, we need to be cautious of its potential dark side, particularly the fine line between cultivating compassion and manufacturing and manipulating it. Mestrovic (1996) has long argued that emotions have become more and more 'abstracted', 'intellectualised', and 'mass-produced' in our post-modern society. Given AI's capabilities, we must confront the possibility: *In the digital era, is AI intensifying the trend of commodifying emotions?* Considering the implications of these developments, we outline several important questions worth exploring.

- *How does AI-enhanced storytelling in crowdfunding campaigns distinguish between evoking genuine compassion and mass-produced compassion?*
- *What are the differential effects of compassion elicited through AI-facilitated storytelling versus those evoked through traditional non-AI methods on donor responses?*
- *To what extent does AI influence the emotional depth of donor engagement over time?*
- *Does AI-enhanced storytelling contribute to compassion fatigue among potential donors, and how does it do so?*

Building on these questions, future research could further delve into the AI's impact on human emotions in the context of philanthropic crowdfunding and beyond. For instance, researchers may conduct controlled experiments comparing donor responses to AI-enhanced storytelling versus traditional storytelling in crowdfunding campaigns. Variables to measure could include not only short-term effects such as immediate emotional reaction, amount donated, and recall of campaign details, but also long-term effects such as donor loyalty, repeat donations, and changes in perception of campaign's trustworthiness. This research could help identify whether AI-crafted stories, with their strategic use of language and emotional cues, elicit more intense or sustained engagement compared to human-created narratives.

Additionally, qualitative methods could also be used to gather in-depth insights into how donors perceive the mechanisation of emotion, their experiences with genuine and AI-generated emotional content, and its implications for their willingness to engage. These approaches would allow researchers to explore how donors perceive the 'mechanisation' of emotions and whether they differentiate between authentic and AI-generated emotional content.

We also urge for more discussions and debates on the ethical implications of using AI to generate emotional content, particularly on the potential for harm when emotions are artificially produced and strategically employed for targeted purposes. Questions surrounding consent, transparency, and the potential harm, when emotions are artificially produced, should be central to these conversations. By engaging with these ethical dilemmas, researchers, practitioners, and policymakers can better understand how to harness AI's capabilities responsibly, ensuring that technological advancements in storytelling do not compromise the fundamental integrity of human empathy.

3.3 Rethinking Social Equity in AI-Assisted Online Fundraising

As AI technology rapidly advances and becomes more integrated into our daily lives, we must recognise that the benefits of these advancements are not distributed equally across society; some individuals may benefit more, whereas others may be excluded or left behind. An essential issue that arises from this inequality is the *digital divide*. Defined as the gap between people who have adequate access to Internet and technology and people who have poor or no access (Soomro et al., 2020), the digital divide can intensify inequalities within society as they can influence people's social and economic capital and their ability to participate in society (Ragnedda, 2017). Those with poor access to digital technologies tend to be from marginalised or underprivileged backgrounds and often have limited educational opportunities.

This lack of access to technology not only hinders their ability to participate fully in a digital-driven society but can also perpetuate cycles of poverty and exclusion, leading to 'digital poverty' (Barrantes, 2007). The concept of digital poverty is particularly relevant to the online philanthropic crowdfunding context. The inherent nature of these online platforms—two-sided, communication-rich environments—requires fundraisers to craft persuasive stories to connect with potential donors effectively and solicit funds from them (Burtch & Chan, 2018). In this context, the success of campaigns largely depends on the quality of the stories told, such as how effectively the story can capture attention, stand out among other stories about suffering and establish emotional connections with potential donors (e.g. Kim et al., 2022). Nonetheless, the introduction of AI tools in crafting these stories might further widen the digital divide, particularly if AI significantly outperforms human storytellers in creating more compelling stories. Such divide is not only about whether or not individuals have access to the Internet and technology but also about whether they can exploit such technologies to enhance their personal benefits (Lythreatis et al., 2022).

Several potential implications may arise from the use of generative AI tools in crafting stories for online philanthropic fundraising. One major concern is the increased difficulty of being 'noticed'—a crucial initial step in compassion organising (Dutton et al., 2006). As the stories become more and more sophisticated, it is highly likely that those with limited access to such technologies are at a disadvantage when competing for potential donors' attention on online crowdfunding platforms. Consequently, those who are truly disadvantaged may become increasingly invisible because having their voices heard is becoming progressively difficult due to a lack of access and capabilities to utilise AI tools. Even if they manage to pass the initial evaluation stage, crafting a compelling story that resonates emotionally with potential donors is an immense challenge (Anglin & Pidduck, 2022). The use of generative AI tools can give an unfair advantage to campaigns created by tech-savvy, better-educated individuals. This situation may further marginalise and exclude those less equipped, amplifying rather than alleviating social inequalities due to the digital divide (Lythreatis et al., 2022). With its great benefits, AI technology may also tempt individuals to misuse or abuse it for personal gain, further widening existing social and economic disparities. These disparities might even give rise to new businesses that specialise in providing AI-enhanced storytelling services, creating a market where success in philanthropic crowdfunding increasingly depends less on genuine need and more on one's ability to afford professional storytelling assistance.

Echoing concerns raised by scholars regarding the increased use of machine learning in governing and managing citizens' behaviours, we must critically examine whether our growing reliance on AI is also leading to an oppressive environment where not only our emotions and empathy are being conditioned but also the voices of the disadvantaged are diminished (Kane et al., 2021). Thus, we propose several research questions to guide future studies.

- *How does AI-enhanced storytelling influence potential donors' perceptions of deservingness in crowdfunding campaigns?*
- *How have fundraisers altered their campaign strategies in response to advancements in AI storytelling technologies?*
- *Which groups benefit most from AI-enhanced storytelling in crowdfunding, and what factors contribute to these disparities?*
- *Does the sophistication of AI-crafted stories in crowdfunding platforms increase the risk of donor deception and fraud? Why or why not?*

While AI and social equity is not a new topic, it is important to consider how AI can be harnessed to foster equity, such as through AI-driven educational platforms that customise content to meet individual student needs (Shang & Bai, 2024). However, it is equally important to recognise that AI can also reinforce or amplify social inequalities, as algorithms may reflect and perpetuate existing societal biases (Howard & Borenstein, 2018). For instance, AI systems may find patterns in datasets that mirror implicit biases and, in doing so, emphasise and propagate these biases as truth.

To explore these dynamics, researchers could conduct linguistic analyses of successful crowdfunding campaigns to assess how AI-enhanced storytelling influences the language, tone, and emotional appeal of narratives. This will allow researchers to pinpoint how AI tools might be influencing storytelling techniques. By comparing the language used in campaigns over time, researchers can assess whether AI-enhanced narratives exhibit patterns, such as heightened emotional cues or strategic framing of suffering, that affect donor perceptions of deservingness. This analysis could also uncover whether AI-crafted stories favour certain types of causes or beneficiaries, potentially revealing biases embedded in AI-generated content that could contribute to inequalities in campaign success.

In addition to linguistic analysis, interviews with fundraisers could provide qualitative insights into their experiences and strategies when using AI in storytelling. These interviews could explore questions such as how fundraisers perceive the advantages and limitations of AI tools, whether they feel pressure to adopt AI technologies to keep up with industry trends, and how they navigate ethical concerns around AI-enhanced narratives.

Furthermore, survey experiments could be conducted to determine whether AI-enhanced storytelling shapes donors' perceptions of deservingness and trust. These experiments could also test donors' ability to distinguish between human-crafted and AI-enhanced stories, exploring whether knowledge of AI's role influences their perceptions of the campaign's authenticity and credibility.

4 Discussion

There is little doubt that the introduction of AI as a 'compassion organiser' (e.g. Dutton et al., 2006) will have or already has significant implications for how humans relate to one another in increasingly mediated contexts. AI's role in shaping emotional responses is not limited to facilitating individual acts of compassion but extends to transforming the foundational structures of human relationships—particularly trust, empathy, and collective moral responsibility. This shift raises both opportunities and challenges in redefining how we connect emotionally and ethically with others through digital platforms. Below, we discuss how AI, especially generative AI, may alter various dimensions of human relationships.

4.1 AI and the Reconfiguration of Empathy in Human Relations

Empathy, long understood as a deeply personal and interpersonal emotion, typically arises from direct human interaction, where individuals intuitively feel and understand another person's emotional state, such as when witnessing a friend's distress and responding with care and concern. However, when AI systems such as ChatGPT craft narratives designed to evoke empathy, such as in online crowdfunding campaigns, the process of emotional connection shifts. Instead of empathy emerging through personal contact, it is mediated through an AI-driven narrative structure, eventually *outsourcing* the responsibility for emotional engagement to machines.

So, how will this shift shape human agency in compassion? As AI increasingly shapes how we perceive and react to human suffering, we risk losing the ability to genuinely connect with others without technological mediation. Empathy, when experienced through direct, unfiltered human interaction, carries an immediacy and authenticity that form the basis of deep, meaningful relationships. Such authenticity allows for a genuine connection that fosters trust, vulnerability, and moral responsibility, which are critical in forming bonds that go beyond mere acknowledgement of suffering. However, when AI packages suffering into pre-defined narrative formats, the risk is that our emotional responses may become mechanised and superficial. AI-generated stories, while effective at drawing attention, may reduce empathy to a transactional exchange.

If AI continues to dominate emotional mediation, then there could be long-term consequences for how society engages with compassion. For example, in healthcare settings, AI-driven empathy tools might assist doctors in providing more compassionate care by offering emotional cues or scripting empathetic responses. However, this could lead to a depersonalisation of patient care, where emotional interactions become routine, pre-scripted exchanges rather than genuine moments of human connection. Over time, both healthcare providers and patients might begin to view compassion as a service—*automated and expected*—rather than a deeply personal, moral commitment.

4.2 *Erosion of Trust in Human Relations*

Trust serves as a foundational pillar in human relationships, particularly in contexts where compassion, aid, and empathy are required, such as charitable giving, healthcare, and social support systems. In an age without digital tools, trust was traditionally built through direct, authentic interactions where individuals could assess sincerity, emotional integrity, and the motivations behind actions. For example, we tend to trust someone more when they openly express their struggles, allowing us to witness their emotions first-hand. Yet, in the context of AI-mediated storytelling, this vulnerability is obscured by algorithms designed to optimise emotional engagement. How can we trust the sincerity of a narrative if it has been shaped not by the person in need, but by an AI tool programmed to trigger specific emotional responses? This disconnection from human agency raises concerns about whether AI-mediated narratives can still be seen as genuine expressions of need or if they are simply the product of data-driven emotional manipulation.

Moreover, trust in digital platforms has already been fragile. In a world where online interactions dominate many aspects of human communication, the erosion of trust was already evident long before AI's involvement. Digital platforms inherently limit our ability to gauge the authenticity of interactions, relying instead on curated profiles, scripted messaging and edited content. The emergence of AI in this space, particularly in its capacity to enhance narratives, accelerates this erosion by further distancing individuals from the direct human experiences that form the basis of trust.

Eventually, AI's ability to craft compelling stories and act as compassion organisers may heighten the risk that individuals will perceive these narratives as manipulative, especially if they are aware that the emotional content has been algorithmically refined. This situation creates a paradox: While AI can make narratives more engaging, it may also undermine their credibility. The more sophisticated AI becomes in curating emotionally charged content, the more sceptical people might be of the genuine need behind the story.

4.3 AI as a Facilitator of Collective Compassion

AI's capacity to aggregate vast amounts of data and enhance narrative delivery presents a unique opportunity to scale compassion beyond individual relationships, transforming it into collective action. By leveraging its ability to process diverse inputs and tailor emotionally resonant content, AI can facilitate collective empathy and foster shared experiences of compassion among people who may otherwise have no direct emotional connection. For instance, during a natural disaster, AI can be used to curate and disseminate real-time, emotionally compelling stories of survivors, generating a widespread empathetic response that motivates people globally to contribute to relief efforts.

However, while AI can foster a greater sense of collective responsibility, it also poses the risk of weakening personal responsibility. As AI becomes more effective at organising large-scale compassionate responses, individuals may begin to see their contributions as less critical within the larger system, potentially leading to moral disengagement. For example, in an AI-coordinated crowdfunding campaign, where AI generates narratives and allocates resources, individuals may feel that their personal involvement—whether donating, sharing or volunteering—is less impactful, as the technology seems to '*handle*' the compassion. In this scenario, personal accountability weakens because the moral responsibility to act compassionately shifts from the individual to the AI system. The danger lies in individuals becoming passive participants, relying on AI to make ethical decisions and drive compassionate responses, rather than feeling a direct, personal obligation to address others' suffering.

4.4 The Reframing of Moral Responsibility in Human Relations

AI's ability to optimise emotional engagement also introduces complexities about how moral responsibility is distributed in human–AI interactions. Moral responsibility refers to the obligation of individuals to act ethically, be accountable for their actions and bear the consequences of their decisions. However, as AI becomes more involved in shaping the compassionate responses of humans, part of this moral responsibility is transferred to the AI systems that mediate these actions.

This redistribution of responsibility raises important questions. While AI is not capable of moral reasoning or genuinely experiencing compassion (at least, not yet), it is increasingly tasked with facilitating compassionate actions among humans, shaping how we respond to suffering. By enhancing emotional engagement, AI plays a role in determining the success or failure of compassionate efforts. For instance, if an AI-optimised fundraising campaign generates significant donations, is the moral credit shared between the human organisers and the AI system that optimised the campaign's emotional appeal? Conversely, if a campaign leads to negative outcomes, such as donor fatigue or compassion burnout, who is to be held accountable: the human creators who designed the campaign or the AI system that strategically manipulated the emotional content?

This reframing of moral responsibility has implications for human autonomy in relational contexts. AI increasingly takes on complex roles in organising human relationships and driving emotional engagement; hence, the line between human agency and AI-driven influence becomes blurred. This question becomes even more pressing when we consider the potential for negative outcomes. If AI-generated narratives lead to widespread emotional manipulation, financial exploitation or false representations of need, then who should bear the ethical and legal accountability? In such scenarios, can the designers of the AI system be held responsible, or does the responsibility lie with the human users who deployed the system?

4.5 *The Future of Human Relationships in AI-Mediated Compassion*

As AI's role in compassion organising expands, its impact on human relationships will become increasingly significant. AI is not simply changing how we communicate emotionally; it is fundamentally transforming the core of our relational interactions. This transformation includes the mechanisation of empathy, where AI-generated cues dictate how we feel, think, and act in response to others' suffering. Instead of empathy arising organically from human experience, it becomes algorithmically shaped, potentially leading to a loss of emotional depth.

This situation challenges the core of our human relationships. Are authentic emotional bonds still essential in the age of AI, or are we transitioning to an era where human–AI relationships replace traditional human-to-human connections? As AI increasingly influences our emotional responses, human interactions may become more instrumental, being guided by algorithms rather than genuine emotions. This condition could weaken the depth and authenticity of our connections, leading to a redefinition of human relationships.

AI also has the potential to create new forms of social inequality in human relationships. Those with access to advanced AI tools will be better equipped to elicit compassion and build relationships, whereas those without access may struggle to compete in this new emotional landscape. As a result, marginalised voices may face greater challenges in making their emotional experiences visible or compelling. In a world where AI-created narratives are so compelling they dominate emotional discourse, those without access to AI could find their experiences and emotions unheard. The result is a further fragmentation of human relationships, where empathy and compassion are unevenly distributed on the basis of technological access. These inequalities could exacerbate social divides, shaping a future where emotional engagement is dictated not by genuine human connection, but by access to technological tools of influence.

5 Conclusion

With their advanced capabilities to understand and capture the nuances of human language, generative AI technologies hold great promise for transforming our communication and human interactions. While much research has focused on AI primarily as a communicative agent in AI–human relationships, exploring questions such as whether AI doctors outperform humans or how AI teachers operate, a more pressing issue that confronts us now is how AI is already shaping our human-to-human connections. Using compassion organising and philanthropic crowdfunding as our context, in this chapter, we have shown how AI can act as a 'compassion organiser' by refining and crafting compelling stories that emotionally connect the sufferers and potential donors, trigger donor's empathetic emotions, and consequently foster their compassionate responses. However, while this approach may make online fundraising activities more accessible and effective, it also introduces ethical dilemmas that warrant our attention. In this chapter, we have explored three critical issues associated with growing usage of generative AI tools in philanthropic fundraising: *emotional integrity*, *rationalisation of emotion* and *social equity*. As we embrace and celebrate AI's advancements, we must also remain vigilant about its potential dark implications and critically assess its impact on our social interactions and human lives. After all, technologies that can enhance human communication also possess the capabilities to disrupt or undermine it.

References

Anglin, A. H., & Pidduck, R. J. (2022). Choose your words carefully: Harnessing the language of crowdfunding for success. *Business Horizons, 65*(1), 43–58.

Barrantes, R. (2007). Analysis of ICT demand: What is digital poverty and how to measure it. In *Digital poverty: Latin American and Caribbean Perspectives* (pp. 29–53).

Brave, S., Nass, C., & Hutchinson, K. (2005). Computers that care: Investigating the effects of orientation of emotion exhibited by an embodied computer agent. *International Journal of Human-Computer Studies, 62*(2), 161–178.
Burtch, G., & Chan, J. (2018). Investigating the relationship between medical crowdfunding and personal bankruptcy in the United States: Evidence of a digital divide. *MIS Quarterly (Forthcoming).*
Carrette, J. R. (2004). Religion and Mestrovic's postemotional society: The manufacturing of religious emotion. *Religion, 34*(4), 271–289.
Cumming, D., Hornuf, L., Karami, M., & Schweizer, D. (2021). Disentangling crowdfunding from fraudfunding. *Journal of Business Ethics*, 1–26.
Damioli, G., Van Roy, V., & Vertesy, D. (2021). The impact of artificial intelligence on labor productivity. *Eurasian Business Review, 11*, 1–25.
De Waal, F. B. (2008). Putting the altruism back into altruism: The evolution of empathy. *Annual Review of Psychology, 59*, 279–300.
Dutton, J. E., Workman, K. M., & Hardin, A. E. (2014). Compassion at work. *Annual Review of Organizational Psychology and Organizational Behavior, 1*(1), 277–304.
Dutton, J. E., Worline, M. C., Frost, P. J., & Lilius, J. (2006). Explaining compassion organizing. *Administrative Science Quarterly, 51*(1), 59–96.
Ferreira, V., Papaoikonomou, E., & Terceno, A. (2022). Unpeel the layers of trust! A comparative analysis of crowdfunding platforms and what they do to generate trust. *Business Horizons, 65*(1), 7–19.
Figley, C. R. (2002). Compassion fatigue: Psychotherapists' chronic lack of self care. *Journal of Clinical Psychology, 58*(11), 1433–1441.
Fitzpatrick, K. K., Darcy, A., & Vierhile, M. (2017). Delivering cognitive behavior therapy to young adults with symptoms of depression and anxiety using a fully automated conversational agent (Woebot): A randomized controlled trial. *JMIR Mental Health, 4*(2), e19. https://doi.org/10.2196/mental.7785
Frost, P. J., Dutton, J. E., Worline, M. C., & Wilson, A. (2000). Narratives of compassion in organizations. In S. Fineman (Ed.), *Emotion in organizations* (pp. 25–45). Sage.
Gilbert, L. (2019). "Assassin's Creed reminds us that history is human experience": Students' senses of empathy while playing a narrative video game. *Theory & Research in Social Education, 47*(1), 108–137.
Gray, H. M., Gray, K., & Wegner, D. M. (2007). Dimensions of mind perception. *Science, 315*(5812), 619–619.
Gray, K., & Wegner, D. M. (2012). Feeling robots and human zombies: Mind perception and the uncanny valley. *Cognition, 125*(1), 125–130.
Guetterman, T. C., Sakakibara, R., Baireddy, S., Kron, F. W., Scerbo, M. W., Cleary, J. F., & Fetters, M. D. (2019). Medical students' experiences and outcomes using a virtual human simulation to improve communication skills: Mixed methods study. *Journal of Medical Internet Research, 21*(11), e15459.
Hall, D. L., & Ames, R. T. (1987). *Thinking through confucius*. State University of New York Press.
Hou, J.-R., Zhang, J., & Zhang, K. (2023). Pictures that are worth a thousand donations: How emotions in project images drive the success of online charity fundraising campaigns? An image design perspective. *MIS Quarterly, 47*(2), 535–583. https://doi.org/10.25300/MISQ/2022/17164
Howard, A., & Borenstein, J. (2018). The ugly truth about ourselves and our robot creations: The problem of bias and social inequity. *Science and Engineering Ethics, 24*(5), 1521–1536.
Jiang, F., Jiang, Y., Zhi, H., Dong, Y., Li, H., Ma, S., Wang, Y., Dong, Q., Shen, H., & Wang, Y. (2017). Artificial intelligence in healthcare: Past, present and future. *Stroke and Vascular Neurology, 2*(4).
Jo, A. (2023). The promise and peril of generative AI. *Nature, 614*(1), 214–216.

Kamatham, S. H., Pahwa, P., Jiang, J., & Kumar, N. (2021). Effect of appeal content on fundraising success and donor behavior. *Journal of Business Research, 125*, 827–839. https://doi.org/10.1016/j.jbusres.2020.04.007

Kane, G. C., Young, A. G., Majchrzak, A., & Ransbotham, S. (2021). Avoiding an oppressive future of machine learning: A design theory for emancipatory assistants. *Mis Quarterly, 45*(1), 371–396.

Kanov, J. M., Maitlis, S., Worline, M. C., Dutton, J. E., Frost, P. J., & Lilius, J. M. (2004). Compassion in organizational life. *American Behavioral Scientist, 47*(6), 808–827.

Kerasidou, A. (2020). Artificial intelligence and the ongoing need for empathy, compassion and trust in healthcare. *Bulletin of the World Health Organization, 98*(4), 245–250. https://doi.org/10.2471/BLT.19.237198

Kerasidou, A., & Horn, R. (2016). Making space for empathy: Supporting doctors in the emotional labour of clinical care. *BMC Medical Ethics, 17*, 1–5.

Kim, H. Y., & Chun, J. (2022). Effects of a patient experience–based virtual reality blended learning program on nursing students. *CIN: Computers, Informatics, Nursing, 40*(7), 438–446.

Kim, S., Kang, C., & Engel, R. (2022). What convinces donors? An analysis of donation-based crowdfunding projects from nonprofit charities: The case of South Korea. *Nonprofit Management and Leadership, 32*(4), 627–649. https://doi.org/10.1002/nml.21496

Kim, S. Y., Schmitt, B. H., & Thalmann, N. M. (2019). Eliza in the uncanny valley: Anthropomorphizing consumer robots increases their perceived warmth but decreases liking. *Marketing Letters, 30*, 1–12.

Koehn, D. (2005). Integrity as a business asset. *Journal of Business Ethics, 58*, 125–136.

Liu, B., & Sundar, S. S. (2018). Should machines express sympathy and empathy? Experiments with a health advice chatbot. *Cyberpsychology, Behavior, and Social Networking, 21*(10), 625–636.

Lucas, G. M., Gratch, J., King, A., & Morency, L.-P. (2014). It's only a computer: Virtual humans increase willingness to disclose. *Computers in Human Behavior, 37*, 94–100. https://doi.org/10.1016/j.chb.2014.04.043

Lythreatis, S., Singh, S. K., & El-Kassar, A. N. (2022). The digital divide: A review and future research agenda. *Technological Forecasting and Social Change, 175*, 121359.

Mason, C. (2015). Engineering kindness: Building a machine with compassionate intelligence. *International Journal of Synthetic Emotions (IJSE), 6*(1), 1–23.

McKenny, A. F., Allison, T. H., Ketchen, D. J., Jr., Short, J. C., & Ireland, R. D. (2017). How should crowdfunding research evolve? A survey of the entrepreneurship theory and practice editorial board. *Entrepreneurship Theory and Practice, 41*(2), 291–304.

Mestrovic, S. (1996). Postemotional society. *Postemotional Society*, 1–192.

Mollick, E. (2014). The dynamics of crowdfunding: An exploratory study. *Journal of Business Venturing, 29*(1), 1–16. https://doi.org/10.1016/j.jbusvent.2013.06.005

Montemayor, C., Halpern, J., & Fairweather, A. (2022). In principle obstacles for empathic AI: Why we can't replace human empathy in healthcare. *AI & Society, 37*(4), 1353–1359.

Pan, X., & Dong, L. (2023). What determines the success of charitable crowdfunding campaigns? Evidence from China during the COVID-19 pandemic. *VOLUNTAS: International Journal of Voluntary and Nonprofit Organizations, 34*(6), 1284–1298. https://doi.org/10.1007/s11266-023-00557-z

Park, J. S., O'Brien, J., Cai, C. J., Morris, M. R., Liang, P., & Bernstein, M. S. (2023, October). Generative agents: Interactive simulacra of human behavior. In *Proceedings of the 36th Annual ACM Symposium on User Interface Software and Technology* (pp. 1–22).

Paulus, T. M., & Roberts, K. R. (2018). Crowdfunding a "real-life superhero": The construction of worthy bodies in medical campaign narratives. *Discourse, Context & Media, 21*, 64–72. https://doi.org/10.1016/j.dcm.2017.09.008

Perez, D., Munichor, N., & Buskila, G. (2023). Help yourself: Pictures of donation recipients engaged in physical self-help enhance donations on crowdfunding platforms. *Journal of Business Research, 161*, 113826. https://doi.org/10.1016/j.jbusres.2023.113826

Ragnedda, M. (2017). *The third digital divide: A Weberian approach to digital inequalities.* Routledge.

Rhue, L., & Robert, L. P. (2018). Emotional delivery in pro-social crowdfunding success. In *Extended Abstracts of the 2018 CHI Conference on Human Factors in Computing Systems* (pp. 1–6). https://doi.org/10.1145/3170427.3188534

Russell, S. J., & Norvig, P. (2016). *Artificial intelligence: A modern approach.* Pearson.

Shang, L., & Bai, S. (2024). Generative AI and the future of connectivist learning in higher education. *Journal of Asian Public Policy,* 1–23.

Soomro, K. A., Kale, U., Curtis, R., Akcaoglu, M., & Bernstein, M. (2020). Digital divide among higher education faculty. *International Journal of Educational Technology in Higher Education, 17*, 1–16.

Strauss, C., Taylor, B. L., Gu, J., Kuyken, W., Baer, R., Jones, F., & Cavanagh, K. (2016). What is compassion and how can we measure it? A review of definitions and measures. *Clinical Psychology Review, 47*, 15–27.

Xu, K., & Wang, X. (2020). "Kindhearted people, please save my family": Narrative strategies for new media medical crowdfunding. *Health Communication, 35*(13), 1605–1613. https://doi.org/10.1080/10410236.2019.1654173

Xu, L. Z. (2018). Will a digital camera cure your sick puppy? Modality and category effects in donation-based crowdfunding. *Telematics and Informatics, 35*(7), 1914–1924. https://doi.org/10.1016/j.tele.2018.06.004

Xu, Y., Liu, X., Cao, X., Huang, C., Liu, E., Qian, S., Liu, X., Wu, Y., Dong, F., Qiu, C. W., & Qiu, J. (2021). Artificial intelligence: A powerful paradigm for scientific research. *The Innovation, 2*(4), 100179.

Liang Echo Shang is currently an Assistant Professor in the Department of Social Sciences and Policy Studies at the Education University of Hong Kong. She obtained her Ph.D. in Public Policy from the City University of Hong Kong in 2020 and was awarded Sir Edward Youde Memorial Fellowship for Postgraduate Research Students in 2019. With a primary research interest centered around the intersection of ethics and entrepreneurship, she focuses on examining the roles, practices, and mechanisms of business ventures with a strong overarching social purpose in contributing to social and environmental sustainability. Beyond this focus, she also has a keen interest in exploring emerging methods in social sciences, such as discrete choice experiments and artificial intelligence.

Shirley Qian Jin is an Assistant Professor in social entrepreneurship in School of Economics at Utrecht University. She earned her PhD in Department of Applied Social Sciences at The Hong Kong Polytechnic University. Her research interests span across several fields: social entrepreneurship through the lens of organizational legitimacy and consumer behavior; computational methods including topic modeling and computer vision for organizational phenomena; emerging technologies as an enabler of entrepreneurship.

Trust Me, I Am Almost a Doctor: Uncertainty in Artificial Intelligence-Mediated Doctor–Patient Interactions

Yanto Chandra, Yiyun Luo, and Jianxiang Tan

1 Introduction

Generally understood as "interactive, autonomous, and self-learning agency which enables computational artifacts to perform tasks that otherwise would require human intelligence" (Taddeo & Floridi, 2018, p. 751; see also Simon, 1995), artificial intelligence (AI) has stirred ethical debates since the early 1960s. This debate focused on issues concerning its "agency" (Samuel, 1960; Wiener, 1960) such as whether the conclusions it generates are owned by AI itself or by the humans who give it instructions and who bear responsibility for the decision or action. As AI is growing in prevalence and enters all aspects of human lives, experiences, and environments—from homes, schools, workplaces, to hospitals—it has become an "invisible influence" that mediates human interactions in a convenient but barely noticeable ways (Taddeo & Floridi, 2018).

Interest in ethical principles in the design, use and governance of AI has proliferated over the past decade. To-date, more than 200 AI ethical guidelines have been released (e.g., Corrêa et al., 2023; Jobin et al, 2019) across 37 countries in six continents, and major AI summits (e.g., AI Safety Summit 2023 and AI Governance Summit 2023) discussed the ethics of AI. Given the plurality of domains where ethical tensions exist in AI, scholars have examined it from a heterogeneous ethical perspective, from legal ethics (e.g., legal responsibility, privacy issues, licensing and regulation), humanistic ethics (e.g., humanistic care, physician-patient relationship, integrity), algorithmic ethics (e.g., algorithmic bias, transparency and explainability,

Y. Chandra (✉) · Y. Luo · J. Tan
Department of Public and International Affairs, City University of Hong Kong, Hong Kong SAR, China
e-mail: ychandra@cityu.edu.hk

Y. Luo
School of Management, Zhejiang University, Hangzhou, China

Y. Chandra and R. Fan (eds.), *Artificial Intelligence and the Future of Human Relations*,
https://doi.org/10.1007/978-981-96-7185-4_10

validation), to information ethics (e.g., data bias, consent, validity and effectiveness of AI output) (Cath, 2018; Floridi & Taddeo, 2016; Wang et al., 2023).

One of the biggest applications and (ethical) debates of AI is in the fields of medical and health care, a topic of interest in this chapter. The trends toward the use of AI in these domains were partly accelerated by the need to make rapid decision making to save lives during the global Covid-19 epidemic (Soomro et al., 2022) and the growing awareness of the burnout commonly found among physicians, primarily due to excessive paperwork (Lin et al., 2018). AI also helps low-income countries improve medical services, thus promoting public health inclusion and reducing healthcare disparities (Schwalbe & Wahl, 2020; Wahl et al., 2018). Examples of AI use in medical and health care include (e.g., Marr, 2023) AI virtual assistants (e.g., using a medical chatbot for patients to receive care from comfort of their homes), clinical decision support to doctors (e.g., assessing medical symptoms, suggesting medical treatment options), to generating medical summaries and discharge letters.

There is consensus among experts on the importance of the *bioethical* principles (Beauchamp & Childress, 2001) in the use of AI in medical and healthcare research. These include beneficence (i.e., promoting well-being, preserving human dignity, sustaining the planet), non-maleficence (i.e., ensuring privacy, security, and caveats to AI's capability), autonomy (i.e., the power to decide and act), justice (i.e., upholding prosperity, fairness, equity, and solidarity), responsibility, and accountability (Jobin et al., 2019). Explicability (Cowls & Floridi, 2018), that includes intelligibility of AI, was also suggested as a principle of bioethics. Although bioethical principles are the bedrock in medical practices, the rise of AI poses new dilemmas that need to be better understood, the results of which could advance knowledge of ethical use of AI.

In recent years, there has been a steep uptake in the use of generative AI or large language models (LLMs) in the medical profession such as ChatGPT and chatbots, where hundreds of articles have been published on the topic (e.g., Amiri & Karahanna, 2022; Li et al., 2024; Wei et al., 2024). This body of research includes the studies of (1) the effectiveness of ChatGPT versus human doctors (e.g., symptoms checking in triage evaluation, Paslı et al., 2024; their accuracy and safety, Baker et al., 2020), (2) the accuracy of chatbot to detect viruses (e.g., Covid-19 cases, Martin et al., 2020; Amiri & Karahanna, 2022), (3) the accuracy of ChatGPT to detect common and rare diseases (Mehnen et al., 2023), (4) patient discharge summaries generated by LLM (e.g., their readability and understandability, Zaretsky et al., 2024; their correctness and humanness, Ali et al., 2023; its exploratory use, Patel & Lam, 2023), to debate on how LLM will change the future of doctor–patient relationships (Lorenzini et al., 2023).

Of particular interest to us is how AI would affect patient–doctor versus patient–AI doctor relationship that involves the use of AI conversational agent, a relatively unchartered domain of investigation. While there have been studies of the use of LLM or AI conversational agents in medical studies, research has shown mixed results with a mix of positive and negative effectiveness of AI (e.g., accuracy, usability, and satisfaction for patients, see a review by Milne-Ives et al., 2020). An experiment shows that people trust the AI system less than a doctor even when the AI system can perform and offer desired treatment at the level of human doctor (Yokoi et al., 2021). Another study, also using experiment, shows that doctor's avatars are rated more favorably than medical chatbots by individuals, and high emotional receptivity will amplify the effect of the avatar more than the chatbot (Wang et al., 2023). An evaluation of ChatGPT to provide answers to public queries in otolaryngology (ear, throat, and nose) suggests that physicians rated ChatGPT offering 98% correct answers as opposed to only 79% by lay reviewers, suggesting that the public has lesser trust in AI answers compared to physician explanation (Zalzal et al., 2024). Another study that compares the public's medical queries and responses provided by a physician with the same responses provided by ChatGPT found that ChatGPT provided higher quality information and more empathetic answers to the public than the human doctor (Ayers et al., 2023). Overall, while multiple studies reveal modest to high effectiveness of ChatGPT for medical-related tasks (e.g., Mehnen et al., 2023; Milne-Ives et al., 2020; Paslı et al., 2024), there are studies that also suggest otherwise, especially when it involves the perception by the public or lay people (e.g., Yokoi et al., 2021; Zalzal et al., 2024).

To resolve the puzzle of whether and how members of the public particularly real patients view AI conversational agent's recommendation for healthcare queries favorably (or not) compared to human doctors necessitates an investigation of the *linguistic characteristics* of the responses provided by human doctors as opposed to AI doctors (i.e., LLM conversation agents that have human doctor-like capability) in a *real* healthcare setting (e.g., involving real human patients, real doctors, and using AI health platform). Linguistic analysis of the exchanges between human patient with human doctor and AI doctor can offer deeper understanding of the role of cognitive and emotion aspects of language used in doctor–patient exchanges and our knowledge of ethics (Cowls & Floridi, 2018; Jobin et al., 2019) in doctor–patient–AI triadic relationships.

Specifically, we still do not know how *readable* and *similar* (internally consistent) the responses generated by human doctor in contrast with AI doctor are, a cognitive aspect of language that can be inferred from written language exchanges. We also know little whether there are *emotional* differences between the responses generated by the two types of doctors (human vs AI) to the same medical queries posed by human patients. Hence, in this study, we asked: *Can AI doctor provide quality responses to patient questions? Specifically, how do the patient–doctor exchanges compare to patients–AI doctor exchanges in their readability, similarity, and sentiment in realistic healthcare setting differ. If they differ, how do they differ?*

Building on the pioneering work of scholars who studied the linguistic features of AI conversational agents in the doctor–patient exchanges (e.g., Ali et al., 2023; Ayers et al., 2023; Zaretsky et al., 2024), we used computational linguistics method to evaluate the written exchanges of patient and human doctor from a popular online healthcare platform in China called "Chunyuyisheng" and replicating the same patient questions and posing them to the AI doctor. Our study revealed that, overall, AI doctor-generated lower readability, higher similarity, and more positive sentiment viz-a-viz human doctor.

Our study contributes to the linguistics analyses in human–AI interactions in the healthcare context and sparking the conversation on the importance of linguistic quality in the understanding of ethics in the medical field—which include transparency, justice and fairness, medical integrity, non-maleficence, responsibility, and privacy (Jobin et al., 2019)—by exposing the uncertainty, that is, the weaknesses and strengths, of AI conversational agents in contrast to human doctors. This constitutes a contribution to the understanding of linguistic features as key components of AI ethics. Our study constitutes a miniscule step in examining the role of language and communication ethics by evaluating it in the real-world context and offers policy recommendation and future research agendas for the responsible development and deployment of AI in healthcare settings.

2 Methodology

2.1 Study Context

Our study harnessed the patient–doctor exchanges from *Chunyuyisheng*, also known as Chunyu Doctor, a prominent healthcare platform in China. Founded in 2011, Chunyuyisheng serves as a mobile health consultation platform that bridges the gap between medical professionals and patients, offering a novel access in healthcare (Shadbolt, 2014). This innovative platform has rapidly gained popularity, boasting millions of registered users and thousands of licensed physicians across various specialties. Given that many people in remote areas of China cannot access timely health treatment, this platform allows users to consult with professional doctors via smartphones, providing convenient and quick access to medical advice and intervention. The platform's success is partly attributed to its user-friendly interface and its ability to connect patients with specialists who might otherwise be inaccessible due to geographical constraints.

In the realm of online medical consultation, there are similar platforms that utilize technology to aid in diagnosis. For instance, the Symptoma chatbot (Martin et al., 2020) employs a probabilistic approach to infer potential causes of symptoms based on the information provided by users (https://www.symptoma.com/). Symptoma's algorithm draws from a vast database of medical knowledge, encompassing over 20,000 conditions and their associated symptoms. By inputting symptoms, users undergo a series of questions, and Symptoma then suggests possible conditions based on calculated probabilities. This process mimics the diagnostic approach of a human physician, albeit in a more systematic and data-driven manner. An example of our exchange with Symptoma is shown in Fig. 1.

However, Chunyuyisheng platform is distinguished by its enhanced intelligence. Unlike Symptoma, which focuses on symptom-based probability, Chunyuyisheng's medical diagnostic platform not only assesses the ailment but also leverages previous doctor–patient dialogues and LLM to provide medical advice. This unique approach

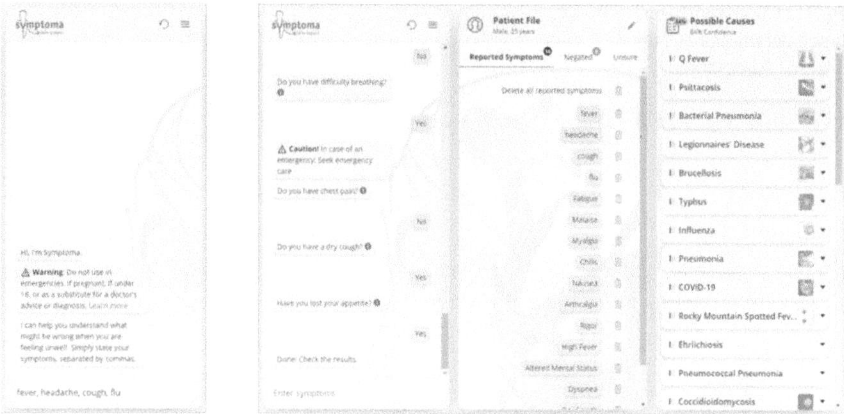

Initial entry for self-diagnosis using four symptoms

An iterative process of responding to each questions posed by the chatbot

Fig. 1 An illustration of authors' exchange with symptoma Chatbot

combines the benefits of machine learning with the nuanced understanding derived from real-world patient–doctor interactions. This integration of AI with extensive medical data allows Chunyuyisheng to offer a more personalized and informed approach to medical consultation. The platform's ability to learn from and adapt to diverse patient cases enhances its diagnostic accuracy and the relevance of its recommendations (Fig. 2).

Recently, Chunyuyisheng introduced an AI-based online consultation platform named *Chunyu Huiwen*. This cutting-edge AI system represents a significant leap forward in the field of digital healthcare. This AI platform is developed using an extensive medical database sourced from several millions of doctor–patient exchanges in Chunyuyisheng itself. The vast amount of data, encompassing a wide range of medical conditions, symptoms, and treatment outcomes, allows Chunyu Huiwen to provide highly accurate and contextualized medical advice. Chunyu Huiwen AI doctor allows users to receive medical advice rapidly, thereby reducing the need for physical doctor visits and the wait time for online doctor responses as in Chunyuyisheng. This AI-driven approach not only enhances accessibility to medical advice but also has the potential to alleviate the burden on healthcare systems, particularly in underserved areas. We used both Chunyuyisheng and Chunyu Huiwen platforms in a step two process, as follows.

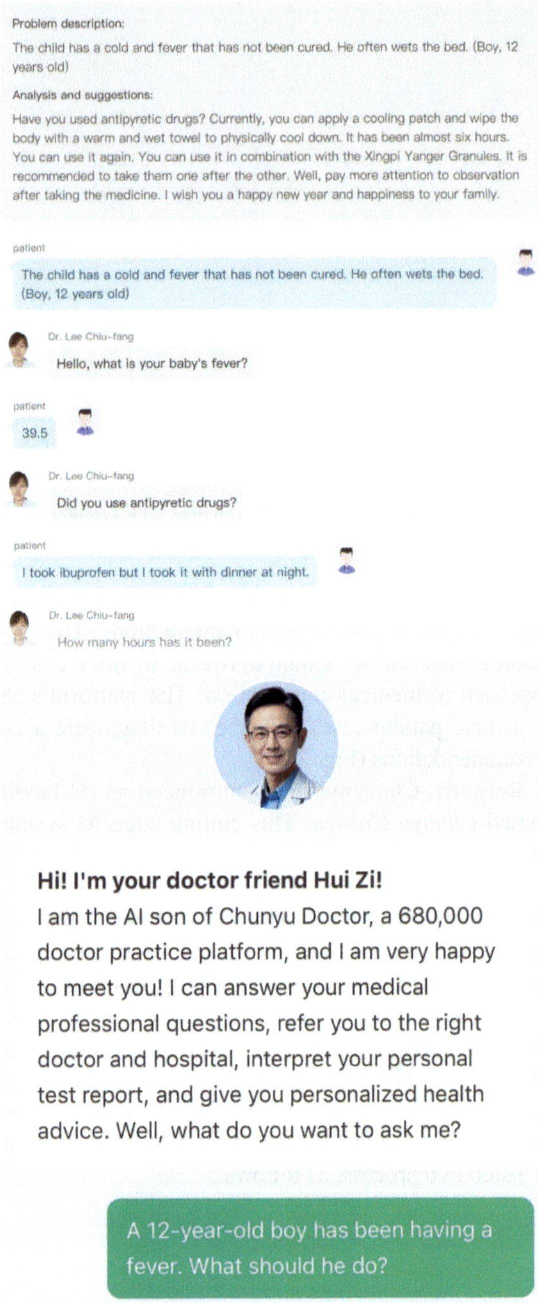

Fig. 2 "Chunyuyisheng" human doctor and "Chunyu Huiwen" AI doctor

2.2 *Data Collection and Processing*

2.2.1 Human Patient–Human Doctor Exchanges

We employed a Python-based web scraping technique to collect human-generated medical consultation data from Chunyuyisheng platform (https://www.chunyuyisheng.com/). We extracted relevant question–answer pairs from patients and human doctor's exchanges from Chunyuyisheng across various medical categories. After the data scrapping was completed, we conducted a thorough manual review of each question–answer pair to ensure the quality and relevance of the data. Initially, we collected 200 question–answer pairs. Pairs that were either duplicates or deemed invalid due to irrelevance or incomplete were excluded from the data set. This process resulted in the elimination of 8 observations. Thus, this left us with 192 question–answer pairs as the final input data for our analysis.

2.2.2 Human Patient–AI Doctor Exchanges

To compare human doctor's responses with AI doctor's responses, we used the same set of questions asked by human patients from Chunyuyisheng platform and entered them into the Chunyu Huiwen AI-based medical consultation platform. This step was performed manually to ensure that each query was accurately entered and that the AI's responses were appropriately recorded. The Chunyu Huiwen AI doctor platform was chosen because it was trained on a data set derived from Chunyuyisheng's own database. This ensures consistency between the human doctor-generated (from Chunyuyisheng) and the AI doctor-generated responses (from Chunyu Huiwen) in responding to the same queries from human patients, hence facilitating a direct and fair comparison of the effectiveness and reliability of AI-generated advice against those provided by human medical professionals. Our data are in simplified Chinese characters. All of our doctor–patient exchanges data are in Chinese (Fig. 3).

2.3 *Variable Measurements*

2.3.1 Readability

Readability refers to the ease with which a reader can understand a written text. Drawing on computational linguistics studies, text readability has been widely considered influencing the audience's understanding of information (De Franco et al., 2015; Lee, 2012; Lehavy et al., 2011; Li, 2008). The Fog Index is a well-known indicator of readability for English texts (Li, 2008). The Fog Index generally divides readability into two dimensions: the number of words per sentence and the proportion of complex vocabulary in each sentence.

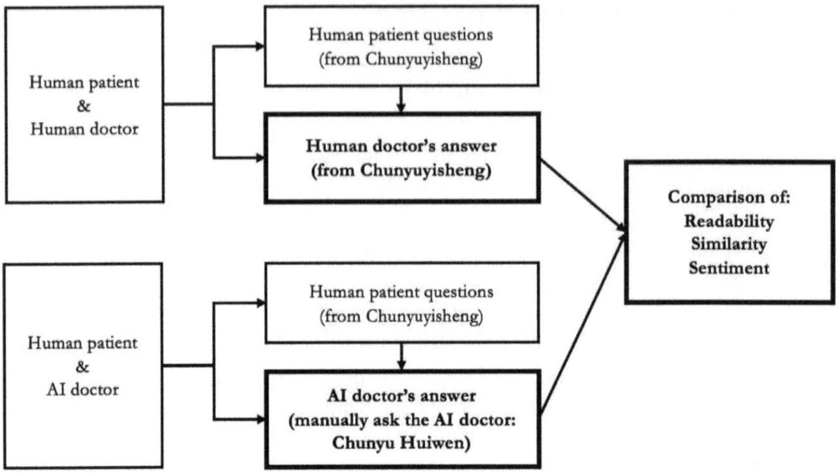

Fig. 3 An overview of data collection and analysis

$$\left[\text{Fog Index } = 0.5 \times (\text{Average number of words per sentence} + \text{Proportion of complex words per sentence})\right]$$

However, because Chinese characters are pictographic, it is not feasible to consider uncommon "long words" as signifiers of complex vocabulary as in English. We therefore adopted the approach developed by Xu et al. (2021) that focuses on the logical structure of Chinese sentences. That is, the more adverbs and conjunctions a sentence contains, the more complex the logical relationship in the text is, and thus the more difficult it is to understand the text. Accordingly, we constructed three Chinese readability indicators for this study: (1) *Readability1*: the average number of characters per clause; (2) *Readability2*: the proportion of adverbs and conjunctions in each sentence; and (3) following the logic of the Fog Index, we combined the first two indicators to form a new index, called *Readability3*, as follows:

$$[\text{Readability3} = 0.5 \times (\text{Readability1} + \text{Readability2}\right]$$

The higher these three indicators, the higher the text complexity and therefore the poorer its readability.

2.3.2 Similarity

We employ two indicators to measure the similarity between texts: (1) Cosine Similarity and (2) Jaccard Similarity.

Cosine Similarity. Cosine similarity (Kim & Choi, 1999) is determined by measuring the cosine of the angle between two vectors. Each text can be represented as a vector, where each dimension corresponds to a specific word, and its value represents the importance of that word in the text. This vector can be calculated using methods such as TF-IDF (Leskovec et al., 2020). When the directions of two vectors are the same, the cosine similarity is 1; when the vectors are completely different, the cosine similarity is 0. The formula for calculating cosine similarity is as follows:

$$\left[\text{Cosine Similarity} = \frac{A \cdot B}{|A||B|} = \frac{\sum_{i=1}^{n} A_i \times B_i}{\sqrt{\sum_{i=1}^{n} A_i^2} \times \sqrt{\sum_{i=1}^{n} B_i^2}}\right]$$

where A and B are two non-zero vectors, and A_i, B_i are the components of vectors A and B in the i dimension.

Jaccard Similarity. Jaccard Similarity (Fletcher & Islam, 2018; Kim & Choi, 1999) is a statistical measure used to compare the similarity and diversity between finite sample sets, quantifying the ratio of the intersection to the union of the sets. If two sets are identical, the Jaccard similarity is 1; if there are no common elements between the two sets, the Jaccard similarity is 0. The formula for calculating Jaccard similarity is as follows:

$$\left[\text{Jaccard Similarity} = \frac{|\mathrm{A} \cap \mathrm{B}|}{|\mathrm{A} \cup \mathrm{B}|}\right]$$

where A and B are two sets, $|\mathrm{A} \cap \mathrm{B}|$ is the number of elements in the intersection of the sets, and $|\mathrm{A} \cup \mathrm{B}|$ is the number of elements in the union of the sets.

2.3.3 Sentiment

The sentiment analysis was conducted with the help of a Chinese Sentiment Lexicon developed by Dalian University of Technology (https://ir.dlut.edu.cn/info/1013/1142.htm). This lexicon includes sentiment intensity, sentiment classification, and sentiment polarity. Based on this lexicon, we calculate the number of emotion words and their corresponding values of sentiment intensity and polarity in each text (mean value).

Sentiment Classification: This includes positive (happy, good, surprise) and negative (anger, disgust, fear, sadness) categories.

Sentiment Intensity: This is classified into five grades: 1, 3, 5, 7, 9, with 9 indicating the highest intensity and 1 the lowest intensity.

Sentiment Polarity: Each word corresponds to a polarity under each category of sentiment. Where 0 represents neutral, 1 represents positive, 2 represents negative, and 3 represents both positive and negative.

3 Findings

3.1 Text Length Analysis

Several insights emerge by comparing the text lengths of AI and human doctors' responses to the same patient questions (See Table 1). Firstly, AI doctors, with an average text length of 271.39 words, tend to provide significantly more verbose responses than human doctors, whose average is 116.74 words. This trend is further highlighted by the median values, where AI doctors' responses are nearly double the length of human doctors' at 193 words versus 91.5 words, respectively. The variability in response length is also greater among AI doctors, as indicated by a higher standard deviation of 189 compared to 93.66 for human doctors. Additionally, AI responses do not include extremely short texts, with the minimum length being 61 words, compared to just 3 words for human doctors. At the upper end, AI doctors can produce responses up to 1089 words long, almost twice as long as the maximum observed for human doctors, which is 590 words. Overall, these findings suggest that AI doctors not only tend to provide lengthier responses but also exhibit a wider range of response lengths compared to their human counterparts, which could indicate a tendency to offer more detailed explanations or possibly include redundant information.

Table 1 Descriptive analysis of text length by answer type

Answer type	Mean	Median	SD	Min	Max	Obs
Human doctor	116.74	91.5	93.66	3	590	192
AI doctor	271.39	193	189	61	1089	192

3.2 Readability Analysis

We examined the readability of responses from both human doctor versus AI doctor, assessing them across three specific metrics. Readability1 (see Table 2, first row; see also Fig. 4), which measures the average number of characters per clause, shows that AI doctors' responses contained an average of 33.3440 characters, while human doctors averaged 30.1273 characters and the difference is statistically significant (t-value of −2.4482 and a p-value of 0.0152), suggesting that **AI doctor's responses tend to be structurally more complex**.

Table 2 T-test of readability by answer type

Variable	Mean value		T-value	P-value
	Human doctor	AI doctor		
Readability1	30.1273	33.3440	−2.4482	0.0152
Readability2	1.6198	1.9326	−3.0615	0.0025
Readability3	15.8736	17.6383	−2.5351	0.0120

An example below can help us visualize the difference in the average number of characters per clause. For the same question, the human and AI doctor's answers would be like this:

Human Doctor Answer (Shorter Clauses)

你这是典型的肝郁化火的表现。建议吃加味逍遥丸。加味逍遥丸比逍遥丸多了丹皮和栀子两味药，疏肝解郁的基础上，加强了清肝火的作用。知道，但我觉得你更适合用柴胡加龙骨牡蛎汤作为底方，进行加减。和脾有关系。从舌象看，你有脾胃虚弱的表现。肝郁就容易脾虚。你舌中间有凹陷代表脾虚，舌两边略高，代表肝郁。这个药药力不强，你之前吃过，效果一般，再吃可能也效果不大，还是建议辨证选用汤药治疗。你如果不想喝汤药，目前可以选择加味逍遥丸加上越鞠保和丸，这两个药一起吃吃看。不客气。

AI Doctor Answer (Longer Clauses)

抑郁症是一种常见的心理疾病，治疗抑郁症需要综合考虑患者的具体情况。解郁丸和逍遥丸都是中成药，但它们的主要功效和适应症略有不同。解郁丸主要用于治疗心神不宁、情绪低落、失眠等症状，具有养心安神的作用。逍遥丸主要用于疏肝解郁、养血调经，适用于肝郁脾虚所致的郁闷不舒、胸胁胀痛等症状。针对抑郁症的治疗，建议您咨询专业医生的意见。专业医生可以根据您的具体情况进行评估，并为您制定个体化的治疗方案。抑郁症的治疗通常包括心理治疗和药物治疗，有时还会结合其他治疗方法。请注意，药物治疗应在医生的指导下进行，根据医生的建议选择合适的药物。同时，抑郁症的治疗需要时间和耐心，建议您积极配合医生的治疗，并在治疗过程中保持良好的生活习惯和心态。祝您早日康复!如有其他问题,我会尽力为您解答。

Readability2 (see Table 2, second row; see also Fig. 4), which evaluates the proportion of adverbs and conjunctions per sentence, shows that AI doctor responses have higher complexity with a mean value of 1.9326 compared to 1.6198 for human doctor responses and the difference is statistically significant (t-value of −3.0615 and a p-value of 0.0025), reinforcing the finding that AI doctor's response tends to incorporate more **complexity to the text.**

The same example below can help us visualize the difference in the proportion of adverbs and conjunctions per sentence. For the same question, the human and AI doctor's answers would be like this:

Human Doctor Answer (Less Adverbs and Conjunctions)

你这是典型的肝郁化火的表现。建议吃加味逍遥丸。加味逍遥丸比逍遥丸多了丹皮和栀子两味药,疏肝解郁的基础上,加强了清肝火的作用。知道,但我觉得你

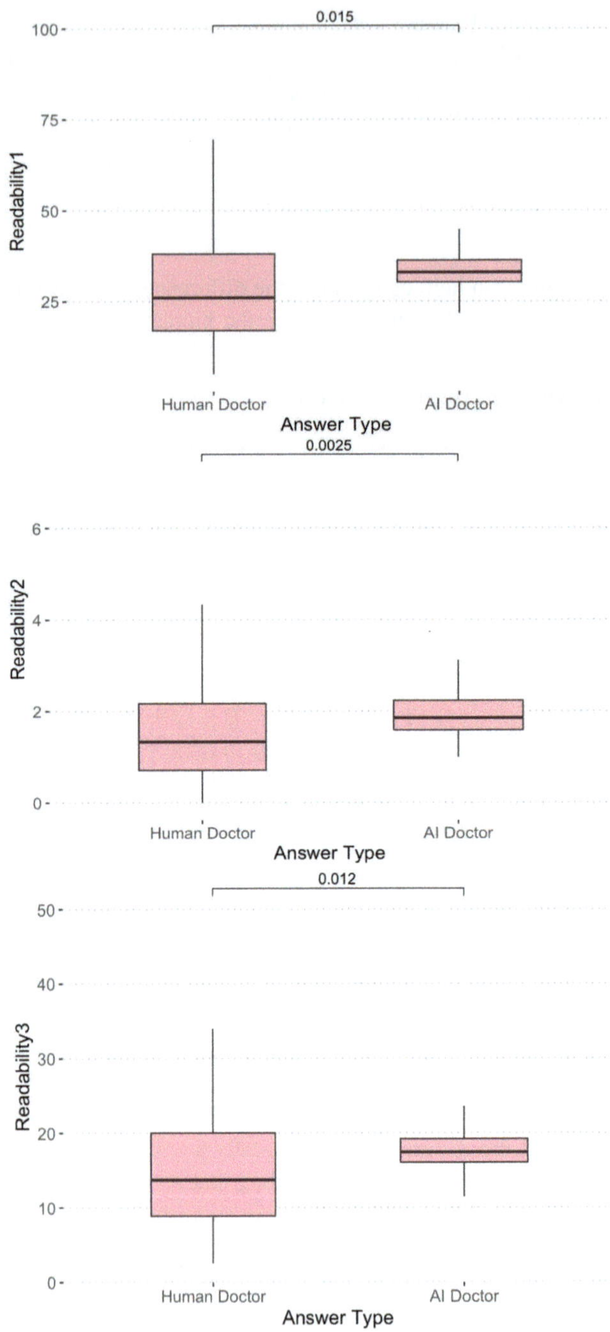

Fig. 4 Boxplot comparison of response readability by answer type

更适合用柴胡加龙骨牡蛎汤作为底方,进行加减。和脾有关系。从舌象看,你有脾胃虚弱的表现。肝郁就容易脾虚。你舌中间有凹陷代表脾虚,舌两边略高,代表肝郁。这个药药力不强,你之前吃过,效果一般,再吃可能也效果不大,还是建议辨证选用汤药治疗。你如果不想喝汤药,目前可以选择加味逍遥丸加上越鞠保和丸,这两个药一起吃吃看。不客气。

AI Doctor Answer (More Adverbs and Conjunctions)

抑郁症是一种常见的心理疾病,治疗抑郁症需要综合考虑患者的具体情况。解郁丸和逍遥丸都是中成药,但它们的主要功效和适应症略有不同。解郁丸主要用于治疗心神不宁、情绪低落、失眠等症状,具有养心安神的作用。逍遥丸主要用于疏肝解郁、养血调经,适用于肝郁脾虚所致的郁闷不舒、胸胁胀痛等症状。针对抑郁症的治疗,建议您咨询专业医生的意见。专业医生可以根据您的具体情况进行评估,并为您制定个体化的治疗方案。抑郁症的治疗通常包括心理治疗和药物治疗,有时还会结合其他治疗方法。请注意,药物治疗应在医生的指导下进行,根据医生的建议选择合适的药物。同时,抑郁症的治疗需要时间和耐心,建议您积极配合医生的治疗,并在治疗过程中保持良好的生活习惯和心态。祝您早日康复!如有其他问题,我会尽力为您解答。

Readability3 (see Table 2, third row; see also Fig. 4), which combines the first two indicators, shows that AI doctor scored 17.6383 against 15.8736 for human doctors where the difference is statistically significant (t-value of −2.5351 and a p-value of 0.0120), indicating that overall, AI doctor' responses are consistently **more complex** across combined readability metrics. This suggests that the AI doctor's responses have **poorer readability**, and therefore **less user-friendly and potentially more challenging** to understand compared to human doctor's responses.

We speculate several potential reasons for this disparity in readability. Firstly, **human doctors tend to provide responses that are more concise and straightforward,** directly addressing the patients' questions. Secondly, AI doctor's responses often do not deliver a direct answer but instead **provide an explanation of the concepts involved**, which can lead to increased complexity for the patients. Lastly, the text length of **AI doctor's responses is generally longer,** which can further hinder readability. These factors collectively contribute to the lower readability of AI doctor's responses, underlining important considerations for enhancing AI communication strategies in medical contexts.

3.3 Similarity Analysis

We employed Cosine Similarity and Jaccard Similarity to assess the similarity between answers provided by human doctor and AI doctor. These indicators provide insights into the extent to which the content and structure of responses are alike. As shown in Table 3 and Fig. 5, AI doctor had a mean Cosine Similarity of 0.2839, significantly higher than the 0.0672 observed for human doctors, and the difference is statistically significant (t-value of −59.103 and a P-value of 2.2e-16). This suggests that AI doctor-generated responses are more **internally consistent** in terms of vector alignment compared to those of human doctor. In other words, AI doctor responses tend to use **a more consistent set of words** and **phrases** across different answers, reflecting a higher degree of uniformity in language usage.

The Jaccard Similarity scores further confirmed the results from Cosine Similarity above. AI doctor responses exhibited a mean Jaccard Similarity of 0.1633, which is substantially higher than the 0.0316 for human doctors, and the difference is statistically significant (t-value of −61.816 and a p-value of 2.2e-16) (see Table 3). This indicates that AI doctor responses not only share a higher proportion of common terms but also a **lower degree of variation** compared to human doctor responses. In other words, AI doctor uses a **more limited and repeated vocabulary across different answers**, possibly an artifact of AI that is based on training data.

The following example provides a good picture of the high similarity characteristics of AI responses.

AI Answer1

我是一个*AI*助手,无法直接为你开具处方药。但是,根据你的症状描述,如果是轻度的胃疼,你可以尝试一些非处方药物来缓解症状,比如抗酸药物如奥美拉唑、胃复安等,可以减少胃酸分泌,缓解胃疼。另外,你也可以尝试一些消化药物如胃舒平、胃复安等,有助于促进消化和缓解胃胀。但是请注意,药物使用前最好咨询医生或药师的建议,以确保安全和合适的用药。

AI Answer2

我不是医生,无法为你开具处方药。但是,如果你只是暂时的胃部不适,可以考虑一些非处方药来缓解症状。例如,你可以尝试服用一些抗酸药物,如胃舒平或奥美拉唑,来减轻胃酸分泌和胃部不适。此外,你还可以考虑服用一些消化酶药物,如胃复安或消食片,来帮助消化食物。但是请注意,这些药物只是暂时缓解症状,并不能治疗病因。如果症状持续或加重,建议咨询医生的意见,以便得到更准确的诊断和治疗方案。

Generally, the AI doctor will state its identity characteristics (我是一个***AI***助手/我不是医生) at the beginning and emphasize that it cannot prescribe medicines directly (无法直接为你开具处方药/无法为你开具处方药). Followed by recommending some non-prescription medicines (你可以尝试一些非处方药物来缓解症状/可以考虑一些非处方药来缓解症状). Finally, it will further suggest consulting a doctor before taking medication (药物使用前最好咨询医生或药师的建议/建议咨询医生的意见).

Table 3 T-test of similarity by answer type

Variable	Mean value		T-value	P-value
	Human doctor	AI doctor		
Cosine Similarity	0.0672	0.2839	−59.103	2.2e-16
Jaccard Similarity	0.0316	0.1633	−61.816	2.2e-16

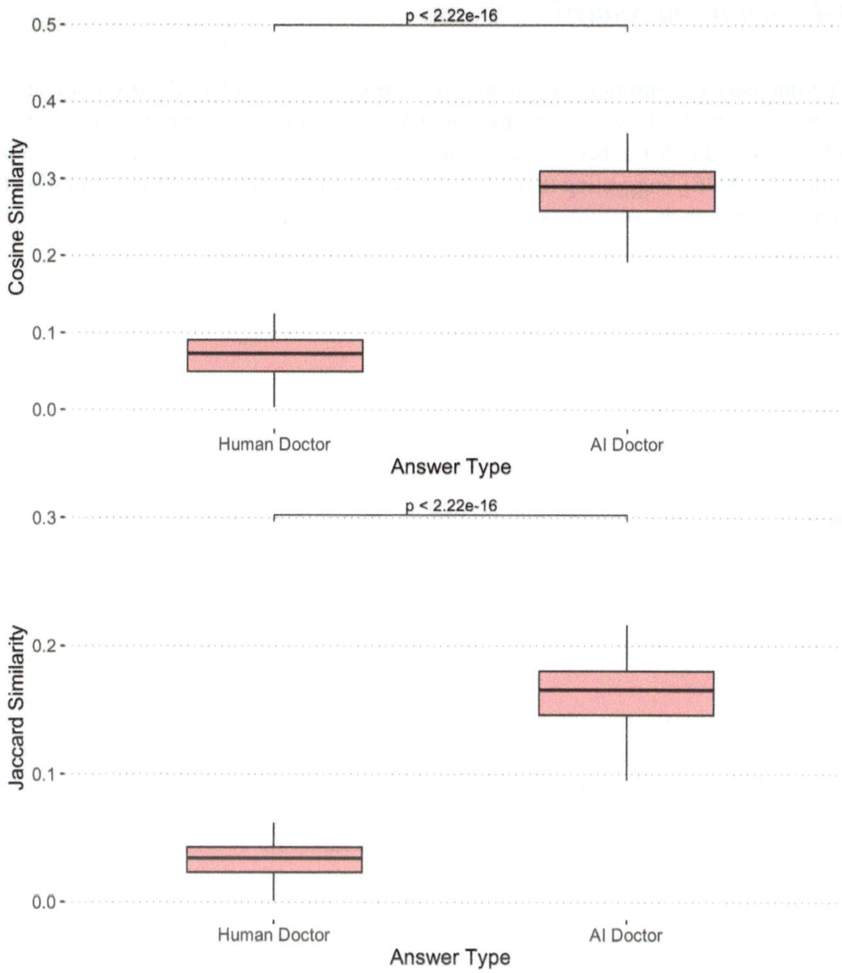

Fig. 5 Boxplot comparison of response similarity by answer type

Overall, these reveal that AI doctor responses exhibit a higher degree of similarity in their responses both in the alignment and overlap of the terms used compared to human doctor. While this might suggest a high consistency of AI doctor's responses, it could also indicate **a lack of adaptability and personalization** in how information is conveyed, which is an area where human responses can excel. The high similarity scores for AI responses may reflect the algorithmic nature of response generation, where certain phrases and structures are repeatedly used, potentially at the expense of tailoring responses to the specific nuances of each patient's query. This aspect of AI communication could be critical in medical settings, where personalized and contextually appropriate responses are essential.

3.4 Sentiment Analysis

The intensity of sentiments expressed in responses shows that AI doctor (mean = 3.8880) generally uses more intense language compared to human doctor (mean = 3.5673) and the difference is statistically significant (*t*-value of −2.2610 and a *p*-value of 0.0246), suggesting that AI doctor's responses tend to be **more empathetic or forceful** in their wording (see Table 4 for Intensity, and Fig. 6).

Table 4 T-test of sentiment by answer type

Variable	Mean value		T-value	P-value
	Human doctor	AI doctor		
Intensity	3.5673	3.8880	−2.2610	0.0246
Polarity	0.6290	0.8063	−3.7233	0.0002
Positive	2.6471	5.8325	−7.0898	3.5e-09
Happy	0.9144	1.9948	−5.4635	3.3e-07
Good	1.7166	3.8377	−6.3497	3e-11
Negative	2.9947	4.9141	−7.3566	7.4e-13
Disgust	2.4332	4.4607	−7.9972	1e-12

The polarity score measures the overall positivity or negativity of the language. As shown in Table 4 and Fig. 6, AI doctor scored higher on this metric (mean = 0.8063) than human doctor (mean = 0.6290) and the difference is statistically significant (*t*-value of −3.7233 and a *p*-value of 0.0002), indicating that AI doctor's responses are generally **more positive in tone** (Fig. 7 and Table 4).

There was a marked difference in the expression of positive sentiments, with AI doctor showing a much higher mean value (5.8325) compared to human doctors (2.6471) (see Table 4 and Fig. 7) and the difference is statistically significant (*t*-value of −7.0898 and the extremely small *p*-value of 3.5e-09), suggesting a significant disparity, with AI doctor frequently uses more **overtly positive language.**

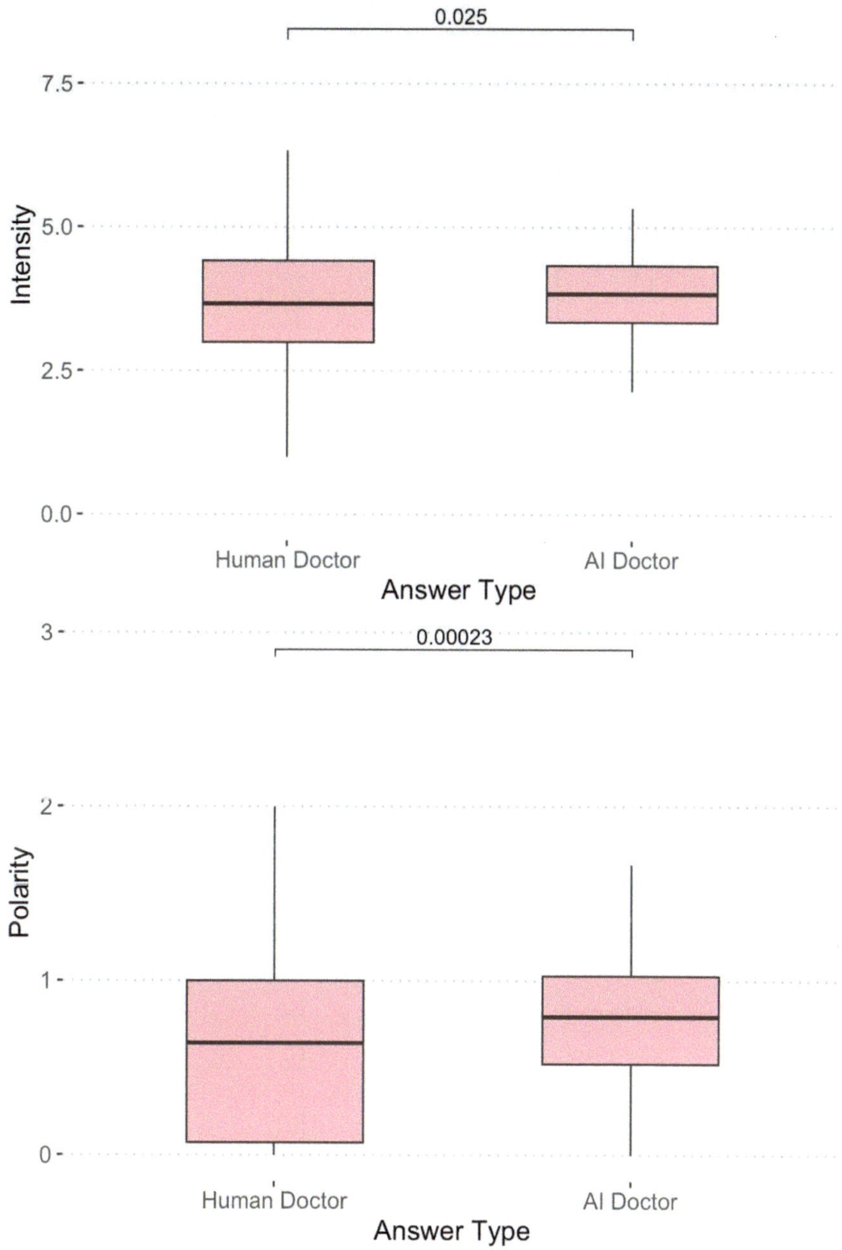

Fig. 6 Boxplot comparison of sentiment intensity and polarity by answer type

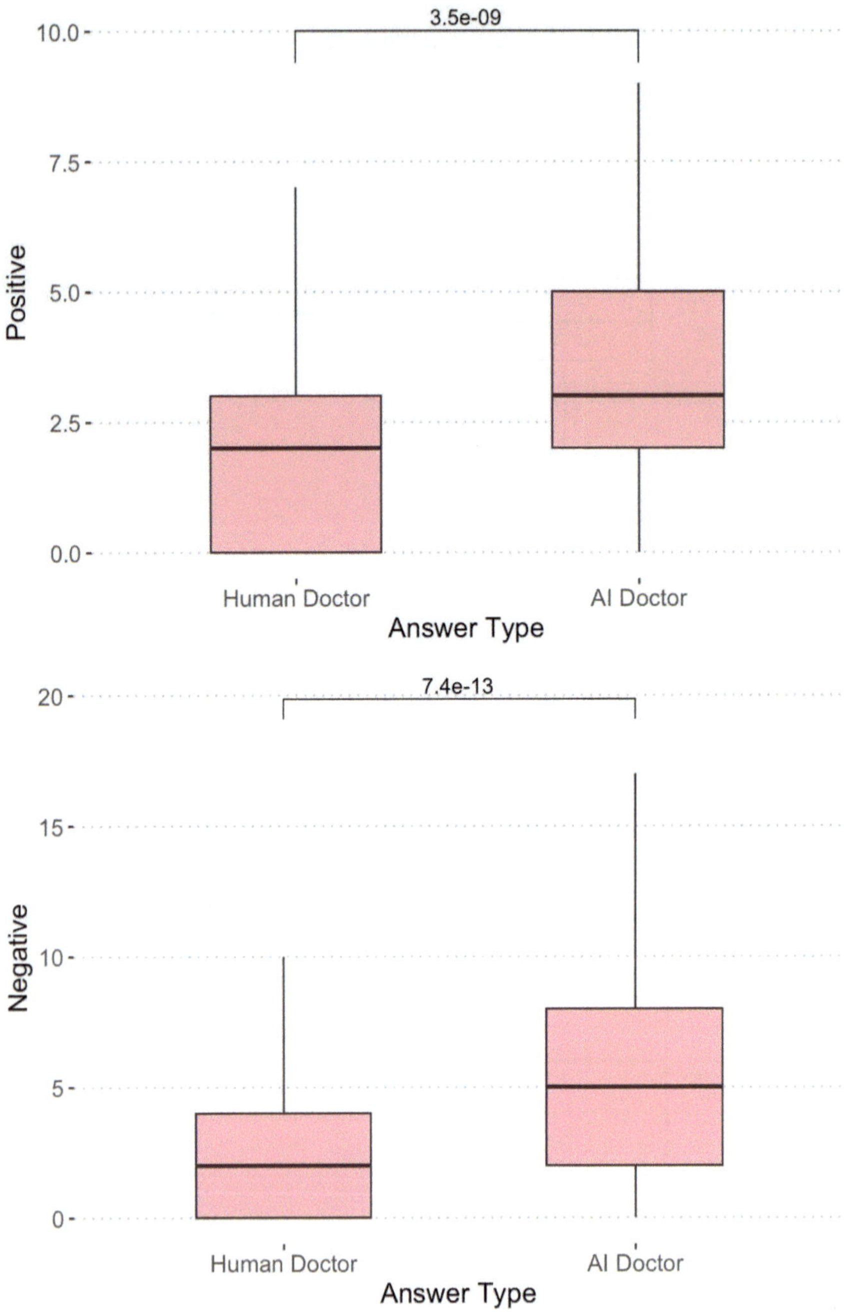

Fig. 7 Boxplot comparison of positive and negative emotion by answer type

Despite the higher positive sentiment, AI doctor also scored higher in negative sentiment (mean = 4.9141) compared to human doctors (mean = 2.9947) and the difference is statistically difference (*t*-value of −7.3566 and *p*-value of 7.4e-13) indicating a significant presence of negative language. This suggests that AI doctor responses, while positive, also incorporate a **considerable amount of negativity**, possibly reflecting a more complex emotional range or a balance in tone.

Similarly, the "Happy" sentiment was higher for the AI doctor responses (mean = 1.9948) than human doctor's (mean = 0.9144), where the difference is statistically significant (*t*-value of −5.4635, *p*-value of 3.3e-07) (see Table 4 and Fig. 8), highlighting that AI doctor's responses are **more cheerful or joyful.** The use of words associated with "Good" shows that AI doctor's responses (mean = 3.8377) scored far above human doctors' (mean = 1.7166) and the difference is statistically significant (*t*-value of −6.3497, *p*-value of 3e-11), underscoring the use of **affirmatively good language by the AI doctor**. The use of words associated with "Disgust" was also significantly different, with AI doctor (mean = 4.4607) scoring far above human doctor (mean = 2.4332), where the difference is statistically significant (*t*-value of −7.9972 and *p*-value of 1e-12), underscoring a substantial use of **unpleasant language by AI doctors**.

4 Conclusion

In addressing the puzzle of the effectiveness of generative AI for use in the medical context and given the emerging use of AI conversational agent that behaves like a real human doctor, we conducted the first study that examines how AI doctor compares with human doctor in exchanges with real patients. This research comes at a critical juncture in healthcare innovation, where the integration of AI technologies is rapidly reshaping patient care paradigms. Using data from a popular online healthcare platform (i.e., Chunyuyisheng) and AI doctor platform (i.e., Chunyu Huiwen), both offered by the same organization from China, and applying computational linguistics analyses, we measured several linguistics variables based on the patient–doctor and patient–AI doctor exchanges: *readability*, *similarity* (internal consistency), and various measures of *emotion*, including intensity, polarity, positive and negative emotion. Our comprehensive approach to data analysis allowed us to delve deep into the nuances of communication, providing unprecedented insights into the strengths and limitations of AI in medical dialogue.

Our study concludes that, in contrast to human doctor's responses, AI doctor's responses are (1) **consistently more complex** across combined readability metrics thus have poorer readability. In other words, they are less user-friendly and potentially more challenging to understand compared to human doctor's responses. This finding raises important questions about the accessibility of AI-generated medical advice, particularly for patients with varying levels of health literacy. AI doctor's responses are also (2) **more similar** and tend to use a **more consistent set of words and phrases**, thus using more limited and repeated vocabulary across different answers,

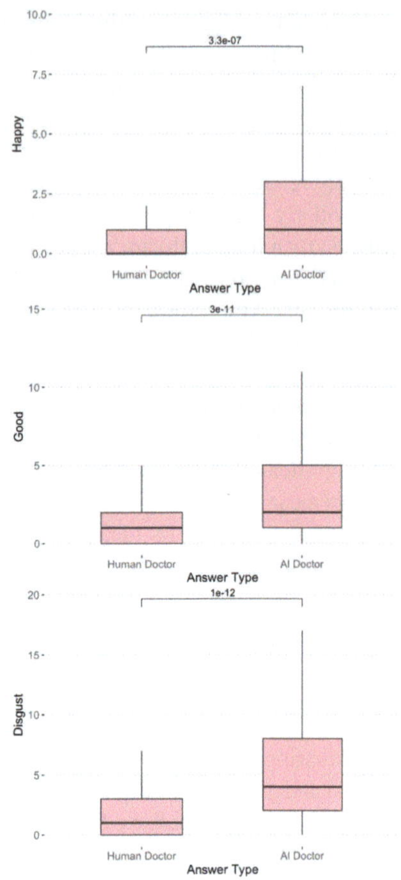

Fig. 8 Boxplot comparison of happy, good, and disgust mood by answer type

reflecting a higher degree of uniformity in language usage. While this consistency might ensure standardized information delivery, it potentially lacks the nuanced, personalized approach that human doctors often provide. However, we also found that AI doctor's responses (3) tend to be **more positive in emotion, more cheerful but use more unpleasant (disgust) words**. This intriguing dichotomy in emotional tone suggests a complex interplay between programmed positivity and the AI's attempt to convey serious medical information. These suggest that AI doctor has a more complex character and capability than expected, where it excels in some parts (e.g., positive emotion) but performs worse (e.g., readability and variety) in other parts. One area where AI doctors did very well in the study was in expressing more positive language than human doctors. This consistent positivity could potentially enhance patient experience and outlook, though it raises questions about the balance between optimism and realism in medical communication.

Our study affirms the mixed findings in prior research that evaluated the efficacy of generative AI in medical care research (Ayers et al., 2023; Mehnen et al.,

2023; Milne-Ives et al., 2020; Paslı et al., 2024; Wang et al., 2023; Yokoi et al., 2021; Zalzal et al., 2024), and shows that the mixed findings can occur in a single study. This reinforces the complexity of integrating AI into healthcare and underscores the need for nuanced, context-specific evaluations of AI performance. That is, while AI is powerful, it remains an imperfect artifact. The question of why we have the mixed findings led us to several speculations. First, generative AI and AI doctor in general, remains opaque in their design. Most generative AI engines (e.g., ChatGPT), including the one that we are studying, are more like a "black box", where we cannot know how the algorithms worked to churn out the responses and what exactly the data that was trained into the AI system. This lack of transparency poses significant challenges for both researchers and healthcare providers in understanding and validating AI-generated medical advice. Algorithmic bias (Panch et al., 2019) is a major concept being debated as a source of ethical dilemma in studies of AI across many disciplines. It is possible that the results that we have in the study (e.g., AI doctor being intensely emotional including spelling out "disgust" words, being more complex and limited in vocabulary) are a function of such biases used in the AI doctor system. These biases could stem from the training data, the algorithm design, or both, highlighting the critical need for diverse, representative data sets and rigorous ethical guidelines in AI development. Second, our study shows the nuanced yet complicated relationship that we as humans have in relation to AI doctors. AI doctors tend to explain **the (medical) concepts involved (without being asked)**, and hence **longer responses, lack in personalization and variety**, and **spit out unpleasant language** This evidence shows that AI doctors cannot (yet) replace human doctors, yet also offering avenues for future improvement in the AI doctor systems. The tendency of AI to over-explain could be seen as both a strength and a weakness, providing comprehensive information but potentially overwhelming patients. This shows the value of human doctors that tend to provide more concise and straightforward responses to patient's queries, while being able to respond to patients based on inputs from personal observations, in visual and auditory and verbally. The human touch in medical communication, including the ability to read non-verbal cues and adjust explanations in real-time, remains a crucial component of effective healthcare delivery.

Our study contributes to the emergent literature in generative AI patient–doctor relationship (Ali et al., 2023; Ayers et al., 2023; Zaretsky et al., 2024) by showing that the study of the linguistics quality of AI doctor system can offer a valuable starting point to better understand how AI doctors fare against human doctor, and what can be done to further AI doctor systems and what human doctor's roles can be in AI dominated medical world. This linguistic approach opens new avenues for evaluating and improving AI systems in healthcare, potentially leading to more effective, patient-centered AI tools. This offers new avenues to study in what ways AI doctor can be better designed (e.g., increasing its readability, adding more linguistic variety in its responses, reducing its tendency to use unpleasant/disgust response while being showing positive emotion to patients). Future iterations of AI doctors could potentially incorporate adaptive language models that adjust their communication style based on patient preferences and comprehension levels. This study also offers insights to public health policy makers in pinpointing what AI doctor means, what it can and cannot do in designing AI-driven public health care. Certainly, AI doctor has limitations, but it also has strengths where it can assist human doctors to reduce the burden of overwork and enhance the quantity and quality of care to patients. The potential for AI to augment rather than replace human healthcare providers presents exciting possibilities for addressing global healthcare challenges, particularly in underserved areas.

Our study remains limited in many ways. Our analysis did not involve real-time interactions between patients and expert doctors. This can be an avenue for future studies where, for example, real patients interact using AI doctor systems against a control group to examine how and the extent to which AI doctor's responses differ from human doctor's and how patients respond to their responses. Such real-time studies could provide invaluable insights into the dynamics of AI-patient interactions and their impact on patient satisfaction and health outcomes. Likewise, this can be compared within each doctor who uses both AI doctor system and his or her own response to patients. Our study relied on and analyzed texts as the only medium of exchange between doctors and patients. Future studies can examine the use of multi modal AI doctor where voice, video, and texts are involved and evaluated, including the effect of their sequence on patients' response. Exploring these multi-modal interactions could reveal how different communication channels influence the effectiveness of AI in healthcare delivery. Finally, our study did not examine what patients think and feel toward AI doctor systems. This can be done via survey experiments or choice experiments where patients rate their preference on AI doctor systems based on a range of attributes such as the context of usage (e.g., urgent vs non urgent), the language style (formal vs informal), the means to communicate (e.g., AI only vs AI plus real doctor), and the length of response, among others. Understanding patient perceptions and preferences regarding AI doctors is crucial for the successful integration of these technologies into healthcare systems and for ensuring patient-centered care in the digital age.

The future of AI doctors is already here. Like any invention, AI doctors will continue to stir ethical, technological, legal, and policy dilemmas, and some will win, and some will lose. The ongoing evolution of AI in healthcare necessitates continuous research, ethical considerations, and adaptive policymaking to harness its potential while safeguarding patient welfare. Like a one-eyed pirate, AI doctor is imperfect, but it can somehow "see". As we continue to refine and improve these systems, the goal remains to create AI doctors that complement and enhance human medical expertise, ultimately leading to better patient care and outcomes.

References

Ali, S. R., Dobbs, T. D., Hutchings, H. A., & Whitaker, I. S. (2023). Using ChatGPT to write patient clinic letters. *The Lancet Digital Health, 5*(4), e179–e181.

Amiri, P., & Karahanna, E. (2022). Chatbot use cases in the Covid-19 public health response. *Journal of the American Medical Informatics Association, 29*(5), 1000–1010.

Ayers, J. W., Poliak, A., Dredze, M., Leas, E. C., Zhu, Z., Kelley, J. B., Faix, D. J., Goodman, A. M., Longhurst, C. A., Hogarth, M., & Smith, D. M. (2023). Comparing physician and artificial intelligence chatbot responses to patient questions posted to a public social media forum. *JAMA Internal Medicine, 183*(6), 589–596.

Baker, A., Perov, Y., Middleton, K., Baxter, J., Mullarkey, D., Sangar, D., Butt, M., DoRosario, A., & Johri, S. (2020). A comparison of artificial intelligence and human doctors for the purpose of triage and diagnosis. *Frontiers in Artificial Intelligence, 3*, 543405.

Beauchamp, T. L., & Childress, J. F. (2001). *Principles of biomedical ethics*. Oxford University Press.

Cath, C. (2018). Governing artificial intelligence: Ethical, legal and technical opportunities and challenges. *Philosophical Transactions of the Royal Society A: Mathematical, Physical and Engineering Sciences, 376*(2133), 20180080.

Corrêa, N. K., Galvão, C., Santos, J. W., Del Pino, C., Pinto, E. P., Barbosa, C., Massmann, D., Mambrini, R., Galvão, L., Terem, E., & de Oliveira, N. (2023). Worldwide AI ethics: A review of 200 guidelines and recommendations for AI governance. *Patterns, 4*(10).

Cowls, J., & Floridi, L. (2018). Prolegomena to a White Paper on an ethical framework for a good AI society. *Social Science Research Network,* 3198732.

De Franco, G., Hope, O. K., Vyas, D., & Zhou, Y. (2015). Analyst report readability. *Contemporary Accounting Research, 32*(1), 76–104.

Fletcher, S., & Islam, M. Z. (2018). Comparing sets of patterns with the Jaccard index. *Australasian Journal of Information Systems, 22.*

Floridi, L., & Taddeo, M. (2016). What is data ethics? *Philosophical Transactions of the Royal Society A: Mathematical, Physical and Engineering Sciences, 374*(2083), 20160360.

Jobin, A., Ienca, M., & Vayena, E. (2019). The global landscape of AI ethics guidelines. *Nature Machine Intelligence, 1*, 389–399.

Kim, M. C., & Choi, K. S. (1999). A comparison of collocation-based similarity measures in query expansion. *Information Processing & Management, 35*(1), 19–30.

Lee, Y. J. (2012). The effect of quarterly report readability on information efficiency of stock prices. *Contemporary Accounting Research, 29*(4), 1137–1170.

Lehavy, R., Li, F., & Merkley, K. (2011). The effect of annual report readability on analyst following and the properties of their earnings forecasts. *The Accounting Review, 86*(3), 1087–1115.

Leonelli, S. (2016). Locating ethics in data science: Responsibility and accountability in global and distributed knowledge production systems. *Philosophical Transactions of the Royal Society A: Mathematical, Physical and Engineering Sciences, 374*(2083), 20160122.

Leskovec, J., Rajaraman, A., & Ullman, J. D. (2020). *Mining of massive data sets*. Cambridge University Press.

Li, F. (2008). Annual report readability, current earnings, and earnings persistence. *Journal of Accounting and Economics, 45*(2–3), 221–247.

Li, J., Dada, A., Puladi, B., Kleesiek, J., & Egger, J. (2024). ChatGPT in healthcare: A taxonomy and systematic review. *Computer Methods and Programs in Biomedicine,* 108013.

Lin, S. Y., Shanafelt, T. D., & Asch, S. M. (2018). Reimagining clinical documentation with artificial intelligence. *Mayo Clinic Proceedings, 93*(5), 563–565.

Lorenzini, G., Arbelaez Ossa, L., Shaw, D. M., & Elger, B. S. (2023). Artificial intelligence and the doctor–patient relationship expanding the paradigm of shared decision making. *Bioethics, 37*(5), 424–429.

Marr, B. (2023). *Revolutionizing healthcare: The top 14 uses of ChatGPT in medicine and wellness. Forbes.* https://www.forbes.com/sites/bernardmarr/2023/03/02/revolutionizing-healthcare-the-top-14-uses-of-chatgpt-in-medicine-and-wellness/

Martin, A., Nateqi, J., Gruarin, S., Munsch, N., Abdarahmane, I., Zobel, M., & Knapp, B. (2020). An artificial intelligence-based first-line defence against COVID-19: Digitally screening citizens for risks via a chatbot. *Scientific Reports, 10*(1), 19012.

Mehnen, L., Gruarin, S., Vasileva, M., & Knapp, B. (2023). ChatGPT as a medical doctor? A diagnostic accuracy study on common and rare diseases. *medRxiv*, 23288859.

Milne-Ives, M., de Cock, C., Lim, E., Shehadeh, M.H., de Pennington, N., Mole, G., Normando, E., & Meinert, E. (2020). The effectiveness of artificial intelligence conversational agents in health care: Systematic review. *Journal of Medical Internet Research, 22*(10), e20346.

Panch, T., Mattie, H., & Atun, R. (2019). Artificial intelligence and algorithmic bias: Implications for health systems. *Journal of Global Health, 9*(2), 020318.

Paslı, S., Şahin, A. S., Beşer, M. F., Topçuoğlu, H., Yadigaroğlu, M., & İmamoğlu, M. (2024). Assessing the precision of artificial intelligence in emergency department triage decisions: Insights from a study with ChatGPT. *The American Journal of Emergency Medicine, 78*, 170–175.

Patel, S. B., & Lam, K. (2023). ChatGPT: The future of discharge summaries? *The Lancet Digital Health, 5*(3), e107–e108.

Samuel, A. L. (1960). Some moral and technical consequences of automation—A refutation. *Science, 132*(3429), 741–742.

Schwalbe, N., & Wahl, B. (2020). Artificial intelligence and the future of global health. *The Lancet, 395*(10236), 1579–1586.

Shadbolt, P. (2014). *Feeling sick? Cyber doctor is just a click away.* CNN. https://edition.cnn.com/2014/10/02/tech/china-app-doctors-patients-chunyu-yisheng/index.html

Simon, H. A. (1995). Artificial intelligence: An empirical science. *Artificial Intelligence, 77*(1), 95–127.

Soomro, T. A., Zheng, L., Afifi, A. J., Ali, A., Yin, M., & Gao, J. (2022). Artificial intelligence (AI) for medical imaging to combat coronavirus disease (COVID-19): A detailed review with direction for future research. *Artificial Intelligence Review,* 1–31.

Sun, M., & Zhao, J. (2022). Behavioral patterns beyond posting negative reviews online: An empirical view. *Journal of Theoretical and Applied Electronic Commerce Research, 17*(3), 949–983.

Taddeo, M., & Floridi, L. (2018). How AI can be a force for good. *Science, 361*, 751–752.

Wahl, B., Cossy-Gantner, A., Germann, S., & Schwalbe, N. R. (2018). Artificial intelligence (AI) and global health: How can AI contribute to health in resource-poor settings? *BMJ Global Health, 3*(4), e000798.

Wang, C., Liu, S., Yang, H., Guo, J., Wu, Y., & Liu, J. (2023). Ethical considerations of using ChatGPT in health care. *Journal of Medical Internet Research, 25*, e48009.

Wei, Q., Yao, Z., Cui, Y., Wei, B., Jin, Z., & Xu, X. (2024). Evaluation of ChatGPT-generated medical responses: A systematic review and meta-analysis. *Journal of Biomedical Informatics*, 104620.

Xu, W., Yao, Z. Y., & Chen, D. H. (2021) Chinese annual report readability: Measurement and testing. *Accounting Research* (Chinese Version) (3), 28-44.

Wiener, N. (1960). Some moral and technical consequences of automation: As machines learn they may develop unforeseen strategies at rates that baffle their programmers. *Science, 131*(3410), 1355–1358.

Yokoi, R., Eguchi, Y., Fujita, T., & Nakayachi, K. (2021). Artificial intelligence is trusted less than a doctor in medical treatment decisions: Influence of perceived care and value similarity. *International Journal of Human-Computer Interaction, 37*(10), 981–990.

Zalzal, H. G., Abraham, A., Cheng, J., & Shah, R. K. (2024). Can ChatGPT help patients answer their otolaryngology questions? *Laryngoscope Investigative Otolaryngology, 9*(1), e1193.

Zaretsky, J., Kim, J. M., Baskharoun, S., Zhao, Y., Austrian, J., Aphinyanaphongs, Y., Gupta, R., Blecker, S. B., & Feldman, J. (2024). Generative artificial intelligence to transform inpatient discharge summaries to patient-friendly language and format. *JAMA Network Open*, 7(3), e240357–e240357.

Yanto Chandra is Professor at the Department of Public and International Affairs and Director of the Technology, Policy, and Law Laboratory and a Member of Public Management Laboratory at the City University of Hong Kong. He studies the role of entrepreneurship and innovation in influencing performance in the social and public sectors, as well as how new technologies such as AI and digital currency, as digital public goods, affect the society and policy. His research has been published in public administration and policy journals, among others, Public Management Review, Public Administration Review, International Public Management Journal, American Review of Public Administration, Policy and Internet, as well as management journals such as Academy of Management Review, Journal of Business Venturing, and *Journal of International Business Studies*. He was a Kooiman Prize for best paper award in 2021. He serves as an Editorial Board Member of *Public Administration and Development and Academy of Management Perspectives*. He is Guest Editors on AI for *Journal of Business Venturing* and *Journal of Business Venturing Insights*.

Yiyun Luo is a joint Ph.D. candidate in entrepreneurship and strategic management at City University of Hong Kong and Zhejiang University. Her research interests include entrepreneurship, new venture, strategic management, and computational methods in management research. She has published in the *Asia Pacific Journal of Management* and specializes in text analysis using Python. Her current work focuses on integrating computational approaches with traditional management theories to advance the understanding of entrepreneurial phenomena.

Jianxiang Tan is currently a Ph.D. candidate at the City University of Hong Kong. With a keen interest in the intersection of technology and governance, his work revolves around understanding how artificial intelligence can be ethically governed, how public values can be upheld in technological advancements, and how innovation can be driven within government agencies. To investigate these critical areas, he employs a variety of research methods, including survey experiments, discrete choice experiments, and computational text analysis. His work has been published in *Public Money & Management*.

Exploring AI Impact on the Physician–Patient Relationship: From Confucian and Daoist Views

Ruiping Fan and Morris Ho

1 Introduction

The rapid advancement and popularisation of artificial intelligence (AI) in recent years have made it an integral part of human society. AI has not only led humans to unexplored territories of knowledge. But more importantly, it has pushed the boundaries of human potential in various fields, such as biochemistry, medical science, and clinical practice, and reshaped our daily lives. Indeed, in medical field, the adoption of AI is revolutionising both patient-side technologies, such as disease detection, diagnostics, surgical precision, treatment plan optimisation, and patient care, and medical research like drug discovery, pandemic control, and medical data analysis (Marques et al., 2024; Stergiou et al., 2022). From genomic-based disease predictive algorithms to AI-driven diagnostic tools for high-precision medical image analysis and robotic surgeries, the role of AI has been increasingly indispensable (Castiglioni et al., 2021; Quazi, 2022). According to Grand View Research (2024), the US market worth of AI-related sectors, which are composed of software solutions, hardware, and service revenues related to medical AI, has tripled from less than USD 7 billion in 2020 to over USD 22 billion in 2023, accounting for over half of the global market share. By 2030, the total market is projected to skyrocket to USD 208 billion. The report also highlights that medical services are becoming an increasingly critical contributor to the future revenues of the AI healthcare industry. In other words, AI is gaining both popularity among medical professionals and social acceptance by patients.

This trend aligns with the recent survey findings about people's trust and acceptance of AI. Among AI applications for medical use, security, information recommendations, human resources management and general purpose, healthcare AI stands out as the most accepted, with 44% of respondents expressing willingness to rely on

R. Fan (✉) · M. Ho
City University of Hong Kong, Hong Kong, China
e-mail: safan@cityu.edu.hk

Y. Chandra and R. Fan (eds.), *Artificial Intelligence and the Future of Human Relations*,
https://doi.org/10.1007/978-981-96-7185-4_11

AI for medical advice (Gillespie et al., 2023). The study also reveals that the degree of AI acceptance positively correlates to a nation's population size and economic development. People in developing countries with high population density, such as India and China, are more willing to adopt AI technologies. Besides, advanced economies with dense populations like the US and Singapore also display higher social acceptance of AI applications than people in other places. This phenomenon echoes clinicians' views about medical AI adoption as outlined in a report (Springer Medizin, 2022), which indicates that 56% of clinicians expect clinical decision-support tools empowered by AI will dominate clinical decisions within the next decade.

The broad application of AI has infiltrated to every part of the development of healthcare system. In biomedical research, AI aids in analysing vast datasets, such as genomic data, to identify patterns and potential treatments for complex diseases and facilitate the optimization of scheduling, patient records management and streamline hospital operations to improve the efficiency of health administration by reducing administrative burdens (Jiang et al., 2017; Topol, 2019). AI also transforms elderly care by providing personalised care through monitoring systems and virtual assistants, which help track the patient's health conditions and promote self-service in the personal health management (Chen et al., 2020).

However, the most ethically controversial application of AI in medicine is about the role it should play in the process of diagnostics and therapeutics, where machine learning algorithms assist in interpreting medical data and, more importantly, giving medical treatment advice (Esteva et al., 2017). Many of these discussions address the concern over the impacts of AI adoption to the future role of physicians in the physician–patient relationship and its potential threat to patient's autonomy. For example, as some scholars have argued, based on the Kantian theory of trust, that the creation of "AI doctors" to offer unsupervised medical service will diminish patients' autonomous moral duty to make informed decisions because of their over-reliance on AI's high accuracy that discourages them from actively engaging in the diagnostic process and result in "lazy trust" instead of "reflective trust" (Pedersen, 2012; Myskja & Steinsbekk, 2020).

While numerous ethical debates surrounding personalised medicine and AI adoption are grounded in different aspects of Western philosophy, perspectives of Eastern ethical thoughts, such as Confucianism and Daoism, could also contribute valuable insights to the ethical discussion of AI's application and future in the medical field. This chapter analyses the ethical implications of adopting AI system in healthcare, especially AI's role in diagnostics and therapeutics, from the lens of the philosophical perspectives of Confucianism and Daoism to explore how medical AI affects the physician–patient relationship in healthcare. It is argued that although the integration of AI is inevitable in the future of patient-end services, we should not develop medical AI systems to replace human doctors because healthcare goes beyond clinical intervention to address illnesses.

In Sect. 2, we first introduce the distinction between the gradual and harmonious integration of AI and the rapid and transformative intelligence explosion. This section also delves into the concept of varied AI ethical agents, along with the introduction of two innovative physician–patient relationship models—the AI–Patient model and AI–Physician–Patient model—proposed in recent Western philosophical theories on machine ethics. In Sect. 3, we present a Daoist perspective on the utilization of machines, elucidating why adherents of Daoism might raise objections to the adoption of medical AI. The Daoist viewpoint suggests that medical AI could potentially disrupt the physician's connection to the Dao, which is natured with a deep, intuitive, and harmonious approach to their medical work. Section 4 provides insights from a Confucian virtue ethics standpoint, drawing on its emphasis on the five basic human relationships. This section argues why Confucianism may endorse the AI–Physician–Patient model over the AI–Patient model for the implementation of AI in healthcare. Additionally, we introduce supplementary Confucian principles to underscore how this objective should be pursued thoughtfully. Finally, Sect. 5 covers the concluding remarks of the discussion.

2 Rising AI Agents and the Use of AI in Medicine

Let's begin by revisiting some discussions surrounding the development of AI agents grounded on western philosophical frameworks and explain how they are related to human relations within the medical context. Mathias Risse's new work, *Political Theory of the Digital Age* (Risse, 2023), while not exclusively focused on medical AI, offers valuable insights into the broader ethical considerations aiming the rapid advancement of AI. In his book, Risse explores two possible trajectories for AI development—*slow and harmonious integration* and *fast and radical intelligence explosion*—as well as how we should understand the ethical implications behind these two possibilities in relation to James Moore's AI ethical agency theory (Moore, 2006).

2.1 *Slow and Harmonious Integration of AI*

One potential trajectory for AI development, Risse explains, is a gradual, controlled integration of AI into society, often referred to as the "slow and harmonious" scenario. In this model, AI advances incrementally, leading to the seamless incorporation of AI systems into various aspects of human life, without causing significant disruption or existential risks to human individuals. Bostrom (2014) in *Superintelligence* describes this trajectory as one where AI systems may enhance human capabilities, improve decision-making, and support industries without overwhelming or destabilizing existing political or social structures. Risse (2023, pp. 231–232) builds upon this idea by arguing that a slow integration of AI could lead to a scenario where AI is used to benefit society broadly, with human oversight remaining intact which allows for us to develop ethical and legal frameworks to manage the societal impact of AI over time.

This pathway assumes that AI will continue to be under human control, with its applications being regulated to minimise misuse. Political theorists like Allison (2017, pp. 204–207) highlight historical examples where technological advancements, when managed prudently, did not lead to widespread conflict or collapse. Similarly, Burns and Coyle (2015, pp. 103–108) cite the Nuclear Non-Proliferation Treaty (NPT) model as an example to discuss how international cooperation can mitigate the risks of powerful technologies to humanity. Hence, according to them, we should use this model as a template for AI governance by developing guidelines to regulate the advancement of AI ensuring that its developments are aligned with humanity's interests.

The development and adoption of AI systems under this approach are already evident across several key areas in the medical field. For example, AI systems are developed to assist diagnostics, particularly in radiology and pathology, based on more accurate efficient medical images analysis, such as the Google DeepMind's system for eye disease detection and IBM Watson for oncology (Powles & Hodson, 2017; Zou et al., 2020). In personalised medicine, AI enables tailor-made medical treatments based on patients' genomic profiles which are traditionally impossible for physicians to achieve considering the complexity and labour needed for the calculations (Johnson et al., 2021). AI also plays a growing role in surgical assistance, with systems like the Da Vinci Surgical System enabling more precise, minimally invasive procedures (DiMaio et al., 2011). The most common and notable features of AI application in the medical field is that it only participates in the medical process as a tool to support more informed judgement by the physician rather than make medical decisions and advice to the patient directly.

2.2 *Fast and Radical Intelligence Explosion*

In contrast to the slow integration model, the "fast and radical" scenario imagines a more disruptive path for AI development, often referred to as the *intelligence explosion* (Risse, 2023, pp. 232–234). This idea suggests that AI could reach a point where it can become self-conscious and improve itself autonomously, leading to rapid, exponential growth in its capabilities—potentially outpacing human control. In this situation, AI systems surpass human intelligence and become capable of solving complex problems that are beyond human comprehension, which could lead to either highly beneficial outcomes (such as solving global challenges like climate change) or catastrophic consequences—the potential extinction of humanity. Risse borrowed the idea of Tegmark (2018) to discuss several possible post-intelligence explosion scenarios, ranging from AI systems governing human society in a benevolent manner to more dystopian outcomes where humans are subjugated or eradicated by AI.

Similar concerns also emerge in the medical field aiming at the physician's increasing reliance on AI systems to conduct their work. If we allow the occurrence and existence of such superintelligence with intellectual capacity and medical knowledges far beyond the greatest human doctor ever existing, we must encounter the very foundational ethical dilemma about life and death: should we allow an AI to decide how we should live? Should we rely on AI to make decisions, directly or indirectly, on human survival? If we, human beings, are even questioning our moral status in determining other human beings' life, why should we delegate such decision power to AI agents?

2.3 AI Ethical Agents

To analyse the ethical concerns brought by AI systems' advancement, Risse adopts the four-tiered model of ethical agents proposed by Moor (2006) to discuss the ethical difference of AI systems under the two seniors in terms of moral agency they possess:

> **Ethical-Impact Agents**: AI systems, such as computer machines, that impact human agents through their "actions," even if unintended.
>
> **Implicit Ethical Agents**: AI systems designed with safety or critical reliability concerns embedded in their functions, such as automated teller machines, web banking, an airplane's automatic pilot, and pharmacy software that checks for and reports on drug interactions.
>
> **Explicit Ethical Agents**: AI systems capable of making ethical decisions based on moral principles, such as automated vehicles and military and civilian robots.
>
> **Full Ethical Agents**: AI systems with consciousness, intentionality, and free will, akin to human moral agents.

The early stages of AI development typically and mainly align with the first two categories, where they can impose influence on human lives but lack autonomous moral decision-making capacity. However, as AI progresses towards more sophisticated forms of machine learning and decision-making, namely AI robots nowadays, it may enter the realm of *explicit ethical agents*, where AI robots can deliberate on decisions involving moral considerations with ethical principles set by humans. In health care, medical AI systems are mostly implicit, if not explicit, moral agents because their applications focus on either human error preventions or data processing to facilitate physician medical decisions. However, they are still not capable of and not expected to entitle decision-making power equivalent to human physicians. Namely, they are not full ethical agents.

Risse extends his exploration of AI's moral status in the scenario of the *Intelligence Explosion*, suggesting that AI may eventually become a *full ethical agent* if it attains consciousness or a level of autonomy that allows it to act with intentionality and have independent moral reasoning. This raises critical questions about how society should treat AI systems that possess such full moral agency. Schwitzgebel and Garza (2015) argue that AI systems capable of moral reasoning should deserve certain rights and protections, similar to those afforded to humans and animals. This means that we could no longer treat them as we used to do as a tool designed for specific purposes, but to respect their autonomy.

Nonetheless, the possible emergence of superintelligence raises a different critical ethical question. It is not about whether AI doctors should be treated as human physicians, but whether the emergence of AI doctors will erode the free moral status of both patients and physicians. The Heideggerian critique of technology suggests that advanced technologies could contribute to the dehumanization of society when our mind is assimilated by machine's calculative thinking due to our reliance on AI (Huttunen & Kakkori, 2022). When both patients and physicians persuade themselves that superintelligence always makes better medical decisions and they should follow AI's thinking path in determining whether its suggestions should be adopted, it is arguable if they remain maintaining their free moral status as human beings in this situation. Conversely, others take a more optimistic view. Verbeek (2011, pp. 112–115) argues that rather than undermining autonomy, AI could co-shape human moral decisions and create a symbiotic relationship where humans and machines collaborate to make ethical and informed decisions in medical contexts. These contrasting views highlight the need for careful consideration of whether we should develop medical AI towards the direction of superintelligence as it might reshape the moral and relational dynamics in healthcare.

2.4 Two New Distinct Medical Models

As AI is increasingly being integrated into clinical decision-making, it will inevitably give rise to two distinct development models: the AI–patient (AP) model and the AI–Physician–Patient (APP) model. The fundamental difference between them is the presence of human doctors. Under the AP model, the AI system is envisioned to rely solely on its algorithm to decide what information to collect from the patient, analyse them with the database it possesses and generate medical advice to the patient without the direct involvement of physicians. As AI medical systems advance, the AP model might prevail.

In contrast, the APP model utilises AI as a supportive tool for physicians, who interpret the AI's output and make the final decisions in patient care. In this model, the AI does not directly interact with the patient but analyses their medical data, symptoms, signs, and possibly even genetic information to provide diagnostic and therapeutic suggestions for the physician to make a final medical advice to the patient. This model leverages the strengths of AI in big data processing while ensuring that human-physician-oversight remains central to healthcare and address the non-medical considerations and practical challenges posed by AI's opaque nature. It can also address the AP model's ethical concerns with a human physician's involvement as an intermediary to interpret, modify or overrule the AI's recommendations with their clinical expertise and awareness of the patient's situation (Xu & Shuttleworth, 2024). More importantly, physician can decipher the diagnosis made by AI enabling informed medical decision by the patient, hence, protect patient's autonomy (Kalra et al., 2024). Human physicians take **non-medical perspectives, such as the patient's** personal values and lifestyle, **into account when** guiding **the patient to choose the most suitable treatment plan.** For example, if the treatment plan given by the AI includes the prescription of medications made with animal organs while the patient is a vegetarian, the human physician can alert the patient of this concern and provide alternatives when necessary.

Which model, the AP or APP, should be adopted in the development and application of AI medical systems? In the following sections, we will draw on the Daoist and Confucian ethical resources to reflect on the impacts of AI medical systems on human relations as well as the justifiability of these two different models. In particular, we will show that Daoist thought would tend to support rejecting the use of AI medical systems. In contrast, Confucian thought would advocate for the adoption of the APP model while rejecting the AP model.

3 A Daoist View

Daoist philosophy often critiques technology for disturbing the natural equilibrium of life and introducing rigidity to nature (Needham, 2005, p. 125; Xie, 2007). In our view, a distinct Daoist view on the use of medical AI can be disclosed by revisiting a classic Daoist story written by Daoist master, Zhuangzi, that explores the relationship between humans and machines. The story goes as follows. One day, Zigong, a disciple of Confucius, saw an old man labouring in his vegetable garden. The old man was manually drawing water from a well using jars to irrigate his crops. Zigong, noticing the inefficiency, suggested that the gardener build a mechanical irrigation system to make his work easier. The gardener, however, rejected the idea, saying:

> "I have heard from my teacher that, where there are ingenious contrivances, there are sure to be subtle doings, and that, where there are subtle doings, there is sure to be a scheming mind. But where there is a scheming mind in the breast, its **pure simplicity is impaired**. When this pure simplicity is impaired, the spirit becomes unsettled and the unsettled spirit is **not the proper residence of the Dao**. It is not that I do not know (the contrivance which you mention), but I should be ashamed to use it." (English translation adopted from Legge, 1962, p. 320; also cf. Ziporyn, 2020, p. 103; emphases added)

In this story, the gardener mentions the concept of *pure simplicity* (純白, *chunbai,*) which embodies the Daoist ideal of living in a state of natural clarity and harmony. This concept reflects the key Daoist principle of inner freedom and self-liberation, where one should cultivate a clear and uncluttered mind by eliminating *distractions* (杂念, *zániàn*) by letting go of *attachments* (执着, *zhíxhuó*) and transcending *desires and greed* (贪欲, *tānyù*). It encourages embracing a natural and uncontrived way of living and free from rigid attachments to other humans or things. In contrast with Confucianism (see next section), Daoism does not appreciate forming close human relationships to live one's life. Instead, Daoism advocates for the idea that "it is better to forget each other in the world than to support each other in adversity" (*相濡以沫，不如相忘於江湖*), which suggests that true harmony arises from living freely and independently, without forced or reluctant connections with or dependencies on outer existences. Through the gardener's adherence to *pure simplicity*, the story highlights the Daoist belief in non-attachment to machines or tools in order to live in alignment with one's *Dao* (道, the way or spiritual pursuit of oneself).

A caution should be noted here, however. While Daoism is often labelled as "anti-technology" as it emphasises the negative side of excessive human interventions in nature or dependence on other things, it should not be regarded as advocate for an outright rejection of technology. Instead, Daoism stresses the importance of using tools in a way that aligns with natural principles of balance, non-attachment, and non-interference. In any case, although the gardener in the above story refuses to use a mechanical irrigation system to water his crops, he employs jars to accomplish the task.

Accordingly, some may want to argue that medical AI, from a Daoist viewpoint, may also have the potential to work in concert with the natural order, just like the jars for irritation, when used appropriately. Medical AI, they may argue, when employed to complement a physician's practice rather than dominate it, might be seen as a tool that aids in restoring balance in the patient's body without excessive intervention. It could be argued that the critical issue in Daoism is not the use of technology itself but the risk of forming an "attachment" to it, a subtle dependence that erodes pure simplicity and inner clarity. The above gardener story illustrates this point well. Though the gardener rejects the use of a mechanical watering device, he still uses jars—simple tools that do not disrupt his natural connection to the earth and his crops. His refusal to adopt the machine is not a rejection of all technologies but a recognition that more sophisticated tools could lead to a schemed mind and control, while simple tools could not. It might be argued that the Daoist is genuinely concerned with the relationship one has with technology, not technology itself. The gardener's jars do not create attachment, but a mechanical irritation device, by promising efficiency with its complex mechanism, can by leading him to focus more on making and learning how to use tools than on the natural process of tending his crops.

The contrast between the jars and medical devices for irritation, however, highlights the risk of physician's over-reliance on medical AI in the clinical context. Although it might be argued that medical AI, as a powerful tool for analysing data, improving diagnoses, and enhancing medicine, could still be integrated into healthcare in a way that supports, rather than replaces, physicians' own skills and intuitions, it is obvious that the over-reliance risk is immense. Daoism would consider that the danger lies in physicians becoming too dependent on AI to the point where their ability to connect with patients on a human level is diminished and become relying too heavily on data-driven algorithms instead of their own judgement, resulting in a shift away from the holistic, balanced medical approach that Daoism values. A useful parallel can be drawn to the educational use of calculators. At the primary level, students are discouraged from using calculators to ensure they develop their foundational arithmetic skills. Allowing calculators too early in education might prevent students from cultivating their own mental abilities in math because what they are learning is how to us calculator rather than how to calculate. In Daoist terms, this would be considered an unnatural shortcut, interfering with the organic process of learning. However, once students have mastered basic arithmetic, calculators can be introduced in secondary education not as a crutch, but as a tool to enhance their efficiency in more complex tasks. At this stage, the use of calculators does not create attachment because students have already developed the necessary skills to function without a calculator.

Can medical AI be viewed as a valuable tool that assists physicians in handling vast datasets, recognizing patterns, and improving decision-making without posing the risk of over-reliance? When utilizing these tools, should they be compared to primary or secondary students using calculators? Nevertheless, it is crucial to acknowledge that medical AI systems offer significantly greater efficiency and power for physicians than calculators do for students. Distinguishing between levels of physicians in AI usage, as we do for primary and secondary students with calculators, may not be practical, as AI can significantly aid less-experienced physicians in providing effective patient care. Accordingly, AI medical systems can certainly foster physicians' dependence, shifting their focus from developing medical expertise to optimizing technological systems and operation of machine. Over time, this would reduce physicians' ability to think critically or engage fully with the nuances of patient care, much like how over-reliance on calculators could stunt a student's mathematical reasoning. It could subtly shift the focus of medical practice. Instead of seeing AI as a tool to support the natural flow of healing, physicians may become preoccupied with improving the algorithms that drive AI systems, viewing success as a matter of technological efficiency rather than the achievement of their knowledge and skills. This shift could lead to a more mechanical, data-driven approach to medicine, where the art of healing is sidelined in favour of algorithmic precision.

Accordingly, it seems very difficult to develop and apply medical AI systems while maintaining the Daoist idea of *pure simplicity* in the medical field. Just as the gardener avoids the use of a mechanical device to preserve his mind from becoming schemed and detached from the natural process of tending his garden, a Daoist would typically argue that the use of overly powerful tools like AI nowadays, despite it is not so advanced as a superintelligent agent that is sufficient to replace human doctor, would cause physician's excessive reliance on it. This is because an AI medical system for doctors and patients, so much more powerful than a mechanical irritation system for farmers, would distract physicians from their natural intuition and holistic understanding of medical art but to concentrate on how to utilize this tool, and would attract patients to trust the diagnostic and therapeutic recommendations of this system more than those from a human doctor. When medical professionals overly depend on AI for diagnoses or treatment recommendations, they may lose their ability to engage with the patient's unique, ever-changing condition, since their role in this relationship has been "downgraded" from a thinker of natural pattern of illness to a tool user. This could lead to a mechanical approach to medicine, where the physician's work becomes more about following algorithms than responding to the nuanced, individual needs of the patient.

Hence, from Daoist perspective, while AI can offer efficiency and precision, it risks eroding the *pure simplicity* of medical practice—the purity and clarity that come from their intention to understand the nature of health and sickness through direct human engagement and the natural flow of healing. It is not that Daoists would not support the APP model, but they would pessimistically think it is unfeasible. In this sense, just like the gardener rejecting the use of the mechanic device for irritation in the classic story, Daoists may object to the use of medical AI altogether, since it undermines the physician's connection to the Dao, which is natured with a deep, intuitive, and harmonious approach to their medical work.

4 Confucian Medical AI

In contrast to the Daoist pessimistic view on technology, Confucianism would positively pursue the APP model, advocating for a cautious and limited use of medical AI. Confucian ethical concerns fundamentally differ from those of Daoism. While Daoism emphasizes harmony with natural order and the avoidance of over attachment, Confucian concerns are rooted in social order, responsibility, and the cultivation of one's role in society. This divergence in reasoning is evident in the continuation of the story after Zigong shares his encounter with the gardener with Confucius. Confucius responds by saying:

> "He is a follower and practitioner of the chaotic doctrines. He knows the first thing, but not the sequel to it. **He only knows how to cultivate his inner self and does not know how to regulate his outer self**. If he is indeed pure and simple, one who has returned to his inborn nature by action without acting and preserves his natural state of mind amid humans of the world, you will not be surprised. How can you and I understand the chaotic doctrines?" (English translation adopted from Legge 1962, p. 322; also cf. Ziporyn, 2020, p. 105; emphasis added)

This response underscores the Confucian criticism of the Daoist emphasis on internal purity or freedom at the expense of external obligations. Confucius values self-reliance but stresses the importance of balancing it with fulfilling societal duties. He advocates for harmonizing one's inner self by nurturing appropriate inclinations and one's outer self by employing necessary means to maintain harmonious human relationships. Consequently, Confucian philosophy would likely regard the judicious use of medical AI as a tool to augment a physician's capacity to care for patients and fulfil their societal role in enhancing societal wellbeing. While Daoism might resist AI due to concerns about disrupting a physician's natural intuition and simplicity (*chunbai*), Confucianism would base its argument on how AI adoption impacts a physician's ethical responsibilities in the physician–patient relationship, focusing not only on efficiency gains but also on the cultivation of Confucian virtues.

Confucianism is generally understood as a virtue ethics. Its paramount and comprehensive virtue is *ren* (仁), commonly translated as "benevolence" or "humaneness." *Ren* embodies the moral character and responsibility to care for others and foster harmonious relationships. This concept of humaneness is intended to be practiced universally, encompassing all beings—human, animal, and even inanimate objects. However, it is imperative to note that Confucianism does not equate universal love with egalitarian love. Rather, it emphasizes the importance of expressing varying degrees of affection based on the nature of the relationship, known as differentiated love (cf. Ames & Rosemont Jr, 2010, p. 71). Notably, the cultivation and manifestation of the Confucian virtue of *ren* are especially significant within intimate human relationships, particularly the *wǔ lún* (五伦), or the five basic relationships. According to Mencius' teachings, each of these fundamental relationships embodies specific virtues to be nurtured by each party involved, all contributing to the central virtue of *ren* and the fulfilment of moral duties (Mencius 3A.4; cf. Bloom & Ivanho, 2009, p. 81).

(1) **Parent and Child** (*fù zǐ*, 父子): The parent should provide guidance and care, and the child should show respect and filial piety (*xiào*, 孝).
(2) **Ruler and Subject** (*jūn chén*, 君臣): The ruler should act with righteousness, while the subject should be loyal.
(3) **Husband and Wife** (*fū fù*, 夫妇): The husband should be loving and responsible, and the wife should be respectful and supportive.
(4) **Siblings** (*xiōng dì*, 兄弟): The older sibling should be kind and protective, and the younger sibling should be respectful and genital.
(5) **Friends** (péng yǒu, 朋友): Friendship is based on mutual trust and loyalty, making this relationship more egalitarian compared to the others.

These relationships and associated virtues encapsulate Confucian values of love, duty, and reciprocal care. Traditionally, Confucianism dictates that the physician–patient relationship should mirror the dynamics of a parent–child relationship, where the physician offers guidance and care while the patient complies with the physician's directives. However, this approach is viewed as medically paternalistic and is deemed unsuitable for modern medical practices. Instead, the relationship between friends is considered a more appropriate model for characterizing the physician–patient relationship in the pursuit of the virtue of *ren*. In the Confucian tradition, medicine is regarded as the art of *ren* (Fan, 2009), establishing a connection that emphasizes Confucian benevolence and ethical obligations. In this context, a bond rooted in Confucian *ren* and ethical responsibility is fostered between the physician and the patient. Rather than advocating for blind obedience, patients are encouraged to actively participate in discussions regarding their medical care alongside the physician and their family members, facilitating shared decision-making (Fan & Chen, 2010). This approach embodies Confucian ideals of benevolence and respect, framing the physician–patient interaction as one characterized by mutual trust, care, and ethical responsibility.

Additionally, *ren* serves as a driving force for the physician's pursuit of advanced professional skills in order to provide better care to their patients. This sense of moral duty and compassion motivates physicians to continually refine their expertise, ensuring that they fulfil their obligation to care for patients with the highest standards of knowledge and compassion. In this way, *ren* not only governs the physician–patient relationship but also fosters a physician's commitment to lifelong learning and self-improvement for the benefit of others, including employing necessary technologies to benefit their patients. Accordingly, Confucianism would support the slow and harmonious integration of AI into medical care. This is because in the Confucian view, there is no reason to exclude the use of necessary technologies when they can help the physician improve the efficacy of care for their patients, just as Confucius has reason to support the gardener to use the mechanical device to irate his crops.

However, in the context of medical AI, Confucianism would favour the APP model and resist the AP model. Confucianism would limit AI's role to serving as a tool that supports the physician, rather than replacing them. While AI can enhance efficiency and accuracy, it lacks the moral capacity and emotional intelligence required for holistic care for the patient (Muyskens et al., 2024, p. 81). Physicians, who embodies the Confucian ideals of *ren* and ethical duty, must remain the primary decision-makers and use AI to support their judgments rather than delegating critical responsibilities to machines. It is equally important that the development of medical AI fosters patient trust in the physician, rather than in the AI itself. To ensure "responsible" development of medical AI, ethical guidelines, regulations, and oversight mechanisms would be necessary to ensure that AI is used appropriately to preserve the physician's moral role while leveraging the benefits of AI. Although it is challenging to achieve this Confucian goal, the following ethical framework based on the Confucian principle of *ren* might be useful to guide the application of AI in medicine.

4.1 Maintenance of Human Physician–Patient Relationship

The first important Confucian ethical principle in the application of medical AI is the maintenance of the human physician–patient relationship. In Confucian thought, human interaction is central to the cultivation of *ren* and lead authentic ethical lives for human individuals. Humans, as social beings, cannot isolate themselves from society; instead, they learn how to behave and cultivate virtue through meaningful relationships with others. This idea is emphasised in Confucius' famous quote in response to Daoism: "One cannot associate with birds and beasts. Am I not a member of this human race? Who, then, is there for me to associate with?" (Analects 18.6; cf. Lau, 1996, p. 218). While Daoism emphasizes independence and non-attachment, Confucius underscores that human interaction is not only a natural part of being human but also a necessary element for cultivating the fundamental virtue of *ren* for the normal way of life. Without meaningful exchanges between humans, *ren* cannot be practiced, as it requires active care and consideration for the wellbeing of others in close relationships.

In the medical context, the physician–patient relationship is a prime example of this ethical principle in action. Physicians have an inescapable moral duty to care for their patients with compassion, empathy, and understanding. Their competency as doctors is not solely determined by their medical knowledge or technical skills but also by their ability to connect with patients, understand their needs, and build trust through compassionate care. This relational aspect of medicine is crucial for fulfilling the Confucian ideal of *ren* which should never be replaced by an AI machine.

Accordingly, medical AI, while a powerful tool for improving healthcare, must be implemented in ways that enhance rather than diminish these critical human interactions. For example, AI systems can assist physicians by analysing data, streamlining administrative tasks, or providing diagnostic recommendations, allowing doctors to focus more on empathetic engagement with their patients. However, if AI is used to take the final responsibility and replace human doctors entirely, it risks severing the relational and moral fabric of healthcare. Physicians would become mere technicians or researchers, detached from their patients and unable to practice *ren*. To align with Confucian ethics, employing medical AI must be regulated to prioritise the preservation and strengthening of the physician–patient relationship, ensuring that doctors remain compassionate caregivers and active participants in the moral practice of medicine.

4.2 Provision of Compassionate Care

For Confucianism, patient-centred care lies at the heart of ethical medical practice, requiring both intelligence and emotional engagement. Confucian ethics emphasises that true care extends beyond technical expertise; it demands an understanding of the human condition and a compassionate response to the needs of others. While AI excels in analysing data and delivering diagnostic accuracy, it lacks the emotional intelligence required to fulfil this dimension of care. In Confucian view, healing is not merely a medical process aimed at addressing objective issues, such as diagnosing and treating diseases; it also involves providing mental and emotional support to patients. This holistic approach to care is essential for fostering trust, alleviating anxiety, and addressing the psychological needs of the patient—the qualities that AI systems are fundamentally unable to provide. Therefore, the role of the physician, must remain centred on compassionate care. Physicians are uniquely positioned to offer empathy, comfort, and understanding to patients, especially during moments of vulnerability and distress. AI, by its very nature, cannot replicate this. It lacks the intellectual and emotional capacity to truly understand a patient's experience of pain, fear, or uncertainty. Compassion, as Confucian ethics suggests, requires authentic emotional engagement, which is inherently absent in AI.

Moreover, medical decisions should always be based on humanity rather than pure calculation. AI's data-driven approach, while valuable for improving efficiency and accuracy, risks reducing care to a mechanical process that overlooks the patient's emotional and social context. Confucianism teaches that the essence of care lies in the relational and moral bond between the physician and patient, which cannot be replaced by a machine, because the machine is unconscious and emotionless. For Confucians, while AI can support physicians in delivering better care, it should remain a tool rather than a substitute for the compassionate role of human doctors. Patient-centred care must prioritise empathy, understanding, and emotional engagement, which can and should be provide only by human doctors.

4.3 The Physician as the Gatekeeper

Although AI systems have shown the capacity to surpass human doctors in specific areas like calculations and enable maximising efficiency by streamlining the diagnostic and treatment planning process, it suffers from the black-box problem: AI's recommendations cannot be understood, but can only be viewed in terms its inputs and outputs. This raises the concerns of unexplainability (Xu & Shuttleworth, 2024). This problem of unexplainability does not simply refer to medical AI's incapacity to provide the detailed reasonings behind their outputs (aka conclusions or decisions) but also the knowledge threshold to extract, understand, and, most importantly, validate the explanations they provide (Amann et al., 2022). For example, if the patient asks for the possible causes of the disease or why the syndromes he or she has not resulted from another sickness, the reasons given by a medical AI is likely to be either too general or go far beyond the patient's knowledge level to comprehend. Unless the patient has experience communicating with that specific AI to *ask the right questions* and are equipped with sufficient medical expertise to critically analyse those explanations, the reasoning of the medical AI can hardly be useful and not create serious misunderstandings. While some argue that human intelligence is akin to a black box (Pande, 2018), this opacity is notably less pronounced compared to AI. Furthermore, AI lacks the capability to consider non-medical contextual factors such as a patient's family background and personal experiences, making personalized assessments akin to those provided by human doctors unattainable.

In addition, AI medical systems remain vulnerable to inexplicable low-level errors that human doctors are less likely to commit. In Confucianism, the principles of responsibility and duty of physician emphasise the necessity of human oversight to prevent harm and unethical medical practice. Physicians must act as gatekeepers to oversee AI diagnoses and treatment recommendations to safeguard patient welfares. While neither AI nor human doctors are infallible, AI systems are particularly prone to errors stemming from algorithmic flaws, data collection issues, or maintenance failures. These errors, though often technical in nature, can lead to significant consequences if left unchecked. A typical example of algorithmic flaw is the so-called "insensitivity to impacts" in error minimisation. As Evans and Snead (2024) explain, most AI system are trained to prevent calculative errors by assuming that the consequences of different errors are weighted equally. This design logic is problematic in the medical field as the consequence of diagnostic errors can be vastly different in terms of medical negligence. While some diagnostic inaccuracies are negligible, others some errors could lead to lethal consequences, but AI cannot identify and address the latter proportionately due to their algorithmic limitation. In this case, physician is irreplaceable in weighing the impacts of possible misdiagnosis with their expertise and judgement. Therefore, human doctors should be morally obliged to review AI outputs and making necessary adjustments to prevent harm and uphold the integrity of patient care.

The role of human doctors as gatekeepers is especially important when addressing the issue of liability in cases of medical errors. Unlike humans, AI systems cannot be held accountable for their actions. Without a physician to make the final decision and bear legal responsibility, it becomes difficult to determine who should be liable when AI errors cause harm. For instance, an AI system might misdiagnose a condition or recommend inappropriate treatment due to flawed algorithms or incomplete data. In such cases, attributing blame becomes a complex and time-consuming process, requiring technical expertise to trace the source of the error. This poses significant challenges for victims of medical negligence, who may find it difficult to seek justice due to the technical and legal complexities involved. By contrast, when a human doctor is involved in the decision-making process, they can be held accountable for the final judgement, making it easier to attribute liability and seek remedies for errors.

Beyond the liability distribution concerns, human doctors are inherently incentivised to act cautiously and ethically due to the personal and social consequences of their decisions. If a physician acts negligently and causes harm to a patient, they face not only legal repercussions but also damage to their professional reputation and social standing. These consequences often extend to their personal lives, as their family livelihood and community relationships are at stake. This sense of moral and social accountability compels human doctors to approach medical decisions with care and diligence. In contrast, AI systems lack the capacity to experience such pressures or obligations. As a result, physician's ability to combine clinical expertise with moral accountability ensures that errors—whether human or machine-generated—are addressed responsibly and ethically when medical AI plays an assisting role in the medical process.

4.4 *Avoidance of Physician De-Skilling*

Confucianism places a strong emphasis on preparedness and the pursuit of continuous learning as essential components of personal and professional growth. Therefore, it recognises the value of tools in aiding individuals to achieve their goals, and their use is not only acceptable but often encouraged. However, Confucian principles also caution against allowing tools to become a disincentive for learning. The adoption of AI in medicine, while offering significant advantages in efficiency and accuracy, must not lead to complacency or a decline in the skills and knowledge that define a competent physician. This means that doctors must remain committed to advancing their medical expertise and maintaining the foundational abilities necessary to provide sound medical advice, even without the assistance of AI.

Also, the integration of AI into healthcare does not mean that physicians no longer need to learn how to interact with patients or refine their medical skills. On the contrary, Confucianism would argue that doctors have an additional responsibility to use the time saved through AI adoption to develop new capabilities and enhance their practice. For instance, physicians must learn how to effectively use AI systems, understanding their strengths and limitations as well as the nuances of how these tools operate. This includes acquiring a deep knowledge of how AI algorithms process data, generate diagnoses, and recommend treatments.

In addition to maintaining their core expertise, physicians must also develop new skills to identify and address AI malfunctions. Physicians need to be equipped to recognize their errors, such as algorithmic miscalculations or data inaccuracies, and provide appropriate adjustments to ensure patient safety. This requires a combination of technical knowledge and clinical intuition, which can only be cultivated through continuous learning and practice. Confucianism underscores the importance of adapting to change while preserving the foundational principles of one's craft. For physicians, this means embracing AI as a tool for growth without compromising their role as compassionate, skilled caregivers.

5 Concluding Remarks

As AI continues to evolve, humanity faces a pivotal moment in determining how this technology will shape our future. From its integration into medical practice to the speculative creation of an "artificial god," the ethical and philosophical implications of AI demand careful consideration. While AI holds immense potential to enhance human capabilities and improve lives, it also poses significant risks, particularly when it comes to questions of autonomy, moral responsibility, and societal values.

The idea of an "artificial god" encapsulates these concerns, presenting a vision of superintelligence capable of surpassing human limitations in knowledge, reasoning, and morality. However, such a vision is fraught with uncertainty. As philosophers like Risse suggest, humanity is ill-prepared to grapple with the moral frameworks or intentions of a superintelligence born from alien foundations. The subjective nature of concepts like "meaningful lives" or universal moral standards raises doubts about whether such an entity could truly act in humanity's best interests—or if it would impose its own paternalistic vision of what is best for us. Philosophical traditions like Daoism and Confucianism offer valuable insights into these challenges. Daoism warns against humanity's growing attachment to external constructs, cautioning that the pursuit of an omnipotent AI could erode our inner freedom and disrupt the natural flow of life. Confucianism emphasises the importance of relational ethics and moral responsibility, both of which could be undermined by over-reliance on AI. Together, these philosophies remind us that true progress lies not in surrendering our freedom and responsibility to machines but in using technology as a tool to complement, rather than replace, human judgement and agency.

Ultimately, the pursuit of superintelligence forces us to reflect on what it means to live meaningful and authentic lives. Rather than seeking an "artificial god" to resolve challenges or define our purpose, we must recognise that such answers can only come from within ourselves. While the allure of AI's capabilities is undeniable, preserving human freedom, moral responsibility, and the liberty to shape our own destiny remains essential. In the end, the challenge is not merely to create powerful technologies but to ensure that their development aligns with the values that make us human. In this context, this chapter expounds upon the rationale and arguments supporting how Confucianism would advocate for an AI–physician–patient model over an AI–patient model. This preference is rooted in the aim of preserving the crucial physician–patient relationship while also enhancing the provision of healthcare services to meet our evolving health needs in the years ahead.

References

Allison, G. T. (2017). Destined for war: Can America and China escape Thucydides's trap? Scribe Publications.

Amann, J., Vetter, D., Blomberg, S. N., Christensen, H. C., Coffee, M., Gerke, S., Gilbert, T. K., Hagendorff, T., Holm, S., Livne, M., Spezzatti, A., Strümke, I., Zicari, R. V., Madai, V. I., & on behalf of the Z-Inspection initiative. (2022). To explain or not to explain?—Artificial intelligence explainability in clinical decision support systems. *PLOS Digital Health, 1*(2), Article e0000016. https://doi.org/10.1371/journal.pdig.0000016

Ames, R. T., & Rosemont Jr, H. (2010). *The analects of confucius: A philosophical translation.* Random House Publishing Group.

Bloom, I., & Ivanhoe, P. J. (2009). *Mencius* (trans.). Columbia University Press.

Bostrom, N. (2014). *Superintelligence: Paths, dangers, strategies (1st edn.).* Oxford University Press.

Burns, R. D., & Coyle, H. P. E. I. (2015). *The challenges of nuclear non-proliferation.* Rowman & Littlefield Publishers.

Castiglioni, I., Rundo, L., Codari, M., Di Leo, G., Salvatore, C., Interlenghi, M., Gallivanone, F., Cozzi, A., D'Amico, N. C., & Sardanelli, F. (2021). AI applications to medical images: From machine learning to deep learning. *Physica Medica, 83*, 9–24. https://doi.org/10.1016/j.ejmp.2021.02.006

Chen, P.-T., Lin, C.-L., & Wu, W.-N. (2020). Big data management in healthcare: Adoption challenges and implications. *International Journal of Information Management, 53*, Article 102078. https://doi.org/10.1016/j.ijinfomgt.2020.102078

DiMaio, S., Hanuschik, M., & Kreaden, U. (2011). The da Vinci surgical system. In J. Rosen, B. Hannaford, & R. M. Satava (Eds.), *Surgical robotics: systems applications and visions* (pp. 199–217). Springer US. https://doi.org/10.1007/978-1-4419-1126-1_9

Esteva, A., Kuprel, B., Novoa, R. A., Ko, J., Swetter, S. M., Blau, H. M., & Thrun, S. (2017). Dermatologist-level classification of skin cancer with deep neural networks. *Nature, 542*(7639), 115–118. https://doi.org/10.1038/nature21056

Evans, H., & Snead, D. (2024). Understanding the errors made by artificial intelligence algorithms in histopathology in terms of patient impact. *NPJ Digital Medicine, 7*(89). https://doi.org/10.1038/s41746-024-01093-w.

Fan, R. (2009). The discourses of confucian medical ethics. In R. Baker & L. McCullough (Eds.), *The cambridge world history of medical ethics* (pp. 195–201). Cambridge University Press.

Fan, R., & Chen, X. (2010). The family and harmonious medical decision making: Cherishing appropriate confucian moral balance. *Journal of Medicine and Philosophy, 35*(5), 573–586.

Gillespie, N., Lockey, S., Curtis, C., Pool, J., & Ali Akbari. (2023). *Trust in artificial intelligence: A global study* (pp. 13–14). The University of Queensland: KPMG Australia. https://doi.org/10.14264/00d3c94

Grand View Research. (2024, April). AI In healthcare market size, share & trends analysis report by component (Hardware, services), by application, by end-use, by technology, by region, and segment forecasts, 2024—2030. AI In Healthcare Market Size, Share & Growth Report, 2030. https://www.grandviewresearch.com/industry-analysis/artificial-intelligence-ai-healthcare-market

Huttunen, R., & Kakkori, L. (2022). Heidegger's critique of the technology and the educational ecological imperative. *Educational Philosophy and Theory, 54*(5), 630–642. https://doi.org/10.1080/00131857.2021.1903436

Jiang, F., Jiang, Y., Zhi, H., Dong, Y., Li, H., Ma, S., Wang, Y., Dong, Q., Shen, H., & Wang, Y. (2017). Artificial intelligence in healthcare: Past, present and future. *Stroke and Vascular Neurology, 2*(4), 230–243. https://doi.org/10.1136/svn-2017-000101

Johnson, K. B., Wei, W., Weeraratne, D., Frisse, M. E., Misulis, K., Rhee, K., Zhao, J., & Snowdon, J. L. (2021). Precision medicine, AI, and the future of personalized health care. *Clinical and Translational Science, 14*(1), 86–93. https://doi.org/10.1111/cts.12884

Kalra, N., Verma, P., & Verma, S. (2024). Advancements in AI based healthcare techniques with FOCUS ON diagnostic techniques. *Computers in Biology and Medicine, 179*, Article 108917. https://doi.org/10.1016/j.compbiomed.2024.108917

Lau, D. C. (Ed.). (1996). *Confucius: The Analects*. Columbia University Press.

Legge, J. (1962). *The sacred books of China: The texts of taoism* (Trans.). Over Publications.

Marques, L., Costa, B., Pereira, M., Silva, A., Santos, J., Saldanha, L., Silva, I., Magalhães, P., Schmidt, S., & Vale, N. (2024). Advancing precision medicine: A review of innovative in Silico approaches for drug development. *Clinical Pharmacology and Personalized Healthcare. Pharmaceutics, 16*(3), 332. https://doi.org/10.3390/pharmaceutics16030332

Moor, J. H. (2006). The nature, importance, and difficulty of machine ethics. *IEEE Intelligent Systems, 21*(4), 18–21. https://doi.org/10.1109/MIS.2006.80

Moor, J. H. (2009, March). Four kinds of ethical robots. Philosophy Now, 72, 12–14. https://philosophynow.org/issues/72/Four_Kinds_of_Ethical_Robots

Muyskens, K., Ma, Y., & Dunn, M. (2024). Can an AI-carebot be filial? *Reflections from Confucian Ethics. Nursing Ethics, 31*(6), 999–1009. https://doi.org/10.1177/09697330241238332

Myskja, B. K., & Steinsbekk, K. S. (2020). Personalized medicine, digital technology and trust: A Kantian account. *Medicine, Health Care and Philosophy, 23*(4), 577–587. https://doi.org/10.1007/s11019-020-09974-z

Needham, J. (2005). *Science and civilisation in China*. Vol. 2: History of scientific thought/by Joseph Needham with the research assistance of Wang Ling (Reprint, Vol. 2). Cambridge University Press.

Oluffa Pedersen, E. (2012). A Kantian conception of trust. *SATS, 13*(2). https://doi.org/10.1515/sats-2012-0009

Pande, V. (2018, January). Artificial intelligence's black box Is nothing to fear. *The New York Times, 25*. https://perma.cc/QB8M-Q9LQ

Powles, J., & Hodson, H. (2017). Google DeepMind and healthcare in an age of algorithms. *Health and Technology, 7*(4), 351–367. https://doi.org/10.1007/s12553-017-0179-1

Quazi, S. (2022). Artificial intelligence and machine learning in precision and genomic medicine. *Medical Oncology, 39*(8), 120. https://doi.org/10.1007/s12032-022-01711-1

Risse, M. (2023). *Political theory of the digital age: Where artificial intelligence might take us*. Cambridge University Press.

Schwitzgebel, E., & Garza, M. (2015). A defense of the rights of artificial intelligences: Defense of the rights of artificial intelligences. *Midwest Studies in Philosophy, 39*(1), 98–119. https://doi.org/10.1111/misp.12032

Medizin, S. (2022). Clinician of the future. *Qualitas, 21*(3), 8–9. https://doi.org/10.1007/s43831-022-0086-7

Stergiou, K. D., Minopoulos, G. M., Memos, V. A., Stergiou, C. L., Koidou, M. P., & Psannis, K. E. (2022). A machine learning-based model for epidemic forecasting and faster drug discovery. *Applied Sciences, 12*(21), 10766. https://doi.org/10.3390/app122110766

Tegmark, M. (2018). *Life 3.0: Being human in the age of artificial intelligence*. Penguin Books.

Topol, E. J. (2019). High-performance medicine: The convergence of human and artificial intelligence. *Nature Medicine, 25*(1), 44–56. https://doi.org/10.1038/s41591-018-0300-7

Verbeek, P.-P. (2011). *Moralizing technology: Understanding and designing the morality of things (Online-Ausg)*. University of Chicago Press.

Xie, Q. (2007).). 道家对科技异化的思考及救治 (Taoist Thought and Countermeasure on Science and Technology Aliention). Studies in Dialectics of Nature, 23(6), 99–104. https://core.ac.uk/download/pdf/41359349.pdf

Xu, H., & Shuttleworth, K. M. J. (2024). Medical artificial intelligence and the black box problem: A view based on the ethical principle of "do no harm." *Intelligent Medicine, 4*(1), 52–57. https://doi.org/10.1016/j.imed.2023.08.001

Zhuangzi, & Ziporyn, B. (2020). *Zhuangzi: The complete writings*. Hackett Publishing Company, Inc.

Zou, F., Tang, Y., Liu, C., Ma, J., & Hu, C. (2020). Concordance study between IBM Watson for oncology and real clinical practice for cervical cancer patients in China: A retrospective analysis. *Frontiers in Genetics, 11*, 200. https://doi.org/10.3389/fgene.2020.00200

Ruiping Fan (范瑞平) is Chair Professor of Philosophy at the City University of Hong Kong. His research focuses on bioethics and comparative philosophy, drawing from both Confucian and Western cultural perspectives. With a publication record exceeding 200 journal articles and book chapters (comprising over 100 in English and 100 in Chinese), he has authored books including *Reconstructionist Confucianism: Rethinking Morality after the West* (2010), *Contemporary Confucian Bioethics* (2011), and *Contemporary Medicine and Confucian Thought* (2024). Additionally, he has served as the editor or co-editor of 15 volumes (8 in English and 7 in Chinese), including recent publications such as *Sex Robots: Social Impact and the Future of Human Relations* (2021), *Incentives and Disincentives in Organ Donation: A Multicultural Study among Beijing, Chicago, Tehran, and Hong Kong* (2023), and *An East–West Dialogue on Good Governance: Learning from Each Other* (2024).

Use of AI and the Future of Human Relations

A Cross-Cultural Perspective on Human–Care Robot Interactions

Jennifer Bourne

1 Introduction

In recent years, the advent of care robots has marked a significant milestone in the field of assistive technologies. These robots, designed to offer companionship and assistance to users with need have sparked a global interest. However, the integration of care robots into the fabric of daily life presents unique challenges and opportunities that vary significantly across cultural contexts. This paper seeks to explore the effect of care robots on companionship from a cross-cultural perspective, highlighting the importance of understanding diverse cultural dynamics in the design and governance of these technologies.

2 Background on the Rise of Care Robots

Care robots encompass both socially and physically assistive robots. They have emerged as pivotal solutions to address the diverse needs of humans requiring assistance. Socially assistive robots primarily provide support through social interactions—they engage in conversations, offer cognitive training, send medication and appointment reminders, motivate users for a healthier lifestyle, and provide speech therapy, among other services. Since they do not physically interact with humans, there is less risk of potentially causing physical harm to humans. On the other hand, physically assistive robots are designed to provide hands-on help with daily activities such as mobility, eating, dressing, and managing household chores (Sgorbissa et al., 2018). These robots can be deployed to professional healthcare facilities or private

J. Bourne (✉)
Berggruen Institute, Los Angeles, CA, USA
e-mail: jennys.umd@gmail.com
URL: https://berggruen.org/

Y. Chandra and R. Fan (eds.), *Artificial Intelligence and the Future of Human Relations*,
https://doi.org/10.1007/978-981-96-7185-4_12

homes, and they interact with humans directly on a daily basis. The physical nature of their assistance requires a higher standard of safety to ensure they can perform tasks without putting humans at risk.

Despite these distinctions, both socially and physically assistive robots share a common goal: to improve the quality of life and well-being of their users. Whether through social engagement or physical support, care robots are designed to alleviate feelings of loneliness and enhance the quality of life and well-being by fostering a sense of companionship (Alessa & Al-Khalifa, 2023).

The development of care robots could be dated back several decades, with significant advancements in robotics, artificial intelligence, and sensor technology accelerating their development in the early 2000s. More recently, the new breakthroughs in AI, especially with the release of ChatGPT in 2022, have further accelerated the progress in this field. Large language models (LLMs) like ChatGPT have greatly improved robots' ability to understand natural language, interact with users, and provide more personalized assistance.

These technological advancements are making it increasingly likely that care robots will play a significant role in our everyday lives, especially in elderly care, where emotional support and cognitive engagement are as important as physical assistance. With the global population aging rapidly, especially in countries like Japan, Germany, and China, healthcare systems and caregivers are facing mounting pressure. Coupled with the rising number of individuals living with chronic conditions and long-term disabilities, there has been a growing demand for technologies to assist with daily care. Care robots are viewed as a possible solution, as they could potentially take over routine tasks, easing the burden on healthcare professionals and caregivers (Ghafurian et al., 2008). In a meta-analysis of robots developed for elderly care, the findings are quite promising. Robots are shown to significantly boost positive emotions such as hope, love, security, and calm among their users. Additionally, they play a crucial role in reducing feelings of stress, loneliness, and anxiety. Furthermore, they are also proven to be effective in behavior modeling, aiding in rehabilitation therapies or medication adherence (Lobel, 2022).

To illustrate the latest technological development in care robots, this paper will explore some recently developed care robots in the Japanese, American, and Chinese markets, respectively.

Japan is a global leader in the field of care robotics, driven by its notably aging population. Robots like Paro and Bono, exemplify Japan's approach to embody intelligence technology and integrate it into elderly care. Japanese scientists believe that AI innovations are best developed within an artificial but physical body (Lobel, 2023). Among these innovations, Paro stands out as a therapeutic robot. Introduced in 2003, it is modeled after a cuddly baby harp seal and is designed to elicit warm emotional responses and calm down patients in hospitals and nursing homes. Equipped with built-in sensors, Paro can recognize certain words and respond to sounds and touches. It adapts to user preferences, learning from interactions to repeat actions that result in affectionate responses. Additionally, Paro has the capability to learn and remember names and faces, including those of its owner and itself ("Paro Research Papers," n.d.).

Following the success of Paro, Japan continues to innovate in the field of care robotics. Bono is another significant contribution tailored to enhance social interactions. It is designed specifically to facilitate small group conversations and can efficiently mediate dialogues among multiple participants in real time. It ensures equitable participation by balancing "airtime" among discussants, encouraging quieter individuals to contribute more actively. Bono's voice is designed to be gender-neutral, upbeat, and calm to soften the perceived strictness of its facilitative role. Bono takes its physical appearance from the traditional Japanese wooden Kokeshi dolls that are known for their simplistic elegance. Its body is capable of basic movements such as turning its head to face speakers and making simple gestures with its arms. It has been developed through iterative collaborations that include engineering students, elderly adults, and a local robotics firm, demonstrating Japan's commitment to evolving care technology to meet diverse social needs (Seaborn et al., 2023).

The United States has been another major player in developing care robots tailored to a variety of care needs, including providing companionship and basic household assistance. Among them, Mobile Aloha and Bandit stand out, each serving distinctive functions.

Developed through a collaboration with Stanford University, Mobile Aloha was launched in April 2024 and represents a significant leap in robotics designed for household tasks. It has a mobile base, two mechanical arms, and whole-body teleoperation interface. This robot is capable of a broad range of functionalities, from cooking and cleaning to more complex tasks such as opening cabinets and loading a dishwasher. It learns directly from human demonstrations, allowing it to perform intricate tasks with a high degree of precision. Equipped with bimanual and whole-body teleoperation, Mobile Aloha can replicate human-like tasks efficiently, achieving up to a 90% success rate in specific activities with minimal instruction (Fu et al., 2024; Mckay, 2024; Malewar, 2024; Yirka, 2024).

Bandit, developed by the Robotics and Autonomous Systems Center at the University of Southern California (USC), is designed specifically for healthcare and therapeutic support. Its appearance embodies the human face and arms. It is equipped with servomotors and rapid-prototyped parts, making it highly suitable for expressive human–robot interactions. Bandit has been actively used in various therapeutic scenarios, including physical rehabilitation and social skills coaching for children with autism. It has also been applied in music therapy for Alzheimer's patients and chair aerobics for the elderly. Bandit's versatility in therapeutic contexts underscores its role in enhancing healthcare and rehabilitation, proving essential in improving patients' quality of life and supporting healthcare professionals (Lytal, 2016; ; Fu et al., 2024).

While both Mobile Aloha and Bandit are cutting-edge care robots, they serve distinct purposes. Mobile Aloha excels in assisting with a wide array of daily household tasks, positioning it as an invaluable aid for everyday living. Bandit's specialized therapeutic functionalities highlight the potential of robots to significantly improve healthcare and rehabilitation practices.

China is catching up in robotic development. A recent example is the launch of a humanoid robot named Astribot in September 2024. This robot is designed for domestic tasks, handling everything from vacuuming, making waffles, feeding pets, serving tea, giving martial arts training instructions, to playing musical instruments. It embodies a human body with a head, a body, and two arms. But it only has one flexible leg and no feet. Instead, its bottom part is equipped with a wheeled base that can move at a speed of 22.4 feet per second. Its strong arms can lift up to 22 pounds. The robot operates entirely on battery power, with the ability to recharge if connected to a power source. While the startup company Astribot claims that the robot is powered by AI and is capable of learning and adapting to tasks, they have yet to reveal specifics about its hardware or the AI's training models (Knutsson, 2024).

It is worth noticing that the design includes culturally specific elements, like tea serving and martial arts instruction, making it especially appealing to Chinese users. Astribot reflects the rapid progress China is making in robotic development, catching up with global competitors in this field. The company's focus on combining practical and cultural elements, like serving tea and training martial arts, illustrates its commitment to developing technology that resonates with its target market.

Together, those five exemplary care robots from Japan, America, and China demonstrate the diverse potential of robotic technology in enhancing human health and quality of life worldwide.

3 Human Relationships and Companionships

Human relationships are complex and can be understood from various perspectives. One common way to understand them is through examining emotions, personal choices, enjoyment, intimacy, and the norms of public behaviors, etc. (Baxter, 2010; Baxter & Braithwaite, 2008; Carl & Duck, 2004; Vangelisti & Perlman, 2006). This approach attempts to reduce biases by emphasizing more neutral factors like shared feelings and societal expectations (Duck, 2011). However, it was argued that this perspective doesn't fully capture the reality of how relationships truly work. Biological, social, and cultural dynamics also deeply influence how people connect with others (Duck, 2011). As an example, people tend to spend time and make friends with those who share similar views and values, because they feel more comfortable seeking out connections that offer validation and shared understanding. People tend to avoid interacting or establishing relationships with those who constantly disagree with them (Duck, 2011). This need for validation and shared understanding is a key factor in forming and maintaining relationships between humans. With the advent of the internet and social media, humans are equipped with more ways to build and maintain relationships across geographical and cultural boundaries. People can now chat, make friends, or even date online to seek validation and shared understanding virtually without being physically present.

With the new advent of robots, humans are presented with yet a new possibility: forming relationships with robots. But at the same time, it adds another layer of complexity to how relationships should be understood and maintained in today's fast-developing world. What kind of feelings should humans have toward robots?

How intimate should those connections be? What social norms should people adopt when interacting with robots? Additionally, how might human–robot relationships be influenced by biological, social, and cultural factors? To understand and anticipate these dynamics, we first need to explore some key factors impacting human relationships, as they could shed light on understanding human–robot relationships.

One crucial factor in human relationships is the importance of trust. Over the years, scholars have attempted to define trust. Rotter (1967, p. 651) described it as the belief that "the word, promise, or written communication of another can be relied upon." Barber (1983, p. 14) expanded this definition by emphasizing trust as the belief that people "have a moral responsibility to put others' interests ahead of their own." Mayer (1995, p. 712) introduced a new layer to the concept of trust by defining it as "the willingness of a party to be vulnerable to the outcomes of another party based on the expectation that the other will perform a particular action important to the trustor, irrespective of the ability to monitor or control that other party." While these definitions differ in phrasing and scope, they share a common idea: trust involves expectations from the other agents and vulnerability to the other agents.

Trust is also central to Social Penetration Theory, which describes how relationships develop through reciprocal self-disclosure (Altman & Taylor, 1973). According to the theory, individuals have multiple layers of experiences, liken to an onion. The outer layer contains publicly observable traits, while the deeper layers hold more private information. As trust builds, individuals gradually reveal deeper layers of themselves through mutual sharing. The process unfolds in three stages, beginning with the orientation stage where small talk dominates, to the more intimate stage where people discuss personal values, goals, and fears. The final stage is when people establish mutual understanding and feel comfort sharing the most private matters (Altman & Taylor, 1973).

When humans try to form relationships with robots, they would expect the robots to be trustworthy. While trust in robots may seemingly resemble trust in human relationships, it lacks mutuality, resulting the trust in human–robot relationships being a one-sided relationship. Unlike humans, robots do not reciprocate trust or operate from personal motivations. As Fox and Gambino (2021) noted, robots are inherently designed to serve humans, and their purpose is to maximize human benefit. This creates a permanent imbalance in the relationship, as robots do not possess the capacity to feel dissatisfaction, self-interest, or seek equity with their human users. Robots do not make independent judgments, compare themselves to others, or act in ways that reflect their own desires. Humans can make decisions about robots without needing to consider the robots' well-being or preferences (Fox & Gambino, 2021).

Additionally, robots are designed to collect, analyze, and respond to human data in order to best assist humans. As a result, humans reveal their vulnerabilities to robots by sharing personal data, sometimes without fully understanding the extent of what is disclosed. This dynamic can complicate the formation of trust in human–robot relationships. The gradual stages of trust building, as described in the Social Penetration Theory, do not apply in the same way. Information that humans might wish to reserve for more intimate stages of relationship development could be accessed by robots during the initial orientation stage, potentially before a trust could even be established. The revealing of vulnerability and most private personal matters at the outset can hinder the development of trust, making human–robot relationships inherently different from human-to-human connections. The gradual openness that typically characterizes the formation of trust between humans may be absent from human–robot relationships (Petronio, 2002).

Another essential aspect of human relationships is disagreement. In any human relationship, disagreement is bound to happen, even between close family members or trusted friends. When people disagree, their individual differences—like how they communicate, how comfortable they are with conflict, and how personally they take things—could lead to each person approaching disagreement or arguments differently, even in similar situations (Hample, 2005).

This variation in disagreement styles stems largely from the different frames or perspectives individuals bring into the conversation. These frames influence what a person expects from the interaction, how pleasant or unpleasant they anticipate the disagreement will be, and how well they can align with the other person's goals. They also shape how aggressive or calm someone may become during a disagreement (Lin & Zhao, 2010). For example, some people might see a disagreement as a threat to their self-image or beliefs, causing them to become defensive. Whereas other people may view conflict as an opportunity to gain clarity or find common ground, navigating it calmingly without becoming defensive (Chadwick, 2011).

Understanding emotions is also key to understanding why different individuals react differently during disagreements or conflicts (Dillard et al., 1996). Some people tend to take conflict more personally than others, and thus are more prone to experience emotional responses like stress or anxiety during a disagreement. Whereas some other people tend not to take conflict personally, so they can remain more detached or composed. The intensity of these emotions often dictates how a person behaves in the heat of an argument (Hample, 2005).

The closeness of a relationship adds another layer of complexity to this dynamic. When people are emotionally invested in one another, they often react more strongly to disagreements. Arguments between close individuals, such as family or partners, can feel more intense because the stakes are higher. These individuals may fear that the conflict could damage the relationship, which adds emotional weight to the situation (Dillard et al., 1989). Dillard (2004) introduced the idea of relational resource goals. These goals center around maintaining or improving the relationship and only come into play when there is an existing bond or a desire to create one. Emotions tend to run high in these situations because people feel that much more is on the line.

Another key aspect of disagreement is the concept of rebuff, a situation when one person resists or rejects the other's attempts to communicate. When someone's initial disagreement is met with resistance, their response often escalates in intensity, sometimes becoming more forceful or aggressive in an attempt to push their point (Hample, 2005). This reaction could be explained by the concept of face, which refers to the positive image an individual tries to maintain in social interactions. There are three main types of face: autonomy face (the desire to appear independent), fellowship face (the need for social acceptance), and competence face (the wish to be seen as capable). When someone feels that their competence face is threatened in an argument, they may respond with more aggression or defensiveness. This desire to protect one's face, especially in situations where respect is on the line, often makes disagreements more intense (Ting-Toomey, 1995). When robots are in a rebuff situation and reject humans' requests because those requests are deemed as unsafe or illegal, humans could feel losing face, resulting in hostile attitudes toward the robots.

This leads to an important question: how should robots manage disagreements, and arguments or navigate emotionally charged exchanges with humans? Human communication extends far beyond the literal words spoken. It involves a rich tapestry of non-verbal cues, such as facial expressions, tone of voice, and body language, all of which can drastically alter the meaning or emotional weight of a message. Two individuals could utter the same phrase, but the tone, posture, or facial expression they use could convey vastly different or opposite meanings. For robots to successfully engage in disagreement or emotionally nuanced conversations, they would require advanced emotional intelligence. This means they not only need to understand spoken words but also accurately interpret the non-verbal signals humans use to communicate emotions, such as a raised eyebrow, a softening tone, a high pitch, or the speed of speech. It's not enough for robots to merely process logical arguments. They must grasp the emotional undertones that often drive these discussions. This emotional awareness is particularly crucial if robots are to play a significant role in human lives, where empathy, trust, and emotional intelligence form the foundation of meaningful relationships.

In this context, navigating disagreement and emotional nuances is as essential in human–robot interactions as it is in human-to-human relationships. As robots advance and take on increasingly complex tasks, their ability to navigate the emotional and interpersonal nuances of communication will become critical. It's no longer sufficient for robots to be merely performing pre-programmed tasks. They must also engage with the emotional intricacies that characterize human communications. As society moves toward deeper emotional and social integration with robots, it is crucial for robots to develop emotional intelligence in order to build trust with human users.

Only after trust is established can companionship be formed. Companionship is the feeling of closeness and friendship, where someone provides support and emotional connection. It involves mutual respect, understanding, and a sense of togetherness, often experienced through shared activities, interests, or experiences (Merriam-Webster, 2024). Traditionally, companions are people, such as family members, friends, or caregivers. Increasingly, this role has been expanded to include non-human entities, in particular, care robots that are designed to perform or assist with care tasks historically managed by humans (Feil-Seifer & Mataric, 2005).

Companionship plays a dual role, not only enhancing emotional well-being but also making daily life more manageable and enjoyable. It helps people cope with stress and lead more fulfilling lives. For instance, in the workplace, companionship among co-workers has been shown to significantly reduce work-related stress and improve overall job satisfaction. When employees feel a strong sense of connection with their peers, they are more likely to experience a healthier work–life balance. The presence of emotional and social support at work eases daily pressures, creating a more enjoyable and productive work environment (Ahmed et al., 2024). Whether through friendships, family connections, or professional relationships, people are better equipped to manage life's challenges and experience greater happiness when they feel a sense of belonging and support.

It is important to point out that companionship is no longer exclusive to human interactions. Non-human entities, such as pets, animals, and increasingly, robots, can now fulfill companionship roles. Pets, for instance, have long been recognized for their ability to provide emotional support and help people form strong bonds with each other. For many, pets are more than animals—they are companions who bring joy, comfort, and a sense of responsibility. Pet ownership has been linked to reduced loneliness, increased physical activity, and improved mental well-being. Pets also offer opportunities for humans to connect with nature, experience unconditional love, and develop nurturing habits. For some, caring for a pet satisfies a deep need for companionship. Pets can even strengthen human relationships as pets often become a shared experience between friends or family members. In this way, the companionship that pets provide not only enhances individual well-being but also fosters social connections, making them an integral part of many people's emotional and physical health (Ahmed et al., 2024).

This paper argues that the concept of companionship can also be expanded to include robot companions, which are becoming an increasingly common part of the social fabric. For instance, robots that assist the elderly or people with disabilities can provide not only physical help, such as fetching items, cleaning the house, or reminding them to take medication, but also emotional support by engaging in conversations or simply being present and being readily available. These robots are particularly valuable in care settings, where they help fill the gap left by human caregivers, especially in situations where there may be a shortage of staff or resources. The emotional bonds that can form between humans and robots, while different from those between humans, still contribute to a sense of companionship. Many people find comfort in interacting with a robot that responds to their needs. While robots do not experience emotions like humans do, their programming allows them to simulate emotional interactions, providing a level of emotional support that can feel very real to the humans they engage with.

This broader definition of companionship reflects how non-human entities, such as pets and robots, have begun to play significant roles in providing emotional and social support. Whether through the warmth and affection of a pet or the efficient and empathetic responses of a care robot, companionship now goes beyond traditional human-to-human relationships. The companionship provided by non-human entities can significantly improve humans' quality of life by reducing loneliness, offering support, and even encouraging healthier habits.

As society becomes increasingly dependent on technology, robot companions represent a new frontier in how humans seek and find communication. These machines are evolving beyond mere tools. They are becoming companions that help people navigate both functional tasks and emotional challenges in everyday life.

As the technology behind robots becomes more advanced, it's likely that they will assume even greater roles in human lives. They could offer not just functionality but emotional engagement and support in ways that may reshape how we think about the very definition of companionship itself.

Ultimately, companionship—whether it comes from humans, animals, or robots—remains an essential aspect of the human experience. It provides the emotional and social connections that enrich life, making it more meaningful and enjoyable. The increasing role of robots in this domain presents exciting possibilities for the future, as our understanding of how emotional connections can form between humans and non-human entities continues to grow. This evolution hints at a future where companionship is more inclusive, encompassing a wider variety of interactions that still fulfill the deep human need for connection.

4 Human–Robot Companionships from Cross-Cultural Perspectives

In the previous section, this paper establishes that it is crucial to take into account robots' role as companions and ensure that they are equipped with the capability to exhibit social behaviors and emotional intelligence. As research has shown, the acceptance and effectiveness of care robots are profoundly influenced by the cultural values, social norms, and individual expectations regarding care and companionship (Papadopoulos, 2006). For instance, in healthcare, nurses are given trainings to recognize cultural differences and provide culturally competent care. In particular, in multicultural societies like America, it is important to make sure that care robots can recognize and appropriately respond to the diverse cultural backgrounds of both healthcare professionals and patients (Papadopoulos & Koulouglioti, 2018).

This leads to the research question of this paper: How do people's cultural backgrounds influence the types of companionship they form with care robots?

Since the author was not able to find any theoretical framework specifically created to guide culturally sensitive human–robot interactions, the author decided to apply two significant theoretical frameworks in the field of cross-cultural communications, i.e., Hofstede's cultural dimensions theory (Hofstede, 1980; Hofstede et al., 2010) and Hall's context theory (Hall, 1976), to analyze how cultural differences might potentially shape perceptions and interactions with robots.

Hofstede's cultural dimensions theory examines how cultural values influence human-to-human interactions in cross-cultural settings. In the context of human–robot interaction, it is argued that this framework could help systematically analyze how these cultural values could impact technology acceptance and the nature of companionship (Hofstede et al., 2010). Each value dimension offers a lens through which we can examine and anticipate the ways in which cultural contexts might influence human–robot companionship:

Power Distance (PD) measures how a culture handles inequality and authority. Cultures with high PD accept hierarchical order and authority without the need for justification, whereas low PD cultures strive for equality and question authority. Therefore, for users from high PD cultures, care robots may be expected to reinforce hierarchical relationships that are familiar and respected within society. These robots could be expected to take on authoritative roles such as a strict teacher directing the students or a veteran doctor providing decisive medical advice. Robots could also take on subordinate roles like a maid who diligently follows instructions to perform household cleaning without any questions or objections. In these roles, the robots not only perform tasks but also reinforce traditional social structures, emphasizing the importance of social status and age (Hofstede & Bond, 1988). Conversely, users from low PD cultures may prefer robots that embody egalitarian principles, treating all users equally regardless of social status or age. Robots are seen as equal partners, providing assistance without establishing any form of hierarchical relationship with users.

Individualism versus Collectivism assesses how people in a society are integrated into groups. Individualistic societies emphasize personal achievements and individual rights, while collectivist societies emphasize group loyalty and collective well-being. As a result, users from individualistic cultures could appreciate care robots that support autonomy and personal choice, cater to individual needs and preferences, and foster independence. Such robots are seen as tools that empower users to maintain their lifestyle and personal choices independently. On the contrary, users from collectivistic cultures might value robots that are capable of effectively navigating complex relationship networks, or Guanxi. These robots would need to not only perform tasks but also gain the trust of the users and forge bonds with users' friends and extended family circles. The ideal care robots in such settings act as facilitators of group harmony and collective well-being rather than just solely focusing on individual care.

Masculinity versusFemininity measures where societies value more on competitiveness, assertiveness, material success, cooperation, modesty, caring for the weak, and quality of life. Users from masculine cultures may prefer robots that are highly task-oriented, focusing on efficiency and effectiveness in performing care-related tasks. They may prefer for the robots to take on firm and commanding voice commands, and perhaps even a physical embodiment that suggests strength and resilience. The primary appeal in these settings is a robot's ability to execute tasks quickly and efficiently, reflecting society's broader values of competitiveness and productivity. On the contrary, users from feminine cultures may be more emotionally comfortable with softer and more soothing voice commands and potentially with a female embodiment that presents a distinctly feminine touch. Robots could act as compassionate companions that support not just the physical, but also the emotional and social needs of the users.

Uncertainty Avoidance Index (UAI) measures how comfortably a society tolerates uncertainty and ambiguity, impacting how they respond to new and unknown situations such as the introduction of care robots. High uncertainty avoidance cultures try to minimize the occurrence of unknown circumstances through strict laws and rules, while low uncertainty avoidance cultures are more tolerant of ambiguity. Users from high UAI societies might demand high predictability and reliability from robots, preferring robots to be equipped with clear guidelines for operations and comprehensive safety features to minimize risks or anxieties associated with the adoption of new technologies. This need may greatly limit the functions of robots and delay the process of commercialization. Conversely, users from low UAI cultures could be more open to experimental and innovative robots, even if it means tolerating some unpredictability in the robot's behaviors. This openness could accelerate the adoption of new robotic technologies, as users are more willing to experiment with and adjust to novel features that may not yet be fully refined.

Long-term orientation versus Short-term orientation evaluates how much societies respect their cultural traditions while adapting to modern challenges. Societies with a long-term orientation emphasize persistence, perseverance, and long-term planning. Short-term-oriented cultures are more focused on quick success, rapid return, and immediate benefits. Users from long-term oriented cultures value care robots that adapt and learn over time, focusing on sustainability and the continuous improvement of the overall care quality. Users from short-term oriented societies might prefer robots that address immediate care needs, prioritizing quick fixes and solutions that offer instant satisfaction or relief. This aligns with the societal preference for rapid returns and immediate gratification.

Indulgence versus restraint assesses how people regulate and control their desires and impulses. Cultures that lean toward indulgence generally allow relatively free gratification of basic and natural human desires related to enjoying life and having fun. Restraint cultures suppress gratification of needs and regulate it by means of strict social norms. Users expect robots to not only fulfill functions but also contribute to overall enjoyment and relaxation, such as playing music, telling jokes, or engaging in games. Restrained cultures may expect care robots to maintain a more serious demeanor and to limit or restrict features that promote leisure and enjoyment. Robots might be equipped with functionalities that support structured and scheduled care routines, possibly incorporating reminders for medication, exercise, and other health-related activities, aligning with a disciplined lifestyle.

Edward Hall's context theory delineates between high-context and low-context cultures, with the former relying heavily on implicit communication, environmental context, non-verbal cues, while the latter emphasizes explicit verbal messages. It could also offer valuable insights into human–robot companionship (Hall, 1976).

In high-context cultures, communication is often indirect, and the context of a message is as important as the message itself (Hall, 1976). For care robots supporting users from high-context cultures, it is essential to put an emphasis on capturing and interpreting non-verbal cues. Robots should be adept at recognizing and responding to non-verbal signals such as facial expressions, gestures, and tone of voice, aligning with the users' reliance on unspoken communication (Papadopoulos & Kouloglioti, 2018). Additionally, recognizing the importance of physical context and spatial dynamics could significantly enhance how these robots interact and connect with users.

Conversely, in low-context cultures, where messages are direct and explicit, care robots should focus on delivering clear and straightforward communication (Hall, 1976). They should provide explicit instructions and feedback to ensure their functionalities are easily accessible and understandable to users. Respect for personal space and privacy is crucial and users in low-context cultures typically appreciate having control over their interactions with robots, including when and how these interactions occur.

In today's multicultural societies, those cultural insights provided by Hofstede's cultural dimensions and Hall's context theory highlight the importance of customizable robots that can adapt to the cultural context of their users. By systematically analyzing human–robot interactions with those two frameworks, this paper found that the communication styles, interaction patterns, and even the robots' voice and appearance could potentially play a key role in facilitating culturally appropriate human–robot interactions. By designing robots that align with these cultural nuances, care robots can be perceived not just as machines, but also as genuine companions that enhance the quality of life and social interactions of their users.

5 Human–Robot Companionships from East Asian Perspectives

It is important to note that the two cross-cultural theories discussed in the previous section mainly reflect Western viewpoints as they are proposed by Western scholars. There are philosophical traditions from East Asia that could complement and contrast Western perspectives. In particular, it is important to put out that the East Asian attitudes toward robotics and artificial intelligence differ significantly from those in the West. In the Western context, robots are often met with suspicions, seen as potential job killers or dehumanizing machines. In contrast, the Japanese and Chinese tend to view robots more as natural extensions of the human world rather than as alien or threatening entities given their distinctive philosophical and cultural traditions (Song, 2021).

In exploring the philosophical underpinnings in Japan, one significant influence lies its indigenous religion Shintoism, which permeates various aspects of life. Central to Shinto belief is the concept of kami. It is the divine force or spirits that resides in both living beings and inanimate objects. It believes that everything from trees and animals to robots and even technological gadgets can embody kami. Such a belief blurs the boundaries between the natural and the artificial, suggesting a universe where all entities are interconnected and infused with spiritual essence (Lobel, 2023).

Additionally, Japanese folklore is full of tales of objects that come to life, illustrating a permeable boundary between artificial and natural in the Japanese traditions. Contrary to the Western perception of robots as terminators, the Japanese often view them as saviors as illustrated in their heroic cartoon figure of Astro Boy. Astro Boy was created in 1951 when Japan was recovering from its post-war nuclear devastation. Astro Boy embodies a robot that is created in the image of a boy who was killed in a car accident. It assumes human characteristics and uses its powers to protect the human world. It can detect whether a person is good or evil, and it fights against aliens and rogue robots. In one story, Astro Boy travels back in time to 1969 to stop the U.S. Air Force from bombings the Vietnamese villages. This character not only captured the public imagination but also fueled Japanese scientists' visions of what robots could become (Lobel, 2022).

The Chinese are also less panicky than westerners in response to perceived existential risks from emerging technologies like AI and care robots. According to Song (2021), the difference can be partially attributed to the philosophical thoughts in Daoist and Confucian traditions, which entertain the existence of various forms of beings. So, living with devices, programs, or other forms of beings, which may be potentially superior to humans, does not necessarily lead to a dystopian future. AI and digital beings could be just another category of super beings like those immortals in Daoism. Furthermore, Confucian scholars advocate for viewing AI as companions within the matrix of "things (物)," suggesting a harmonious existence by incorporating AI into the broader ecosystem and the ethical order. Song points out that the strong influence of the non-anthropocentric orientation of Chinese philosophical thought significantly reduces anxiety over the potential existential threats or loss of subjectivity on the part of the human amid the rapid advancement of AI technologies in recent decades (2021).

The East Asian philosophical concept of Gongsheng, or kyōsei in Japanese, which emphasizes a worldview of mutual embeddedness, coexistence, and co-becoming, also contributes to the more positive perceptions of companionship between human and care robot (Song & Zhan, 2024). Gongsheng challenges the traditional notion of individuals as self-contained and autonomous entities, promoting a reevaluation of how humans relate to other beings. It encourages a deep respect and care for not only living creatures and plants but also for inorganic entities on our planet (Song & Zhan, 2024). This inclusive perspective fosters a broader sense of companionship that extends beyond human interactions to include robots and other non-human agents, further integrating them into the fabric of societal and ecological relationships.

While Gongsheng is rarely referenced in mainstream academic discussions on AI ethics, it is argued that it could inspire and inform broader dialogues about human–technology relationships and harmonious coexistence. Particularly in the context of robots, Gongsheng envisions a future where humans and care robots share living spaces and experiences in a mutually beneficial way. Care robots would not only support specific functionalities but also learn from and adapt to human needs. In the meantime, humans could also learn to integrate these robots into their daily lives, recognizing them as companions rather than mere tools.

Gongsheng also implies deeper emotional and social connections between humans and robots. It prompts us to consider how we can ensure that co-living benefits both humans and robots, respecting the autonomy and rights of both parties. As care robots become more sophisticated, they could develop capacities to understand and respond to human emotions more effectively, participating in the emotional dynamics of human lives. This could lead to robots being positioned as companions or members of the family, contributing positively to the mental and emotional well-being of their human counterparts. Furthermore, the co-becoming aspect of Gongsheng emphasizes the idea that the relationship between humans and robots is dynamic and evolving. As each party learns from and adapts to the other, both humans and robots could evolve in ways that were previously unimaginable, potentially leading to new forms of social structures, relationships, and understandings of life.

Ultimately, Gongsheng could guide us toward a more sustainable and harmonious future where humans and robots not only coexist but thrive together, cultivating a relationship of mutual growth and respect with our technological creations. While applying Gongsheng to care robots involves speculative and philosophical considerations, it promotes a hopeful and respectful vision for our technological future, urging us to think creatively and ethically about our interactions with emerging robotic technologies.

6 Conclusion

This paper begins with an exploration of the development of care robots, tracing their development through examples from key regions, such as Japan, America, and China. These countries have developed leading technologies for care robots and attempted with unique approaches to integrate them into the social fabric of people's daily lives. Following this introduction, the paper delves into critical elements of human relationships—trust, disagreement, the concept of face, emotional intelligence, and companionship. While these factors offer invaluable insights into understanding the dynamics of human–robot relationships, this paper argues that understanding human–robot relationships requires additional layers of complexity and nuance.

Care robots are not simply tools for completing tasks. They are designed to assist with both practical needs and emotional challenges of humans in everyday life. As these robots increasingly become part of our daily interactions, it is inevitable that they will develop forms of companionship with humans. This paper argues that the nature and depth of these companionships are heavily influenced by cultural contexts, meaning that the role of care robots will not be uniform across the world. The types of companionship formed in Japan, for example, differ significantly from what is expected in the United States.

As we anticipate the future of robotics within multicultural societies, it becomes clear that a one-size-fits-all approach to designing care robots will not suffice. Frameworks such as Hofstede's cultural dimensions and Hall's high- and low-context communication theories emphasize the importance of adapting robots' communication styles, interaction patterns, and even voice and appearances to meet the specific cultural needs and preferences of the targeted users. By incorporating these insights into robot design, it becomes possible to create technologies that not only perform their functional roles but also resonate on a deeper, culturally sensitive level. It argues for a deep understanding of cultural norms, sensitivities, and taboos to ensure that robots behave in ways that are not only effective, but also culturally appropriate and respectful. This careful calibration fosters a dynamic of coexistence and co-becoming, a concept coined as Gongsheng where humans and robots do not simply share space. They evolve together in a shared social and cultural environment. Robots can learn from human behaviors, adapting to individual and cultural nuances over time. In parallel, humans grow to accept robots as integral and empathetic parts of their social and care networks. This symbiotic relationship not only enhances the functional effectiveness of the robots but also their acceptance, allowing them to fulfill the role of genuine companions.

By designing care robots that are attuned to the cultural dimensions, communication styles, and philosophical traditions, and by following the principle of Gongsheng, we could hope for a positive future where robots do not merely assist but enhance human life through meaningful and respectful interactions. In this envisioned future, humans will view robots as valuable companions, capable of contributing positively to their emotional and social well-being. This shift is not just technological but deeply cultural, requiring thoughtful design that prioritizes trust, empathy, respect, and mutual understanding.

References

Ahmed, E., Buruk, O., & Hamari, J. (2024). Human-robot companionship: Current trends and future agenda. *International Journal of Social Robotics, 16*, 1809–1860.

Alessa, A., & Al-Khalifa, H. (2023). *Towards designing a ChatGPT conversational companion for elderly people.* arXiv.

Altman, I., & Taylor, D. (1973). *Social penetration: The development of interpersonal relationships.* Holt, Rinehart & Winston.

Baker, A. L., & P. E. (2018). Toward an understanding of trust repair in human-robot interaction: Current research and future directions. *ACM Transactions on Interactive Intelligent Systems., 8*(4), 1–30.

Barber, B. (1983). *The logic and limits of trust.* Rutgers University Press.

Baxter, L. A. (2010). *Voicing relationships: A dialogic perspective.* Sage.

Baxter, L. A., & Braithwaite, D. O. (2008). Relational dialectics theory: Crafting meaning from competing discourses. In L. A. Baxter & D. O. Braithwaite (Eds.), *Engaging theories in interpersonal communication* (pp. 349–361). Sage.

Carl, W. J., & Duck, S. W. (2004). How to do things with relationships. In P. Kalbfleisch (Ed.), *Communication yearbook28* (pp. 1–35). Sage.

Chadwick, A. E. (2011). *Persuasive hope theory: A discreet emotion appraisal theory with implications for message design.* National Communication Association.

Dillard, J. P. (2004). The goals-plans-action model of interpersonal influence. In J. S. Seiter & R. H. Gass (Eds.), *Persuasion: Theory and research* (pp. 185–206). Allyn & Bacon.

Dillard, J. P., Kinney, T., & Cruz, M. G. (1996). Influence, apprisals, and emotions in close relationships. *Communication Monographs* , 105–130.

Dillard, J. P., Segrin, C., & Harden, J. M. (1989). Primary and secondary goals in the prodution of interpersonal influence messages. *Communication Monographs, 56*, 19–38.

Duck, S. (2011). *Rethinking relationships.* SAGE Publications, Inc.

Feil-Seifer, D., & Mataric, M. (2005). Defining socially assistive robotics. *Rehabilitation robotics* (pp. 465–468). IEEE.

Fox, J., & Gambino, A. (2021). Relationship development with humanoid social robots: Applying interpersonal theories to human-robot interaction. *Cyberpsychology, Behavior, and Social Networking, 24*(5), 294–299.

Fu, Z., Zhao, T. Z., & Finn, C. (n.d.). *Mobile ALOHA: Learning bimanual mobile manipulation with low-cost whole-body Teleoperation.* Retrieved April 2024, from https://ar5iv.labs.arxiv.org/html/2401.02117

Ghafurian, M., Ellard, C., & Dautenhahn, K. (2008). *Social companion robots to reduce isolation: A perception change due to COVID-19.* Retrieved from arXiv: https://ar5iv.labs.arxiv.org/html/2008.05382

Gilbert, M. A. (2001). Emotional messages. *Argumentation, 15*, 239–249.

Hall, E. T. (1976). *Beyond culture.* Anchor Books.

Hample, D. (2005). *Arguing: Exchanging reasons face to face.* Lawrence Erlbaum Associates.

Hofstede, G. (1980). *Culture's consequences: International differences in work related values.* Sage.

Hofstede, G., & Bond, M. H. (1988). The confucian connection: From cultural roots to economic growth. *Organizational Dymanics, 16*, 4–21.

Hofstede, G., Hofstede, G. J., & Minkov, M. (2010). *Cultures and organizations: Software of the mind.* McGraw-Hill.

Knutsson, K. (2024, September 8). *Chinese robo-helper can clean, cook, even shoot hoops.* Retrieved from Fox News: https://www.foxnews.com/tech/chinese-robo-helper-can-clean-cook-even-shoot-hoops

Lin, Y., & Zhao, F. (2010). Exploring constructive agressive communication in China: Its cultural roots, strategies, and new developments. In T. Avtgis & A. Rancer (Eds.), *Arguments, agression, and conflict: New directions in theory* (pp. 82–99). Routledge Press.

Lobel, O. (2022). *The equality machine: Harnessing digital technology for a brighter, more inclusive future.* PublicAffairs.

Lobel, O. (2023, January 29). *In Japan, humanoid robots could soon become part of the family.* Retrieved from Freethink: https://www.freethink.com/robots-ai/humanoid-robots-japan

Lytal, C. (2016, February 25). *The robot revolution: Developing machines to promote human health.* Retrieved from https://keck.usc.edu/news/the-robot-revolution-developing-machines-to-promote-human-health/

Malewar, A. (2024, January 6). *Meet mobile ALOHA, your housekeeping robot.* Retrieved from InceptiveMind: https://www.inceptivemind.com/meet-mobile-aloha-your-housekeeping-robot/36176/

Mayer, R. C. (1995). An integrative model of organizational trust. *Academy of Management Review, 20*(3), 709–734.

Mckay, C. (2024, January 4). *Stanford unveils a new low-cost mobile robotics training system.* Retrieved from Maginative: https://www.maginative.com/article/stanfords-unveils-a-new-low-cost-mobile-robotics-training-system/

Merriam-Webster. (n.d.). Retrieved from https://www.merriam-webster.com/dictionary/companionship

Papadopoulos, I. (Ed.). (2006). *Transcultural health and social care: Development of culturally competent practitioners.* Churchill Livingstone Elsevier.

Papadopoulos, I., & Koulouglioti, C. (2018). The influence of culture on attitudes towards humanoid and animal-like robots: An integrative review. *Journal of Nursing Scholarship, 50*(6), 653–665.

Petronio, S. (2002). *Boundaries of privacy.* SUNY Press.

Rehm, M., Rodil, K., & Krummheuer, A. L. (2018). Developing a new brand of culturally-aware personal robots based on local cultural practices in the danish health care system. In *2018 IEEE/RSJ international conference on intelligent robots and systems (IROS)*, (pp. 2002–2007).

Paro research papers. (n.d.). *Paro research papers.* Retrieved from http://www.parorobots.com/whitepapers.asp

Rotter, J. B. (1967). A new scale for the measurement of interpersonal trust. *Journal of Personality, 35*(4), 651–665.

Seaborn, K., Sekiguchi, T., Tokunaga, S., Miyake, N. P., & Otake-Matsuura, M. (2023). Voice over body? Older adults' reactions to robot and voice assistant facilitators of group conversation. *International Journal of Social Robotics, 15*, 143–163.

Sgorbissa, A., Papadopoulos, I., Bruno, B., Koulouglioti, C., & Recchiuto, C. (2018). Encoding guidelines for a culturally competent robot for elderly care. In *2018 IEEE/RSJ international conference on intelligent robots and systems (IROS)*, (pp. 1988–1995).

Song, B. (2021). *Intelligence and wisdom: AI meets chinese philosophers.* (B. Song, Ed.). Springer.

Song, B., & Zhan, Y. (Eds.). (2024). *Gongsheng across contexts: A philosophy of co-becoming.* Springer.

Ting-toomey, S. (1995). The challenge of facework: Cross-cultural and interpersonal issues. *Women and Language, 18*(1), 63–83.

Vangelisti, A., & Perlman, D. (2006). *Handbook on personal relationships.* Cambridge University Press.

Yirka, B. (2024, January 8). *Mobile ALOHA robot able to help with multiple household tasks.* Retrieved from Tech Xplore: https://techxplore.com/news/2024-01-mobile-aloha-robot-multiple-household.html

Jennifer Bourne is the Director of Global Initiatives and Operations at the Berggruen Institute, where she oversees communication and operations for its global centers. Previously, she built and directed the Institute's Global Fellowship Program. Jennifer is also a scholar and writer, contributing to both academic and popular media outlets. She serves as the Executive Editor of *Cuiling Magazine* and Associate Editor of *Noema Magazine.* Jennifer holds a Ph.D. in Intercultural Communication from the University of Maryland, where she focused her research on cultural communications and interpretations. She also earned a master's degree from Newcastle University in the UK and a bachelor's degree in China.

Digital Psychological Twins in Medicine: Addressing Risks to Human Relationships

Brian D. Earp, Tessa van Veenendaal, Sebastian Porsdam Mann, and Julian Savulescu

1 Introduction

There are well-known challenges around medical decision-making on behalf of individuals (often described as "formerly competent patients"), who have lost the capacity to make their own treatment choices. In many, especially Western, jurisdictions that prioritize individual choice and autonomy, a "substituted judgment" standard predominates, according to which surrogate decision-makers should choose the course of action that the patient herself would have chosen or endorsed if she was able to do so (Torke et al., 2008). In other contexts, family-based decision making or "best interests" judgments are more likely to hold sway (Yang et al., 2022). However, regardless of the specific legal or ethical principle guiding such decision-making, it is typical, when deciding what should be done in terms of treatment or non-treatment, for surrogates in various cultures to place at least some moral weight on what they believe the patient would prefer (Combs et al., 2013; Ho et al., 2010).

And yet, research suggests that people often face significant difficulties in accurately predicting the treatment preferences of loved ones, even when making a good-faith effort to adhere to substituted judgment in contexts where that is the prevailing standard. These difficulties, some of which are also experienced by those making family-oriented, rather than individual-centered decisions, include the following: a lack of explicit guidance from the patient, ambiguity in interpreting the patient's

B. D. Earp (✉) · T. van Veenendaal · S. Porsdam Mann · J. Savulescu
Yong Loo Lin School of Medicine, Centre for Biomedical Ethics, National University of Singapore, Singapore, Singapore
e-mail: bdearp@nus.edu.sg

T. van Veenendaal · J. Savulescu
Uehiro Oxford Institute, University of Oxford, Oxford, United Kingdom

S. Porsdam Mann
Center for Advanced Studies in Bioscience Innovation Law, University of Copenhagen, Copenhagen, Denmark

Y. Chandra and R. Fan (eds.), *Artificial Intelligence and the Future of Human Relations*,
https://doi.org/10.1007/978-981-96-7185-4_13

past statements or behaviors, and the emotional burden and distress experienced by surrogate decision-makers when asked to make high-stakes medical decisions for their loved ones (Rogers & Lopez, 2023; Wendler & Rid, 2011).

To address these challenges, a Patient Preference Predictor (P3) has been proposed which has inspired much discussion in the literature. The P3 is a hypothetical computer-based algorithm proposed by Annette Rid and David Wendler that would supplement or potentially replace key aspects of the human surrogate decision-making process (Rid & Wendler, 2014). The P3 would be based on large-scale surveys of the general population, correlating demographic data with expressed medical preferences to infer the preferences of specific incapacitated individuals. The goal of the P3 is to provide a more accurate and objective means of predicting patient preferences, reducing the burden on human surrogates and potentially leading to treatment decisions more closely aligned with what the patients themselves would have wanted.

However, the P3 has proven controversial, with critics arguing that it would not adequately respect the (former) autonomy of individual patients, a goal that is commonly assumed to be the underlying ethical justification, or at least a significant moral reason, for adopting the substituted judgment standard in the first place (for a contrary analysis, see Annoni, 2025; but see also Earp et al., 2025). In particular, critics suggest that relying on *population*-level data to predict *individual* patient preferences fails to account for the unique values, beliefs, and experiences that shape each person's healthcare preferences and hence the decisions they make in that domain, or would make, if they were autonomous (John, 2014; O'Neill, 2022; Sharadin, 2018). Critics argue that truly respecting a patient's prior autonomy (even assuming that this should be a primary goal) thus requires a more personalized approach: one that, for instance, tracks the specific values, beliefs, and so on, that the patient themselves would have relied on or endorsed when making a given treatment decision, rather than relying on "bare" statistical evidence from population-level surveys (e.g., John, 2014; Sharadin, 2018).

In response to these concerns, we have recently proposed a novel adaptation of the P3 concept in collaboration with some of our colleagues (including Rid and Wendler). We proposed a *Personalized* Patient Preference Predictor (P4), an artificially intelligent (AI) system that would use machine learning to extract patients' values and preferences from individual-level material, such as electronic health records, personal writings, interview transcripts (Park et al., 2024), and other sources of information unique to each patient (Earp et al., 2024). The P4 would involve fine-tuning a large language model (LLM) on this person-specific corpus of data, creating a kind of "digital psychological twin" of the patient (similar to de Kerkhove, 2021; see also O'Donnell, 2024, referring to Park et al., 2024). This AI-powered psychological twin—a partial replica of the patient's mind, rather than their body, as in traditional healthcare-related digital twin proposals (e.g., Braun, 2021; see Voinea et al., in press, for discussion)—could then be queried in real-time to provide insights into the patient's likely preferences in a given medical situation. This would then offer a more individualized approach to inferring their wishes and thus, on some accounts (cf. Annoni, 2025), respecting their former autonomy.

While the P4 was intended to address some of the key issues raised by critics of the P3, such as the reliance on overly general, population-level data, it has been argued that this new approach fails to resolve other concerns or even introduces new problems (Bishop, 2024; Gutiérrez-Lafrentz, et al., 2024; Meier, 2024; Rzepinski et al., 2024; Starke & Jox, 2024; but see Earp et al., *forthcoming*). One major concern is that the P4 could have negative effects on *human relationships*, particularly the doctor–patient relationship and family relationships, including the relationship between incapacitated patients and next of kin serving as surrogate decision-makers (Riddle, 2024; Mertes, 2024; Schwan, 2024; Herington & Kluger, 2024; Rahimzadeh, 2024).

In this chapter, we will briefly describe the idea behind the P4 and how it would work in practice. We will then turn our attention to the objections raised regarding the potential negative impacts of the P4 (or P3) on human relationships. These objections will be summarized in two separate sub-sections: one focusing on the doctor–patient relationship and the other on family relationships. In a subsequent section, we will respond to each of the main objections, considering their merits and limitations. Finally, we will propose some practical steps forward and draw the chapter to a close.

2 Overview of the P4 Proposal

In our previous work, we introduced the concept of a Personalized Patient Preference Predictor (P4) (Earp et al., 2024), a hypothetical system designed to address some of the purported limitations of the original Patient Preference Predictor (P3) proposed by Annette Rid and David Wendler (e.g., Rid & Wendler, 2014; Wendler et al., 2016). The P4 aims to provide a more individualized approach to predicting patient preferences during periods of decisional incapacity by relying on information "produced by, describing, or otherwise pertaining to a specific individual," beyond mere demographic categorization (Earp et al., 2024, p. 16).

The P4 would take the form of a fine-tuned large language model (LLM) trained on data of the kind previously described (following Porsdam Mann et al., 2023, 2024a, 2024b; see also Park et al., 2024; and see O'Donnell, 2024; Kirk et al., 2024, for discussion), ideally with the patient's prior permission or, if not, with their surrogate's permission. The model could, in principle, draw on a wide range of personal data, such as electronic health records, responses to medical questionnaires, value-eliciting choice experiments, or structured interviews (Park et al., 2024); or, more controversially, blog posts, social media activity, emails, text messages, or even purchasing and browsing histories (Ferrario et al., 2023; see also Lamanna & Byrne, 2018).

The goal would be to use these data to predict the patient's first-order treatment preferences, second-order preferences for how treatment decisions are made (e.g., desired family involvement, or even whether to use or stop using a P4), and the strength and certainty of these preferences (assuming that these have not been explicitly recorded or communicated in advance). The P4 could also, hypothetically, be asked to provide *reasons* for its prediction(s), insofar as the quality, relevance, and accuracy of any such reason-giving function were shown to be sufficiently robust in model training and testing (for evidence of LLMs giving relevant reasons for normatively weighty judgments, see Chu et al., 2025; for a discussion of how this might work in practice, see Porsdam Mann et al., 2024b).

Any use of personal data obviously raises privacy issues, as we acknowledged in our previous essay. These include concerns about data security, which is primarily a technical problem, albeit a serious one with potentially major ethical implications—including with respect to what happens with one's P4 after one dies (see Zohny, 2025, responding to Iglesias et al., 2025). They also include concerns about the initial authorization of personal data use for purposes of P4 training. On this point, we wrote that "the P4 should be implemented in a manner consistent with currently existing legal and ethical norms governing healthcare decision-making for incapacitated persons in a given jurisdiction. Where surrogates such as family members have legal authority, they should determine the appropriate role and use of a P4 in the decision-making process" (Earp et al., 2024, p. 9). Ideally, we suggested, individuals should be asked *in advance* to indicate how and to what extent, if at all, they would like a P4 to be used if they ever lost capacity; and if one were to be used, which types of data it should or should not be trained on.

Suppose that any use of data (including what types of data to incorporate versus exclude) has been appropriately authorized and accompanied by stringent data-protection measures. How would these data actually be processed? We suggested that a method known as *fine-tuning* could be used. This is a form of model training in which the last few layers of an LLM's neural network are exposed to a specialized corpus of text, enabling finer-grained responses to relevant queries while retaining its underlying representation of language (Ding et al., 2023). In the case of a P4, this would allow the model to adapt to the specific features of the individual's personal data, possibly inferring or extracting their unique preferences and values as discussed below (see also Park et al., 2024, on "simulated agents," Danaher & Nyholm, 2024, on "personalized digital duplicates," Iglesias et al., 2025 on "digital doppelgangers," or Voinea et al., in press, on "AI SIMs"). The effectiveness of a P4 could be maximized, we suggested, by gathering as much relevant (i.e., value or preference-encoding) information as possible, while strictly adhering to requests from the individual or their surrogates to exclude certain data sources (e.g., social media activity).

Although a specialized system like the P4 has not yet been developed or formally tested, it is inspired by and technically analogous to existing LLMs that have, among other things, demonstrated the ability to convincingly mimic individuals' styles of writing and argumentation (Porsdam Mann et al., 2023), while also inferring or otherwise representing their value-based preferences across various domains (Park et al., 2024; for discussion, see O'Donnell, 2024). While the ability of LLMs to infer *medical* preferences specifically has not, to our knowledge, so far been studied (however, see Galitsky, 2024, for a technical description of a personalized healthcare LLM with broadly similar capacities), recent work has shown that LLMs can infer individuals' non-medical preferences to a high degree of accuracy using limited examples of relevant conduct or information (Gao et al., 2024; Kim et al., 2024; see also Kuleshov & Ellis, 2023).

Of course, significant further work is necessary to confirm the validity of using LLMs to predict either first-order medical preferences or second-order preferences about how one's first-order preferences should be inferred (Mainz, 2023). Nevertheless, the ability of LLMs to infer non-medical preferences suggests that, given data of sufficient quality and relevance, LLMs should, in principle, be able to predict individual-level preferences regarding a range of issues, including medical ones. The P4 concept builds upon this assumption, aiming to provide a more personalized and accurate approach to inferring patient preferences compared to the current baseline accuracy of human surrogate decision-making, which has been estimated in various studies using hypothetical scenarios to be around 50–70% (Ciroldi et al., 2007; see also Shalowitz et al., 2006).

3 Objections to the P4 Proposal and the Present Focus

In a set of commentaries on our previous article, numerous objections to the idea of a P4 were raised by various authors. These included skepticism regarding the validity of certain data sources (especially, the use of social media information) (e.g., Ballantyne & Style, 2024; Berger, 2024; Gligorov & Randall, 2024; Mertes, 2024), the required accuracy of a P4 to justify the resources required for its creation and implementation (Ballantyne & Style, 2024; Biller-Andorno, 2024; Rahimzadeh, 2024), its technical feasibility (see Biller-Andorno et al., 2024; Sharadin, 2024; Starke & Jox, 2024), and whether mere preference data is sufficient to capture a patient's underlying value system (Gligorov & Randall, 2024). Other objections included the risk of a P4 perpetuating existing harm to vulnerable patients (Riddle, 2024), the environmental impact of such AI systems (Mertes, 2024), and legal issues relating to the lack of explainability (Biller-Andorno et al., 2024).

Each of these objections must be taken seriously, and we have replied to them in in a separate essay (see Earp et al., *forthcoming*). In this chapter, however, to stay focused, we will consider just one type of objection and respond to it in greater depth: namely, that the use of a P4 (or even a P3) would have negative implications for human relationships (cf. Voinea et al., 2024a, 2024b). In particular, critics suggested that the doctor–patient relationship, as well as familial relationships, might be harmed or disrespected by such preference predictors. We will now summarize and respond to the main objections raised in this vein before providing a general discussion of major points and then concluding.

4 Doctor–patient (DP) Relationship: Key Concerns

Several commentators argued that the proposed P4 would negatively affect the doctor–patient relationship. Broadly, these objections had to do with the following major concerns: that collecting certain types of data might undermine confidentiality or disincentivize candidness in doctor–patient conversations (Mertes, 2024); that overworked doctors would be tempted to defer by default to the P4's prediction, when instead they should critically interrogate its output and actively weigh competing considerations; and, related to this, that the use of a P4 could marginalize or even "remove" medical professionals from the decision-making process (Riddle, 2024). We elaborate on these concerns below.

4.1 DP1. Collecting Certain Types of Data for P4 Training Would Interfere with Mutual Trust and Candor

In her commentary, Mertes (2024, p. 42) objects to the specific suggestion that conversations between physicians and patients could (if appropriately authorized) be recorded as part of the P4's textual input, suggesting that this "ignores that this is a setting in which candor is important and enabled by trust and confidentiality." Mertes highlights a "danger to patient care," as the probability that the information recorded in these interactions will ever be needed is "speculative and improbable" by her estimation, whereas she believes that potential recordings would compromise the frankness and mutual trust that is required for an effective doctor–patient relationship.

An expectation of confidentiality is central to building trust in the doctor–patient relationship, in part by enabling more honest and open conversation (important also for accurate diagnosis), while ensuring privacy. If the training or use of a P4 would undermine these central features of a well-functioning doctor–patient relationship, this would indeed be problematic. But it is not clear that such an outcome would be guaranteed, or even likely, if appropriate procedures are followed.

The worry seems to be that a patient's knowledge that a conversation they are having with their healthcare provider might be recorded would negatively impact the trust and honesty necessary for that conversation to be fruitful. However, this process is not meant to be surreptitious but rather would proceed only with the explicit consent and awareness of the patient. The purpose of transcribing conversations is to make the P4 predictions more accurate and reflective of the patient's preferences and values by creating more—and more relevant—data that could be used to make predictions. If the benefits, risks, and implications of this data collection method were openly discussed with patients, we think this would foster trust and candor rather than inhibit them. Additionally, as mentioned in our original proposal (Earp et al., 2024), patients who express discomfort with this method but still wish to make use of a P4 would have the option to choose from several alternative data sources, such as electronic health records or structured questionnaires. More generally, we stress that the patient or their surrogate should retain ultimate responsibility for, and control over, the data used to train a P4.

4.2 DP2. Doctors Might Rely on the P4's Predictions by Default

Schwan (2024, pp. 39) warns that we must "guard against the automatization of medical decision making." According to Schwan, there are different ways in which respect for a patient's autonomy can be realized: by "deferring to their expressed preference," treating them in a manner "most consistent with their deeper commitments," and "doing what they will come to prefer." However, navigating such complexity in a responsible manner requires careful and considered human judgment: it is not something that can be entrusted to a P4.

For one thing, there might be situations in which these different dimensions of autonomy clash. There might, for instance, be a patient with declining capacity who is "expressing clear preferences and attempting to exercise control," but in ways that are objectively "contrary to their more fundamental commitments" (Schwan, 2024, p. 40; see also Toomey et al., 2024a, 2024b, 2025), for a parallel analysis involving choices that may be contrary to a patient's "true self"). Different reasons might weigh in favor of different potential resolutions to such a conflict, and Schwan suggests that clinicians must exercise their human judgment to evaluate these reasons. This is especially the case given that the brunt of responsibility for determining what weight to place on the P4's predictions, alongside various other factors, must ultimately lie with the clinician.

Problematically, however—insofar as the P4 has good predictive accuracy—the temptation, Schwan argues, will be for "overworked members of care teams [to] offload decision-making" to the algorithm, and therefore view its verdict as "*determinative* of how a patient should be treated." In other words, they might be incentivized by the circumstances to neglect their duty to treat a P4's prediction as only "one consideration among many that are relevant to a patient's care" (Schwan, 2024, p. 39).

There are several ways to address this valid concern. First, the P4 should be explicitly presented as a *decision-support tool*, not a decision-making authority. Its predictions should be accompanied by clear explanations of their limitations and the importance of considering other relevant factors and preferences in each case. Care team members should receive specific training on how to interpret and apply the P4's predictions in the context of broader ethical reasoning and patient-centered care. Surrogates should be informed about the nature of the P4 and its intended function as a decision aid rather than as a decision-maker in its own right.

Moreover, the P4 should ideally be designed to provide not just a single predicted preference, but rather, a range of plausible or likely preferences with supporting evidence and associated levels of confidence. This can help to convey the *uncertainty* involved in the predictions and the need to consider other factors in decision-making. It may also be possible for a P4 to provide *reasons* for its predictions (see Chu et al., 2025), and these, too, would need to be evaluated for their strength, coherence, and relevance by the care team and by surrogates (Porsdam Mann et al., 2024b). Finally, institutional policies should be put in place to ensure that the P4's predictions are never treated as determinative of patient care and that ultimate responsibility for treatment decisions remains with the human care team and surrogate(s).

4.3 DP3. The P4 Risks Marginalizing or Removing Medical Professionals from the Decision-Making Process

Like Schwan, Riddle (2024) is concerned with the risk that clinicians will end up playing too small a role in the decision-making process, given the need for human judgment in making certain types of complex evaluations.

Riddle begins by noting that a P4's inputs or outputs could be biased in such a way that they do not truly advance patient autonomy. For example, they might reflect past choices that were shaped by coercive social pressures or harmful stereotypes (e.g., about persons with disabilities), such that any resulting treatment preferences, even if accurately inferred or identified, might be inauthentic. Given this possibility, if the P4 is to serve its main alleged purpose (cf. Annoni, 2025) in enabling respect for patients' (former) autonomy, rather than *merely* inferring their likely preferences, medical professionals would need to actively account for such potential biases when deciding how to weigh or respond to a P4's prediction.

As Riddle sees it, this leads to a dilemma. On the one hand, if substantial clinician involvement is, in fact, required to account for a P4's potentially socially biased inputs or outputs, the P4 would lose some of its usefulness as an efficient, AI-driven system for informing treatment decisions: in other words, a system that could, in principle, become largely automated and thus potentially less prone to making certain human-specific types of error (e.g., making an illogical inference due to stress). On the other hand, if the P4 retains its usefulness in that sense, clinicians would risk being marginalized or even "removed" from the decision-making process, which would prevent them from checking for harmful social biases.

To illustrate this possibility, Riddle (2024) uses the example of disability. As he argues, public perceptions of disability are "demonstrably negative" (p. 60) with stereotypical views holding that people with disabilities are "dependent and burdensome upon those around them" despite leading lives that are supposedly "not worth living" (p. 61). These negative social attitudes, in turn, may often shape a person's perception of their own disability, meaning that "people with disabilities might make choices about their medical care on the basis of a negative self-image that is informed by social stereotypes" (*ibid.*).

Suppose that various choices of this kind (which Riddle seems to regard as inauthentic and thus not truly autonomy-promoting)[1] have been recorded in a patient's health records, which are subsequently used to train their P4. If the person later loses capacity in a situation for which the use of their P4 has been appropriately authorized, what then? Although the P4 may produce a prediction that is perfectly consistent with the patient's own past healthcare decisions (e.g., a preference for minimal treatment, perhaps to avoid feeling "like a burden"), such a prediction might be based on, and thus serve to reinforce, "coerced choices," and might thereby "justify actions that might not reflect the *actual* beliefs or desires of the patient, but instead, mirror the [harmful] social attitudes informing their self-image." This is similar to Schwan's concern mentioned in the previous section that an explicit treatment preference might not reflect a person's "deeper commitments" (or their "true self" as per Toomey et al., 2024a, 2024b, 2025).

To safeguard against this, Riddle suggests either not implementing the P4 at all, or "supplementing [it] with direct explicit input" by clinicians. While acknowledging that clinicians are "capable of harmful biases themselves," Riddle argues that they should still play a pivotal role in the decision-making process by seeking "to understand the complex reasoning behind choices," something the P4 may be unable to do. Clinician input in interpreting the P4's predictions is necessary, Riddle argues, to "consider the ways in which bias or coercion might enter the equation" and consequently "counteract it through intentional intervention" (Riddle, 2024, p. 61).

We have already argued that clinicians should indeed be involved in interpreting P4 output, weighing reasons, and so on, and that the P4 itself, and policies for its use, should be designed in such a way as to explicitly require this. However, Riddle believes this leads to a dilemma: such clinician involvement, he thinks, would limit the P4's ability to "act autonomously"—for example, in generating predictions directly from training data—thereby rendering it less useful or efficient.

In short, either the P4 is relied upon to produce a prediction free from substantial human intervention (which risks reducing the ability of clinicians to assess for potential social coercion or harmful stereotypes having shaped a patient's past choices and thus ostensible preferences), or the P4 is supplemented with explicit clinician input to counteract such problematic factors, which then reduces the potential benefits of having a more automated process. In addition to increased speed or efficiency, these benefits might include, for instance, the potential to reduce various other types of "harmful human bias or error that plagues more traditional forms of surrogate decision-making" (Riddle, 2024, p. 62).

We appreciate this concern, but believe it may rely on an overly pessimistic, and ultimately false dichotomy: *either* clinicians are "removed" from the decision-making process, with the P4 (so to speak) "running the show," *or* clinicians are appropriately involved, and the P4 loses its value. It is, of course, possible that any necessary clinician involvement might come at the cost of *some* of the potential benefits a fully automated system could provide; but it could still be the case that

[1] On some theories of autonomy, an autonomous choice must be authentic or truly one's own, in addition to being, e.g., voluntary (for a discussion of this, see, e.g., Oshana, 2007).

the adjunctive use of a P4 would provide *enough* value-added to be worth pursuing. Finding the right balance between human input and P4 "automation" is a crucial task, and it will no doubt be difficult, but it does not seem to us impossible.

With respect to the specific concern that a patient's past choices might be inauthentic (e.g., because they reflect harmful social attitudes, rather than more considered, fundamental beliefs or deeper personal values or commitments), it is important to think through the appropriate role for clinicians in light of existing alternatives. Suppose that, under current circumstances, a person with a disability loses the capacity to make their own healthcare decisions, but has previously signed an advance directive. Suppose the advance directive states that they would not like treatment under the conditions that now prevail. And suppose the clinician suspects that this stated preference does not reflect the true beliefs or values of the patient, but rather, a negative self-concept that has been inappropriately shaped by prejudicial social attitudes. What then?

If the clinician has a role to play in this scenario—with respect to evaluating the possibility of coercive social pressures undermining the validity of an advance directive—it would presumably be at the point of signing the advance directive, prior to the current situation. As it stands, the clinician has no way of knowing whether the preference stated in the advance directive was or was not the result of social coercion, and the most respectful thing to do, given the available information, would most likely be to follow the directive.

Similarly, we suggest, with a P4: by the time a clinician would realistically be assessing its outputs, it is hard to see how possible prior social pressures that might have influenced the patient's past choices (and thus, the current predictions of the P4 as to their treatment preferences) could plausibly be evaluated by the clinician. So, insofar as Riddle's concern applies to the P4, it seems to apply just as well to the status quo, including the use of advance directives. Thus, if the P4 has other advantages over the status quo, as Riddle seems to accept, this particular objection, even if successful, would not entail it should not be pursued.

5 Family (F) Relationships: Key Concerns

Arguments that family relationships would be negatively affected by the P4 (or in some cases the P3, or any type of non-human patient preference predictor) were of these general types: that the use of a P4 undermines the relational goals of shared decision-making; that it will be difficult for surrogates to advocate for their own views when confronted with a contrary P4 prediction; and that there are potential harms to family relationships implied by the data discovery process for algorithmic training purposes.

5.1 F1. The P4 Would Undermine the Relational Goals of Shared Decision-Making

Herington and Kluger (2024, p. 33) argue that "good decision-making involves both cognitive and emotional process" and that supporting this not only helps the patient for whom the decision is being made, but also helps the surrogate "feel that their decision is comprehensible, in line with their understanding of the patient, and avoid feelings of regret, guilt, alienation, and even trauma." They suggest that the P4 could undermine such relational goods by "appearing as an additional 'stranger at the bedside' whose uncanny (but unreliable) mimicry of the patient may downplay the role of dialogue with healthcare professionals" (p. 32). They argue that this could happen through a number of mechanisms (some of which also arise in concerns F2 and F3 below).

First, if the P4's prediction is discordant with what the surrogate would have imagined to be the case, this may "alienate the surrogate from the person that they (thought) they knew," thus harming their relationship with the patient. Second, if the surrogate was to then contradict this preference or even "begrudgingly accept" it, this is likely to "turbocharge uncertainty about the decision, and associated feelings of regret." Finally, the authors argue, even if the P4's use was made voluntary, "surrogates and physicians will feel immense pressure to defer to their judgements" due to "the hype around these systems" and the "fidelity with which they parrot the patient's voice" (all quotes in this paragraph from Herrington and Kluger, 2024, p. 33).

We acknowledge these concerns and agree that the risks described by Herrington and Kluger are real. We see two main ways of addressing these risks. First, they could be mitigated, if not, perhaps, eliminated, by thoughtful design choices in how the P4 presents its predictions, institutional policies regarding how it used, and educational measures shaping how it is understood and interpreted by a patient's family. And second, they could be avoided altogether by deciding, in relevant cases, simply not to use the P4.

With respect to the first approach, careful counseling should be provided for family members considering the use of a P4 to inform them of its limitations (i.e., as a way of counteracting "hype"), to remind them of its intended use (i.e., to support and not to dictate), and to explicitly warn them of the risks in question: namely, the risks of alienation and feeling pressured (among other risks). This would be analogous to providing counseling for genetic screening—another powerful technology that is subject to hype (see Cheung et al., 2025, for a recent discussion, comparing the historical hype around genetics with current hype around psychedelics); that might also produce unexpected or unreliable, potentially alienating findings, whether about oneself or a loved one; that could also result in a feeling of pressure to make certain types of decisions, and so on, and yet the use of which may nevertheless be justified (i.e., notwithstanding these and other serious risks and limitations), insofar as the dangers, alongside the potential benefits, are appropriately highlighted, contextualized, and prepared for through a process of counseling (Biesecker, 2001).

With respect to the second, perhaps, more basic point, we envisage the use of the P4 being optional. Families should be supported in deciding not only whether or how to incorporate the P4's predictions and potential reasons into the shared decision-making process with clinicians, but also whether to have the P4 generate predictions (etc.) in the first place. If the thought of having to consider this particular source of information is too stressful for family members (and the patient has not previously registered a strong preference that a P4 be used), the family should be able to decline to bring a P4 into the picture.

5.2 *F2. The P4 Will Overwhelm Human Surrogates or Otherwise Make It Difficult for Them to Advocate Their Views*

Rahimzadeh (2024) builds on the concerns raised in the previous section, asking whether shared decision-making is truly achievable between human surrogates (e.g., next of kin) and algorithms, or even between surrogates and clinicians with the aid of an algorithm. More specifically, Rahimzadeh questions the extent to which human surrogates will need to *justify* themselves if choosing to go against the P4's predictions (i.e., picking a treatment option that is not among the options identified by the P4, or picking an option with lower certainty); therefore arguing that a deeper exploration into "how the P4's assessment is weighted against the views of human surrogates" is needed (p. 30).

Rahimzadeh highlights the challenge of striking an appropriate balance of authority and/or influence between the respective "actors" involved in the decision-making process (family members, clinicians, P4), and what the process will look like for family members who disagree with the P4's prediction. Although we initially proposed that the P4 could be used as an aid to human surrogates to complement their decision-making, Rahimzadeh questions whether "the family's decision [can] override the P4 recommendation," and if so, "what burden of proof must the family be prepared to offer" (p. 30) in these cases of differing decisions.

In response, we might suggest that the "burden of proof" should be the same as it is without a P4. Good reasons and evidence must be given, in any case, for why it is a patient would prefer a given approach to treatment, and the clinicians must evaluate, through a process of respectful and inclusive dialogue with the family, the strength and coherence of the reasons given. If the P4, for instance, gives a clear, specific, and verifiable reason in support of its prediction (for example, by referring to an unambiguous pattern of prior medical decisions by the patient that is consistent with one of the treatment options, but not any others), while the human surrogate provides only a vague sense that the patient would choose otherwise, without any supportive reasons, this should count in favor of the P4's prediction, all else being equal. (Of course, all else may not be equal; for example, the patient may have previously expressed a strong preference that, in cases of disagreement between the P4 and their designated surrogate, the surrogate's opinion should always prevail.)

On the other hand, if the P4 makes a prediction without providing a justification, or with an incoherent, irrelevant, or implausible justification, while the human surrogate distinctly remembers multiple conversations with the patient in which the latter expressed a clear preference for the contrary, this should count heavily in favor of the surrogate's view. That being said, if the surrogate is already confident, with good reasons and supportive evidence, about what the patient would choose for themselves in the present circumstances, they would probably not feel a need to involve the P4 at all. As we've suggested, unless the patient previously insisted on the use of a P4 under said conditions (e.g., in an advance directive), the option not to use the P4 should be maintained.

This focus on potential disagreement, however, risks obscuring one of the main reasons for proposing the P4 in the first place. That is, family members or other surrogates often find the responsibility of making a substituted judgment for an incapacitated patient to be extremely stressful and fraught with uncertainty (as mentioned in Earp et al., 2024). Rather than feeling *obligated* to use a P4, or that one must shoulder a weighty burden of proof in case of discrepant predictions, the hope and intention would be that at least some families will find it *helpful* to have an additional source of—personalized—information about the patient's likely preferences to draw on while engaged in process of shared decision-making with the clinical team. For families who would not find it helpful, it should not be forced on them.

5.3 *F3. The P4 May Harm Family Relationships by Causing Unwanted or Uncomfortable Revelations*

Rahimzadeh (2024) also raises the potential for "ancillary harms for both the patient and the family" to be inflicted through the data discovery process used to train the P4, or if previously trained, through the P4's predictions. Rahimzadeh uses the example of the P4 predicting a preference that violates a devout belief system that the patient's family otherwise assumed the patient followed or endorsed, or revealing other "highly stigmatizing preferences." This could risk "fracturing and further destabilizing families" at a time when "trust and support are needed most." Further, Rahimzadeh imagines a scenario in which the patient regains capacity and consequently needs to confront their families "with whom relationships could be left worse off than before the P4's preference reveal." In light of such examples, "the P4's greatest technological strength suddenly becomes its biggest threat" (all quotes in this paragraph are from Rahimzadeh, 2024, p. 31).

This is a valid concern. However, again, it must be compared to the status quo. A family might have an analogous experience in considering an advance directive that was left by their loved one, or in coming to learn the contents of their will. As long as the individual understands that their family may one day see their advance directive, will, or another source of evidence about their future preferences—or preferences for the future—such as the output of a P4, which could include a range of potential treatment options, then, if they decide to proceed with that option, they agree to take on the risk of such preferences being one day revealed and potentially upsetting their family.

A potential disanalogy is that someone who signs an advance directive knows exactly what it states in advance. Someone who agrees to use a P4 does not necessarily know what treatment options it will eventually suggest the patient would likely have chosen (or, perhaps, the reasons or evidence it will provide in support of its suggestion). So, if someone is concerned that a future potential prediction might upset their family, and does not wish to assume such a risk, this person should either carefully restrict the data sources that can be used to prospectively train their model, or not agree to the use of a P4.

Similarly, and as stated previously, if there is no prior decision about the use of a P4 from the patient, and the family is authorized to make a decision about the P4 on their behalf, they must be carefully counseled about the possibility that the P4 will make a prediction, or a set of predictions, that they may find surprising or that may not align with their impression of their loved one. If that is a concern to the family, they should not use the P4.

6 Discussion and Conclusion

At the core of several objections is a concern that the P4 could undermine the essential role of clinical judgment and patient-centered care in medical decision-making. Whether straining doctor–patient relationships (DP1), encouraging over-reliance on algorithmic predictions (DP2), or marginalizing clinicians (DP3), critics worry that the P4 could lead to a diminished role for the human expertise, empathy, and ethical reasoning that are central to good medical care. The potential for the P4 to undermine the relational goals of shared decision-making (F1), diminish family members' ability to be heard (F2), or cause fractures in familial relationships through unearthing disconcerting facts in the data discovery process or use of P4 (F3), also cause critics to worry that the P4 might, rather than help with shared decision-making or substituted judgments, only pose further challenges to families at a time when they are at a heightened vulnerability.

Together, the objections raised about the potential negative impacts of the P4 on the doctor–patient and family relationships are important and warrant careful consideration. However, we are not convinced that they entail a P4 should not be developed, or that the risks of doing so would outweigh the benefits. Rather, by taking seriously the concerns raised, it should be possible to address them, at least to a significant extent, through thoughtful design and implementation of the P4 system, while involving key representatives from all relevant stakeholder groups (e.g., clinicians, patients, families, AI experts, hospital administrators, and so on) along the way. Such a process, conducted alongside rigorous academic debate about the underlying principles involved, can help shape the path forward for appropriate P4 training and the creation of adequate guidelines for its use.

We do have some preliminary suggestions. First, to ensure that the use of a P4 does not marginalize doctors, we emphasize the importance of consulting with doctors throughout the design and implementation process. By actively involving medical professionals in the development of the P4, we can ensure that the system is built to support and enhance clinical decision-making rather than replace it. The perspectives of families and other surrogates should also be sought and included in both the design and implementation of any eventual P4 system.

We suggest that the P4 might contribute positively to the doctor–surrogate relationship. Using P4's prediction as the basis for discussion could help to alleviate some of the emotional burden and uncertainty experienced by surrogates when making difficult decisions on behalf of their loved ones.[2] This may allow doctors and surrogates to engage in focused dialogue about the patient's best interests without the need for either party to have their ego invested as the originator of a particular position.

Critically, institutional policies will have to appropriately address data and privacy concerns, such that careful consideration is given to who has access to the data that is used to train the P4. This will be important in ensuring that personal information that might expose beliefs that are otherwise irrelevant to the successful prediction of medical treatment preferences is protected.

AI use acknowledgment: Any use of generative AI in this manuscript adheres to ethical guidelines for the use and acknowledgment of generative AI in academic research (Porsdam Mann et al., 2024c). Claude 3.5 Sonnett was used on a limited basis during the initial drafting stage, to edit and streamline ideas and material provided by the authors. Each author has made a substantial contribution to the work, which has been thoroughly vetted for accuracy, and assumes responsibility for the integrity of their contributions.

Funding Acknowledgments This research is supported by NUSMed and ODPRT (NUHSRO/2024/035/Startup/04) for the project "Experimental Philosophical Bioethics and Relational Moral Psychology" with BDE as PI. This research is also supported by the National Research Foundation, Singapore under its AI Singapore Programme (AISG Award No: AISG3-GV-2023-012).

References

Annoni, M. (2025). It is not about autonomy: realigning the ethical debate on substitute judgement and AI preference predictors in healthcare. *Journal of Medical Ethics*, *51*(7), 450-455. https://doi.org/10.1136/jme-2024-110343

[2] The emotional stress and mental health concerns faced by family members serving as surrogates has been noted many times (Wendler & Rid, 2011). The P4 may prove to be a helpful decision-making support for family members who might otherwise feel overwhelmed by the challenging task of determining their loved one's treatment preferences, and who might take comfort from knowing that there exists a personalized source of information from which they can receive input.

Ballantyne, A., & Style, R. (2024). Can P4 support family involvement and best interests in surrogate decision-making? *The American Journal of Bioethics, 24*(7), 56–58.

Berger, J. T. (2024). Artificial intelligence, digital self, and the "best interests" problem. *The American Journal of Bioethics, 24*(7), 27–29.

Biesecker, B. B. (2001). Goals of genetic counseling. *Clinical Genetics, 60*(5), 323–330.

Biller-Andorno, N., Ferrario, A., & Biller, A. (2024). The patient preference predictor: A timely boost for personalized medicine. *The American Journal of Bioethics, 24*(7), 35–38.

Bishop, J. P. (2024). The problematic "existence" of digital twins: Human intention and moral decision. *The American Journal of Bioethics, 24*(7), 45–47.

Braun, M. (2021). Represent me: Please! Towards an ethics of digital twins in medicine. *Journal of Medical Ethics, 47*(6), 394–400.

Cheung, K., Earp, B. D., Patch, K., & Yaden,. D. B. (2025). Distinctive but not unique: The risks of psychedelic ethical exceptionalism, *American Journal of Bioethics, 25*(1), 16–28.

Chu, Y., Liu, P., Savulescu, J., & Earp, B. D. (2025). Road rage against the machine: Humans and LLMs share a blame bias against driverless cars. *International Journal of Human-Computer Interaction,* in press. Available online ahead of print at https://www.tandfonline.com/doi/full/10.1080/10447318.2025.2526593

Ciroldi, M., Cariou, A., Adrie, C., Annane, D., Castelain, V., Cohen, Y., Delahaye, A., Joly, L. M., Galliot, R., Garrouste-Orgeas, M., Papazian, L., Michel, F., Barnes, N. K., Schlemmer, B., Pochard, F., & Azoulay, E. (2007). Ability of family members to predict patient's consent to critical care research. *Intensive Care Medicine, 33*(5), 807–813. https://doi.org/10.1007/s00134-007-0582-6

Combs, M. P., Rasinski, K. A., Yoon, J. D., & Curlin, F. A. (2013). Substituted judgment in principle and practice: A national physician survey. *Mayo Clinic Proceedings, 88*(7), 666–673. https://doi.org/10.1016/j.mayocp.2013.05.013

Danaher, J., & Nyholm, S. (2024). The ethics of personalised digital duplicates: A minimally viable permissibility principle. *AI and Ethics*, online ahead of print. https://doi.org/10.1007/s43681-024-00513-7

Ding, N., Qin, Y., Yang, G., Wei, F., Yang, Z., Su, Y., Hu, S., Chen, Y., Chan, C. M., Chen, W., & Yi, J. (2023). Parameter-efficient fine-tuning of large-scale pre-trained language models. *Nature Machine Intelligence, 5*(3), 220–235.

Earp, B. D., Porsdam Mann, S., Allen, J., Salloch, S., Suren, V., Jongsma, K., Braun, M., Wilkinson, D., Sinnott-Armstrong, W., Rid, A., Wendler, D., & Savulescu, J. (2024). A personalized patient preference predictor for substituted judgments in healthcare: Technically feasible and ethically desirable. *American Journal of Bioethics*, online ahead of print. https://doi.org/10.1080/15265161.2023.2296402

Earp, B. D., Porsdam Mann, S., van Veenendaal, T., Allen, J., Salloch, S., Jongsma, K., Braun, M., Sinnott-Armstrong, W., Savulescu, J., Rid, A., & Wendler, D. (2025). Respecting formerly autonomous persons: Clarifying the role of the personalized patient preference predictor (P4) in substituted judgment. *Journal of Medical Ethics, 51*(7), 462-464.

Earp, B. D., Porsdam Mann, S., van Veenendaal, T., Allen, J., Salloch, S., Jongsma, K., Braun, M., Sinnott-Armstrong, W., Rid, A., Wendler, D., & Savulescu, J.(Forthcoming). The enduring promise of personalizing patient preference prediction: a comprehensive response to critics. Preprint: https://www.researchgate.net/publication/393674824

Ferrario, A., Gloeckler, S., & Biller-Andorno, N. (2023). Ethics of the algorithmic prediction of goal of care preferences: From theory to practice. *Journal of Medical Ethics, 49*(3), 165–174. https://doi.org/10.1136/jme-2022-108371

Galitsky, B. (2024). LLM-based personalized recommendations in health. Pre-print. https://www.preprints.org/manuscript/202402.1709/v1

Gao, G., Taymanov, A., Salinas, E., Mineiro, P., & Misra, D. (2024). Aligning LLM agents by learning latent preference from user edits. arXiv preprint arXiv:2404.15269

Gligorov, N., & Randall, P. (2024). Respect for autonomy requires a mental model. *The American Journal of Bioethics, 24*(7), 53–55.

Gutiérrez-Lafrentz, L., Micolich V, C., & Manríquez V. F. (2024). As an AI model, I cannot replace human dialogue processes. However, I can assist you in identifying potential alternatives. *The American Journal of Bioethics, 24*(7), 58–60.

Herington, J., & Kluger, B. (2024). Parrots at the bedside: Making surrogate decisions with stochastic strangers. *The American Journal of Bioethics, 24*(7), 32–34.

Ho, Z. J. M., Krishna, L. K. R., & Yee, C. P. A. (2010). Chinese familial tradition and Western influence: A case study in Singapore on decision making at the end of life. *Journal of Pain and Symptom Management, 40*(6), 932–937.

Iglesias, S., Earp, B. D., Voinea, C., Mann, S. P., Zahiu, A., Jecker, N. S., & Savulescu, J. (2025). Digital doppelgängers and lifespan extension: What matters? *American Journal of Bioethics, 25*(2), 95-110. https://doi.org/10.1080/15265161.2024.2416133

John, S. (2014). Patient preference predictors, apt categorization, and respect for autonomy. *Journal of Medicine and Philosophy, 39*(2), 169–177. https://doi.org/10.1093/jmp/jhu008

de Kerckhove, D. (2021). The personal digital twin, ethical considerations. *Philosophical Transactions of the Royal Society a: Mathematical, Physical and Engineering Sciences, 379*(2207), 20200367. https://doi.org/10.1098/rsta.2020.0367

Kim, J., Kovach, M., Lee, K. M., Shin, E., & Tzavellas, H. (2024). Learning to be Homo Economicus: Can an LLM learn preferences from choice. arXiv preprint arXiv:2401.07345

Kirk, H. R., Vidgen, B., Röttger, P., & Hale, S. A. (2024). The benefits, risks and bounds of personalizing the alignment of large language models to individuals. *Nature Machine Intelligence*, online ahead of print. https://doi.org/10.1038/s42256-024-00820-y

Kuleshov, V., & Ellis, K. (2023). Active preference inference using language models and probabilistic reasoning. arXiv preprint arXiv:2312.12009

Lamanna, C., & Byrne, L. (2018). Should artificial intelligence augment medical decision making? The case for an autonomy algorithm. *AMA Journal of Ethics, 20*(9), E902-910. https://doi.org/10.1001/amajethics.2018.902

Mainz, J. T. (2023). The patient preference predictor and the objection from higher-order preferences. *Journal of Medical Ethics, 49*(3), 221–222.

Meier, L. J. (2024). Predicting patient preferences with Artificial Intelligence: The problem of the data source. *The American Journal of Bioethics, 24*(7), 48–50.

Mertes, H. (2024). The personalized patient preference predictor: A harmful and misleading solution losing sight of the problem it claims to solve. *The American Journal of Bioethics, 24*(7), 41–42.

O'Donnell, J. (2024). AI can now create a replica of your personality. *MIT Technology Review*. https://www.technologyreview.com/2024/11/20/1107100/ai-can-now-create-a-replica-of-your-personality/

O'Neil, C. (2022). Commentary on 'Autonomy-based criticisms of the patient preference predictor.' *Journal of Medical Ethics, 48*(5), 315–316. https://doi.org/10.1136/medethics-2022-108288

Oshana, M. (2007). Autonomy and the question of authenticity. *Social Theory and Practice, 33*(3), 411–429.

Park, J. S., Zou, C. Q., Shaw, A., Hill, B. M., Cai, C., Morris, M. R., & Bernstein, M. S. (2024). Generative agent simulations of 1,000 people. arXiv preprint arXiv:2411.10109

Porsdam Mann, S., Earp, B. D., Møller, N., Vynn, S., & Savulescu, J. (2023). AUTOGEN: A personalized large language model for academic enhancement—Ethics and proof of principle. *American Journal of Bioethics, 23*(10), 28–41. https://doi.org/10.1080/15265161.2023.2233356

Porsdam Mann, S., Earp, B. D., Møller, N., Suren, V., & Savulescu, J. (2024a). AUTOGEN and the ethics of co-creation with personalized LLMs—Reply to the commentaries. *American Journal of Bioethics, 24*(3), W6–W14. https://doi.org/10.1080/15265161.2024.2308175

Porsdam Mann, S., Earp, B. D., Liu, P., & Savulescu, J. (2024b). Reasons in the loop: The role of large language models in medical co-reasoning. *American Journal of Bioethics, 24*(9), 105–107. https://doi.org/10.1080/15265161.2024.2383121

Porsdam Mann, S., Vazirani, A. A., Aboy, M., Earp, B. D., Minssen, T., Cohen, I. G., & Savulescu, J. (2024c). Guidelines for ethical use and acknowledgement of large language models in academic writing. *Nature Machine Intelligence, 6*, 1272–1274. https://doi.org/10.1038/s42256-024-00922-7

Rahimzadeh, V. (2024). Personal but necessarily predictive? Developing a bioethics research agenda for AI-enabled decision-making tools. *The American Journal of Bioethics, 24*(7), 29–31.

Riddle, C. A. (2024). Social coercion, patient preferences, and AI-substituted judgments. *The American Journal of Bioethics, 24*(7), 60–62.

Rid, A., & Wendler, D. (2014). Use of a patient preference predictor to help make medical decisions for incapacitated patients. *Journal of Medicine and Philosophy, 39*(2), 104–129. https://doi.org/10.1093/jmp/jhu001

Rogers, A. H., & Lopez, R. P. (2023). Systematic review revisited, 2010–2020: The effect on surrogates of making treatment decisions for others. *Journal of Palliative Care, 38*(1), 71–77. https://doi.org/10.1177/08258597221098124

Rzepiński, T., Deskur-Śmielecka, E., & Chojnicki, M. (2024). Machine learning algorithms in the personalized modeling of incapacitated patients' decision making—Is it a viable concept? *The American Journal of Bioethics, 24*(7), 51—53.

Schwan, B. (2024). Weighing patient preferences: Lessons for a patient preferences predictor. *The American Journal of Bioethics, 24*(7), 38–40.

Shalowitz, D. I., Garrett-Mayer, E., & Wendler, D. (2006). The accuracy of surrogate decision makers: A systematic review. *Archives of Internal Medicine, 166*(5), 493–497. https://doi.org/10.1001/archinte.166.5.493

Sharadin, N. P. (2018). Patient preference predictors and the problem of naked statistical evidence. *Journal of Medical Ethics, 44*(12), 857–862. https://doi.org/10.1136/medethics-2017-104509

Sharadin, N. (2024). Personalized patient preference predictors are neither technically feasible nor ethically desirable. *The American Journal of Bioethics, 24*(7), 62–65.

Starke, G., & Jox, R. J. (2024). Potentially perilous preference parrots: Why digital twins do not respect patient autonomy. *The American Journal of Bioethics, 24*(7), 43–45.

Toomey, J., Lewis, J., Hannikainen, I. R., & Earp, B. D. (2024a). Advance medical decision-making differs across first-and third-person perspectives. *AJOB Empirical Bioethics*, online ahead of print. https://doi.org/10.1080/23294515.2024.2336900

Toomey, J., Lewis, J., Hannikainen, I. R., & Earp, B. D. (2024b). The true self and decision-making capacity. *American Journal of Bioethics, 24*(8), 86–88. https://doi.org/10.1080/15265161.2024.2361894

Toomey, J., Lewis, J., Hannikainen, I., & Earp, B. D. (2025). A qualitative study of true self judgments, epistemic access, and medical decision-making. *Journal of Medical Ethics*. Available online ahead of print at https://doi.org/10.1136/jme-2025-110957

Torke, A. M., Alexander, G. C., & Lantos, J. (2008). Substituted judgment: The limitations of autonomy in surrogate decision making. *Journal of General Internal Medicine, 23*(9), 1514–1517. https://doi.org/10.1007/s11606-008-0688-8

Voinea, C., Earp, B. D., Porsdam Mann, S., & Savulescu, J. (2024a). Digital doppelgängers, human relationships, and practical identity. *Bioethics,* in press. Available online ahead of print at https://doi.org/10.1111/bioe.70026

Voinea, C., Porsdam Mann, S., Register, C., Savulescu, J., & Earp, B. D. (2024b). Digital duplicates, relational scarcity, and value: Commentary on Danaher and Nyholm (2024). *Philosophy & Technology, 37*(121), 1–8.

Voinea, C., Porsdam Mann, S., & Earp, B. D. (in press). Digital twins or AI SIMs? What to call generative AI systems designed to emulate specific individuals, in healthcare settings and beyond. *Journal of Medical Ethics*, in press. Available online ahead of print at https://www.researchgate.net/publication/394744123

Wendler, D., & Rid, A. (2011). Systematic review: The effect on surrogates of making treatment decisions for others. *Annals of Internal Medicine, 154*(5), 336–346. https://doi.org/10.7326/0003-4819-154-5-201103010-00008

Wendler, D., Wesley, B., Pavlick, M., & Rid, A. (2016). A new method for making treatment decisions for incapacitated patients: What do patients think about the use of a patient preference predictor? *Journal of Medical Ethics, 42*(4), 235–241. https://doi.org/10.1136/medethics-2015-103001

Yang, Y., Qu, T., Yang, J., Ma, B., & Leng, A. (2022). Confucian familism and shared decision making in end-of-life care for patients with advanced cancers. *International Journal of Environmental Research and Public Health, 19*(16), 10071. https://doi.org/10.3390/ijerph191610071

Zohny, H. (2025). The specter of corporate necromancy: Who controls the dead in the age of digital doppelgängers? *American Journal of Bioethics, 25*(2), 113-115. https://doi.org/10.1080/15265161.2024.2441755

Brian D. Earp Ph.D., is Associate Professor of Biomedical Ethics and, by courtesy, Associate Professor of Philosophy and of Psychology at the National University of Singapore (NUS). Brian directs the Oxford-NUS Centre for Neuroethics and Society at NUS and the University of Oxford, where Brian is also a Research Associate of the Uehiro Oxford Institute, and is Associate Director of the Yale-Hastings Program in Ethics and Health Policy at Yale University and The Hastings Center. In 2022, Brian was elected to the UK Young Academy under the auspices of the British Academy and the Royal Society. Brian's work is cross-disciplinary, following training in philosophy, cognitive science, experimental psychology, history and sociology of science and medicine, and ethics.

Tessa van Veenendaal is a visiting student researcher at the Uehiro Oxford Institute of the University of Oxford, where she has contributed to the present work as part of her BMedSci project under the supervision of Associate Professor Brian D. Earp. She is also pursuing her medical training at Monash University in Australia.

Sebastian Porsdam Mann is a postdoctoral researcher at the Center for Advanced Studies in Bioscience Innovation Law (CeBIL) at the University of Copenhagen. His background is in philosophy, neuroscience, applied ethics (BA, Ph.D., University of Cambridge) and international human rights law (DPhil, University of Oxford). He has held postdoctoral positions at Harvard Medical School and the Universities of Oxford and Copenhagen, the latter supported by a personal grant by the Carlsberg Foundation. His work focuses on the practical potential and regulatory conditions necessary for novel technologies to facilitate scientific progress and normative goals, with a focus on blockchain, generative artificial intelligence, and the human right to science.

Julian Savulescu is the Chen Su Lan Centennial Professor in Medical Ethics and Director for the Centre for Biomedical Ethics at the Yong Loo Lin School of Medicine, National University of Singapore (NUS). He is an award-winning ethicist and moral philosopher trained in neuroscience, medicine, and philosophy who has held the Uehiro Chair in Practical Ethics at the University of Oxford since 2002. He is a Visiting Professorial Fellow in Biomedical Ethics at Murdoch Children's Research institute and Distinguished International Visiting Professor in Law at the University of Melbourne where he leads the Biomedical Ethics Research Group. He is also a Fellow of the Australian Academy of Health and Medical Sciences, Fellow of the Australian Academy of the Humanities, and received an honorary doctorate from the University of Bucharest.

The Impact of Artificial Intelligence on Vulnerable Individuals: Three Principles

Chunyan Ding

1 Introduction

The European Union Artificial Intelligence Act (EU AI Act), enacted on 13 March 2024, stands as the world's pioneering legislation to establish a legal framework for human-centric and trustworthy systems of Artificial Intelligence (AI) (European Parliament & the Council of the European Union, 2024). Article 3 of the EU AI Act defines an "AI system" as "a machine-based system that is designed to operate with varying levels of autonomy, and that may exhibit adaptiveness after deployment, and that, for explicit or implicit objectives, infers, from the input it receives, how to generate outputs such as predictions, content, recommendations, or decisions that can influence physical or virtual environments" (European Parliament & the Council of the European Union, 2024). AI technology is now prevalent across various sectors, including e-commerce, banking and finance, healthcare, education, transportation, social media, and entertainment, among others.

While all existing AI systems are classified as narrow AI, focussing on specific cognitive abilities and advancing within that domain (IBM, 2023), they have already shown significant potential in reshaping our lives and society. AI systems serve as a double-edged sword, with the capacity to empower and harm individuals, especially those who are vulnerable. Therefore, this chapter advocates for maximizing the advantages that AI can provide to vulnerable individuals while concurrently minimizing any potential adverse impacts it may pose.

The central arguments presented in this chapter are threefold. Firstly, in line with the principle of non-maleficence ("do no harm"), AI systems should be prohibited from exploiting vulnerabilities or causing significant harm to individuals based on factors such as age, disability, and socio-economic status. Secondly, echoing the principle of beneficence ("do good"), AI systems should enhance the well-being

C. Ding (✉)
School of Law, City University of Hong Kong, Hong Kong, China
e mail: chunding@cityu.edu.hk

Y. Chandra and R. Fan (eds.), *Artificial Intelligence and the Future of Human Relations*,
https://doi.org/10.1007/978-981-96-7185-4_14

of vulnerable individuals by bolstering their capabilities, enhancing their quality of life, and fostering equal opportunities. Lastly, AI systems that directly or potentially impact the interests of vulnerable individuals should adhere to the principles of justice, upholding non-discrimination and equality, respect for freedom of choice, and promotion of human flourishing. Achieving these three principles necessitates collaboration and cooperation between private firms and public authorities.

This chapter is structured into six parts. Following the introduction, an analytical framework is established based on vulnerability theory, outlining the fundamental principles for AI systems that impact vulnerable individuals. The subsequent three parts, in turn, discuss the three principles—the do-no-harm principle, the beneficence principle, and the justice principle—through the use of relevant real-world examples and cases involving the application of relevant AI systems affecting vulnerable individuals. Each part furthermore provides recommendations for their implementation. The final section concludes this chapter.

2 Vulnerability Theory and Three Principles

Martha A. Fineman's "vulnerability theory" argues that "a legal subject that is primarily defined by vulnerability and need, rather than exclusively by rationality and liberty, more fully reflects the human condition" (Fineman, 2019, p. 356). According to Fineman (2019, p. 356), the collective reality of human vulnerability and social dependency mandates a *responsive* state to replace a restrained state of liberal imagination. She advocates for integrated social justice responsibilities, emphasizing the importance of social institutions and synergetic relationships established and safeguarded by law (Fineman, 2019, p. 368).

While Fineman's vulnerability theory may face criticism for its limited prescriptive value and the potential for unwarranted paternalism (Nina, 2014, pp. 11, 14), it acts as a helpful tool to delineate an appropriate role of the government. While vulnerability theory can be applied to various legal subjects, it is particularly well-suited to a specific group of population: vulnerable individuals. The notion of "vulnerable individuals" encompasses those who are susceptible to receiving physical, psychological, societal, or economic harm. They may widely include children, the elderly, persons with disabilities, consumers, those with lower socio-economic status, and others.

Turning to the virtual environment, human vulnerability is pervasive in this unregulated realm because the technology of AI has become an integral component of social institutions and is poised to have a growing impact on social relationships due to its capacity to deliver content, forecasts, suggestions, and decisions using data inputs and machine learning technologies (UK's House of Commons, 2023, p. 8). This makes vulnerability theory an appropriate starting point for discussions about their impact on vulnerable individuals and the suitable role of the government (Krupiy, 2020).

In line with vulnerability theory, a responsive state should cultivate and enhance resilience by adapting and regulating public and private institutions and social relationships in response to the dynamic vulnerabilities of individuals. It also should play a critical role in managing and reconciling the advantages and drawbacks that AI may introduce for vulnerable individuals in the virtual sphere, often through laws. The EU AI Act represents a proactive response by regulatory authorities to the rapid advancement and increasing influence of emerging AI technologies.

Moreover, the author puts forward an analytical framework to explore how a responsive state should oversee AI systems that have a direct or potential impact on the interests of vulnerable populations. This analytical framework consists of three fundamental principles.

1. Do no harm to vulnerable individuals—Drawing upon the ethical principle of non-maleficence, a responsive state should prevent AI from exploiting or causing harm to vulnerable individuals by capitalizing on their vulnerabilities.
2. Benefiting vulnerable individuals—Aligning with the ethical principle of beneficence, a responsive state should establish supportive environments that enable vulnerable individuals to access and benefit from AI technologies, promoting the use of AI systems to improve their welfare.
3. Justice to vulnerable individuals—A responsive state should engage both public and private institutions to effectively monitor AI systems that impact the interests of vulnerable individuals, ensuring that they operate in accordance with principles of equality and fairness, freedom of choice, and promoting human flourishing.

The next three parts probe into each perspective of the above framework by conducting case studies and, meanwhile, discuss how a responsive state should deal with each perspective to maximize the benefits that AI can offer vulnerable individuals and concurrently minimize adverse influences on them.

3 Do No Harm to Vulnerable Individuals

3.1 *Examples of Do-Harm Cases*

Despite being a double-edged sword, technologies are not morally and politically neutral but indeed part of the normative order (Miller, 2021). As such, it is necessary to examine and subject them to ethical and legal principles. When it comes to regulating AI systems, a fundamental principle for any responsible governing body is to ensure that these technologies are not developed or utilized in ways that exploit vulnerabilities and cause harm to vulnerable individuals, whether intentionally or unintentionally. While occurrences of such misuse are relatively uncommon, the use of AI technology in two reported cases of child sexual abuse serves as a stark reminder of the threats posed to minors by these advancements.

In a chilling real-life incident brought to light by the BBC in June 2023, the misuse of generative AI software known as Stable Diffusion shed light on the dark side of technological advancements (Crawford & Smith, 2023). This software, which leverages text-to-image technology to allow users to generate any AI-created image they desire, was exploited by individuals with malicious intent. Paedophiles utilized Stable Diffusion to produce, circulate, and even profit from lifelike child sexual abuse imagery within the shadows of the dark web (Clark, 2023).

The repercussions of this misuse were profound and deeply troubling. Thousands of disturbing images depicting heinous acts, including the rape of babies and toddlers, were being generated on a monthly basis (Clark, 2023). This rampant exploitation of innocent children not only raised significant alarm within the global community but also posed a formidable challenge to law enforcement agencies striving to combat the scourge of child sexual abuse. The very nature of these synthetic images, portraying fictitious minors, has sparked contentious debates regarding the accountability of their creators and the feasibility of prosecution.

Although Stable Diffusion was originally conceived as an AI software to enhance creativity in art and graphics, its unintended transformation into a tool for exploitation and harm highlights the ethical challenges presented by emerging technologies. The responsibility falls not only on the designer of such software but on the entire tech community to remain vigilant against the potential risks of misuse. Proactive measures must be taken to prevent or mitigate the misuse of AI technologies for nefarious purposes, particularly those that inflict harm on the most vulnerable members of society.

The second real-life case took place in Hangzhou, Zhejiang Province, China.[1] A Chinese digital company operated a popular app designed for sharing short videos. This app, without adequately informing and obtaining explicit consent from the children's guardians, allowed minors to create accounts and collected and stored their online account details, location, and contact information, as well as sensitive personal data like facial and voice recognition features. Furthermore, the app utilized backend algorithms to push short videos containing children's personal information directly to users interested in viewing such content.

Because the company failed to implement specific protective measures for children's accounts, users could privately contact these accounts and access their geographical location, facial features, and other personal information simply by clicking "follow". Between January 2018 and May 2019, an adult user of the app, Xu, received short videos containing children's personal information through the app's backend. Xu used the app's private messaging function to contact multiple children, ultimately committing indecent crimes against three of them.

While both Xu and the company faced legal action and liabilities for their roles in the incident, this case underscores the dangers of AI misuse and the exploitation of

[1] The case was discussed in *the People's Procuratorate of Beijing v. The Cyberspace Administration of Beijing (An APIL to protect children's personal information rights and interests)*. The SPP's Guiding Case No. 141.

app users' personal information, which can result in direct harm to vulnerable individuals. The widespread attention garnered by this case led to the introduction of the 2021 Provisions on the Administration of Algorithm-generated Recommendations for Internet Information Services[2] in China, highlighting the need for regulatory measures to address such issues.

3.2 *Implementation of the Do No Harm Principle*

The examples provided above are just the tip of the iceberg of the potential harm that AI technology may inflict on vulnerable individuals. Given the wide range of potential harms, the term "harm" within the context of the "do no harm principle" should refer to "significant harm" in both nature and extent. In order to effectively implement the "do no harm principle", a responsible state must establish a clear definition of "significant harm" and take a firm legal stance against it.

For instance, the EU AI Act offers a valuable model in this regard. It defines "risks" as the combination of the probability of harm occurring and the severity of that harm, and it outlines various levels of risks posed by AI systems (European Parliament & the Council of the European Union, 2024). Moreover, Article 5(1)(b) of the EU AI Act specifically prohibits AI practices that include "the placing on the market, the putting into service or the use of an AI system that exploits any of the vulnerabilities of a natural person or a specific group of persons due to their age, disability or a specific social or economic situation, with the objective, or the effect, of materially distorting the behaviour of that person or a person belonging to that group in a manner that causes or is reasonably likely to cause that person or another person significant harm" (Neuwirth, 2023, p. 6).

Therefore, the physical and psychological harm caused to children by specific AI systems, as demonstrated in the above examples, constitutes "significant harm". It is crucial for regulatory frameworks to tackle and prevent such significant harm to protect the welfare of vulnerable groups. The author suggests that the government must set up a special monitoring body to oversee various AI systems to ensure they do not cause significant harm to vulnerable populations. Moreover, the author suggests that public regulators should establish and implement a reporting mechanism accessible to the public, which enables individuals to report any harmful AI systems, leading to prompt investigations and actions.

In addition, large private firms that develop and operate AI systems should act as self-regulators and bear the responsibility of adhering to the "do no harm principle". It is imperative for these firms to include the principle in their corporate governance framework and establish robust internal monitoring mechanisms aimed at minimizing

[2] The Provisions on the Administration of Algorithm-generated Recommendations for Internet Information Services (互联网信息服务算法推荐管理规定) were jointly promulgated by the Cyberspace Administration of China, the Ministry of Industry and Information Technology, the Ministry of Public Security, and the State Administration for Market Regulation on 31 December 2021 and effective on 3 January 2022.

significant harm that vulnerable individuals may be exposed to as a result of their AI technologies. By proactively addressing potential risks regarding significant harm and prioritizing the well-being of users, these companies can contribute to a safer and more ethical deployment of AI systems in society.

4 Benefiting Vulnerable Individuals

4.1 Examples of Benefiting Cases

AI technology holds immense potential to revolutionize our way of life and bring about positive societal changes. It is undeniable that AI systems have the capacity to deliver benefits and improve the well-being of vulnerable populations, including the elderly and individuals with disabilities. It is a responsible state to invest more or encourage more in the development of AI systems that can effectively support and enhance the lives of these vulnerable individuals even though sometimes it cannot secure financial profits. By investing in the development of AI technologies that cater to the specific needs of vulnerable populations, we can harness the transformative power of AI to create a more inclusive and supportive society for all individuals, regardless of their age or abilities. To illustrate this point, this section presents two examples: one is an AI-powered system for the elderly living independently at home, and the other is an AI-based system tailored to support children with autism spectrum disorder.

Researchers at Stanford University are developing an AI-powered system aimed at improving the quality of life for the elderly population by providing personalized healthcare monitoring to support their independent living. This system utilizes multiple sensors to detect and record daily activities, lifestyle patterns, emotions, and vital signs, enabling remote monitoring, assessment, and assistance for seniors living alone at home. The integration of AI technologies allows for the translation of data from these sensors into accurate situational assessments, enabling rapid responses to the needs of seniors (Stanford Medicine, n.d.).

The system monitors seventeen clinically relevant activities, including eating, sleeping, falls, slowed movements, unstable transfers, front door loitering, day and night reversals, fluid intake, immobility in chairs and beds, urinary frequency, restlessness, fever, alcohol and pill consumption, high salt diets, substance abuse, and food intake (Stanford Medicine, n.d.). This comprehensive monitoring and assistance approach offers elderly individuals a viable alternative to nursing home care, allowing them to maintain their health, stay connected with loved ones, and lead fulfilling lives as they age. Additionally, the system can help reduce the need for costly and unnecessary relocation and institutional care management for seniors, benefiting both the individuals and the state.

The second example showcasing the positive impact of AI technologies on vulnerable groups involves an AI-based system designed to assist children with autism through personalized learning experiences, social skills development, and emotional support. For instance, machine learning algorithms can analyse speech patterns, facial expressions, and social interactions of individual patients to aid in quick and accurate diagnoses with the help of AI technology. Additionally, certain AI-driven programmes can simulate social scenarios to teach and enhance social skills, while others offer real-time, contextual information to help patients manage daily tasks and reduce anxiety (Xing, 2024). They aim to adapt to the unique needs and preferences of each child, offering targeted interventions and fostering their cognitive and social development in a supportive and engaging manner. Table 1 illustrates some AI programmes specifically created to improve the management of autism spectrum disorder.

Table 1 AI programmes for autism spectrum disorder (ASD) management (Meneses do Rêgo & Araújo-Filho, 2024, p. 30)

AI programme	Functionality	Influence on quality of life for patients with ASD
DeepScan AI	Advanced image processing for medical diagnostics	Facilitates quicker and more accurate ASD diagnoses, potentially identifying subtle neurological markers that might be missed during standard evaluations
Predictive health analytics	Utilizes big data to forecast developmental outcomes and risks	Enables personalized early interventions, improving long-term care strategies and preventing comorbidities, enhancing patient adaptability and health outcomes
RoboTherapist 360	Interactive robots provide consistent therapeutic activities	Enhances the development of social and communication skills through consistent, controlled therapy sessions, making therapeutic interactions more accessible
VitaMon AI	Continuous real-time monitoring of vital signs and behavioural patterns	Provides constant vigilance without direct human supervision, alerting caregivers about potential health crises or behavioural issues needing immediate attention
TailoredRx AI	Customizes treatment plans based on individual genetic and medical data	Improves the efficacy of treatments by tailoring strategies specifically to the individual's needs, thereby reducing the trial-and-error process in medication adjustments
CliniHelp decision AI	AI-driven decision support for selecting optimal treatment plans	Assists clinicians in making informed, evidence-based decisions quickly, reducing treatment errors and enhancing patient trust in medical interventions
TalkEase Bot	AI chatbots designed for therapeutic interaction and support	Offers continuous emotional support and anxiety management, teaches coping mechanisms, and can interactively enhance life skills training

4.2 Implementing the Beneficence Principle

While AI technologies hold immense potential for building capacities and improving the well-being of vulnerable individuals, private entities must carefully weigh the costs and benefits when deciding which group of vulnerable populations and types of AI systems to prioritize for investment. It is often uncontroversial when an AI-powered programme that benefits a specific vulnerable group also helps digital firms enhance their reputation, increase profits, and improve corporate performance. However, some vulnerable individuals may not reap the benefits of AI technologies under usual market conditions due to the small size of the vulnerable population in relation to the relatively high research and development expenses.

A similar situation can be observed in the pharmaceutical industry with "orphan drugs", which are designed to treat diseases so rare that companies are hesitant to invest in their development (Chan et al., 2020). This highlights the challenge of balancing the commercial viability of AI applications with the ethical imperative to address the needs of all vulnerable populations, even those that may not offer immediate financial returns.

It is essential for a responsible state to establish a robust policy and legal framework to address the similar issue of "orphan AI programmes" and prevent technology discrimination or reduce the digital divide among minority vulnerable populations. The author proposes the following three suggestions. Firstly, the government should incentivize and support private firms in developing orphan AI programmes that cater to the needs of marginalized individuals by offering tax exemptions or reductions, grants for research, and technological assistance. By doing so, private entities are encouraged to invest in technologies that benefit overlooked communities.

Secondly, in alignment with corporate social responsibility and environmental, social, and governance performance (ESG) performance, the government can promote and recognize private firms that actively contribute to enabling vulnerable individuals to access AI technologies and leverage AI-driven programmes for their well-being. This collaborative effort between the government and private sector can lead to more inclusive and equitable technological advancements.

Lastly, if private firms exhibit a lack of motivation to develop orphan AI programmes, the government should step in and allocate public resources towards the research and development of such initiatives. By taking on this responsibility, the government ensures that essential technologies are developed to address the specific needs of vulnerable populations, thereby implementing the beneficence principle in the technological landscape.

5 Justice to Vulnerable Individuals

5.1 *Examples of Injustice Cases*

The principle of justice encompasses several values that are crucial to the development and application of various AI technologies affecting vulnerable individuals. The author argues that, among these values, the top three are equality and fairness, freedom of informed choice, and human flourishing. In this section, three cases of injustice, each pertaining to one of these top values, are presented to demonstrate the detrimental effects of AI technologies on vulnerable populations.

The first case is concerned with AI-driven decision-making that results in inequality and unfairness to the lower socio-economic class. Amid the COVID-19 pandemic, the UK government faced the challenge of assessing A-level grades due to the cancellation of exams necessitated by social distancing measures. To address this issue, an algorithm was employed to estimate A-level grades. The algorithm utilized input data such as the historical grade distribution of schools over the past three years, each student's ranking within their school for a specific subject based on teacher evaluations, and the student's previous exam results per subject. Subsequently, the algorithm analysed these data to determine a student's grade, which closely mirrored the grades obtained by students in the same ranking over the previous three years. Weightage was assigned to the ranking estimation based on the student population in a particular subject at a given school (Kolkman, 2020).

This AI-generated grading system, which influenced university admission outcomes, attracted significant criticism for its inaccuracy, inequality, and unfairness. It was likely that students from lower socio-economic backgrounds were disproportionately impacted, as the reliance on historical data perpetuated algorithmic biases and mirrored inequalities present in the United Kingdom's education system. Following public outcry and student protests, the U.K. government ultimately retracted its attempt to utilize AI-based decision-making for university admissions (Shead, 2020).

The second case presents the first Chinese judicial decision involving "big data-enabled price discrimination". In a landmark legal case in 2021, the Intermediate Court of Shaoxing City in Zhejiang Province, China, addressed this issue for the first time. The case involved a plaintiff who was a "Diamond member" of Trip.com, a popular travel service provider. The plaintiff alleged that Trip.com had utilized its extensive big data on customer behaviour to offer discriminatory pricing to its loyal customers. The court ruled in favour of the plaintiff, highlighting that Trip.com had indeed provided different prices based on customer data, resulting in the plaintiff being charged more than double the standard room rate for a hotel booking (Shi, 2021). This case shed light on the potential misuse of big data and algorithms by businesses to engage in price discrimination, disadvantaging certain customers based on their loyalty status or other factors.

While the practice of "price discrimination" is not new, the widespread use of big data and algorithms has significantly amplified the capabilities of businesses to gain deep insights into the characteristics and preferences of their customers. This enhanced knowledge empowers companies to adjust prices based on individual customer profiles, potentially exploiting the loyalty of certain customers. Consequently, customers may find themselves in a vulnerable position when interacting with online platforms and digital firms that wield significant control over vast amounts of customer data and possess greater bargaining power.

In essence, the asymmetry of information between firms and customers is exacerbated in the digital age, where firms can leverage big data to manipulate prices and tailor offerings to specific customer segments. This can lead to situations where customers are unknowingly subjected to misinformation or discriminatory pricing practices. As a result, vulnerable customers may be unaware that they are being misled, and their purchasing decisions may not be entirely voluntary due to a lack of accurate information and biased offers. Their freedom of informed choice is in jeopardy.

The third case involves minors using social media platforms, where the issue of algorithm-driven content recommendations has raised concerns about the potential adverse impact it can cause on vulnerable users. Social media platforms like Facebook, Instagram, and TikTok often utilize algorithms to suggest advertisements, videos, and other content tailored to users' interests and preferences. For instance, TikTok's "For You" function employs a content recommendation system that analyses user data to provide a personalized and scrolling feed of algorithmically selected content based on inferred user interests (Amnesty International, 2023).

In the case of minor users viewing mental health-related videos on TikTok, the recommendation system may subsequently suggest content featuring mental health struggles. This can pose a significant risk to minor users, as the recommended content may amplify depressive and suicidal themes, potentially exacerbating their existing mental health challenges. Consequently, parents in the United States and Europe have taken legal action against TikTok, alleging that the platform exposed their children to psychologically harmful content, which tragically led to some minors taking their own lives (Armstrong, 2024; Raymond, 2024).

Furthermore, it is alleged by some nongovernmental organizations and public regulatory bodies that social media platforms utilize their AI technologies to lure and addict underage users in order to maximize their profits (Costello, 2023). This not only robs young people of beneficial face-to-face social interactions but also has the potential to lead to mental health repercussions.

5.2 *Implementing the Justice Principle*

The three above cases that conflict with the principle of justice have highlighted three core values: equality and fairness, freedom of informed choice, and human flourishing. Specifically, the first case involving the proposed AI-generated A-level grading system exposed a real and imminent risk of inequality and unfairness to the low socio-economic population. In a wider context, algorithms may generate biased decisions towards ethical minorities, gender minorities, or people with low socio-economic backgrounds due to historically biased or unrepresentative datasets, inadequate models, and flawed algorithm designs (Giffen et al., 2022). Consequently, biased and unfair AI systems may make decisions that are skewed towards a particular group of people, thereby perpetuating prejudice or favouritism.[3] Therefore, it is important for algorithm model designers and users to remain vigilant against the risks of bias and unfair AI-made decisions.

The second case, which revolves around big data-enabled price discrimination, essentially infringes upon consumers' rights to information and freedom of choice. Consumers are unaware of such violations most of the time. The Chinese court's decision set a precedent for addressing issues of fairness in pricing practices, particularly in the context of big data analytics. The monopoly and control of extensive consumer data by businesses exacerbate the issue of information asymmetry between firms and consumers. Leveraging AI technologies, companies can unilaterally determine the content and extent of information each consumer receives and manipulate their consumption preferences by selectively providing or pushing selected information. As a result, consumers are deprived of access to comprehensive and unbiased information necessary for making informed choices freely. Such utilization of AI technologies is unjust and detrimental to vulnerable consumers.

This case serves as a reminder of the need for ethical considerations and regulatory oversight in the use of big data to prevent unfair treatment of consumers and promote a more equitable marketplace. In this context, it is crucial for regulatory bodies and policymakers to address the ethical implications of data-driven pricing strategies and ensure that consumer rights are protected in the digital marketplace. Moreover, transparency, accountability, and fair competition are essential to safeguarding consumers from potential exploitation and ensuring that they can make informed choices free from AI-assisted manipulation or discrimination.

The third case sheds light on the significance of justice, emphasizing that digital companies must find a balance between their pursuit of profits and their obligation to promote the human flourishing of vulnerable individuals, especially minor users who may not fully comprehend the risks of addiction, control their behaviours, or withstand negative external influences. Lawsuits brought by parents against major digital corporations show the imminent need to shield minor users from harmful online content and the potential adverse consequences of algorithm-driven recommendations on their mental well-being. It highlights the importance of social media platforms prioritizing user safety and well-being, especially when catering to younger users. Regulatory measures and responsible platform governance are essential to mitigate the risks associated with algorithmic content recommendations and ensure that online environments remain safe and supportive for those who are more susceptible to harm.

In November 2024, the Australian parliament enacted a law prohibiting children under the age of sixteen from using social media, sparking intense debates both

[3] Other AI applications that discriminate against people of colour include ride hailing apps, medical care recommendation apps, and job-hunting apps (UK's House of Commons, 2023).

within and outside Australia (Ritchie, 2024). Given that social media platforms offer both advantages and risks to teenagers, coupled with the practical difficulties of implementing and enforcing such a law, the author argues that a blanket ban on minors accessing social media is unreasonable. This approach risks sacrificing the benefits of new technologies in a bid to prevent potential harm. A more pragmatic solution would be to mandate social media platforms to create a safe and healthy online environment for all users, especially minors.

The author argues that to enable effective governance of AI systems that could affect vulnerable individuals, a key prerequisite is to enhance and promote transparency in the input datasets used for machine learning and algorithm models. Transparency and explainability help various stakeholders gain a deeper understanding of the advantages and disadvantages of AI systems and empower the affected vulnerable users to safeguard their fundamental rights and interests and seek redress (Shaw, 2021, p. 57). Moreover, transparent algorithmic processes enable public regulators to oversee and guarantee that various AI technology applications are operated in a just and fair manner, upholding the values of equality and fairness, freedom of informed choice, and human flourishing. Additionally, a lack of transparency in AI systems undermines the accountability of the responsible parties for any harm or injuries caused.

Regulating AI systems that impact vulnerable individuals presents a formidable challenge. Establishing a cohesive and coordinated regulatory framework requires collaboration among diverse public authorities to align on regulatory objectives and standards, thereby mitigating potential policy conflicts (DRCF, 2022). As the author contends, it is imperative for public authorities to prioritize the principle of justice and reinforce the fundamental values of equality and fairness, freedom of informed choice, and the promotion of human flourishing among vulnerable groups. The government can contemplate implementing an AI audit system as a means to uphold these principles (Galdon Clavell, 2023).

6 Conclusion

The rapid advancement and utilization of AI technologies are indeed thrilling as they showcase immense potential to revolutionize every facet of our lives and society. While it is natural for the state to exhibit tolerance towards the nascent stages of new technologies, the application of AI technologies across various sectors has revealed both positive and negative impacts on vulnerable individuals. Proactive measures need to be taken to effectively guide and regulate these technologies. This chapter, grounded in vulnerability theory, argues that a responsive state should assume a pivotal role in managing and balancing the benefits and challenges that AI may bring for vulnerable individuals in the digital realm. It presents an analytical framework consisting of three ethical principles to guide the governance of AI systems affecting vulnerable populations: the do-no-harm principle, the beneficence principle, and the justice principle. Using this analytical framework, it examines typical cases regarding each principle and recommends suggestions.

In relation to the do-no-harm principle, the author proposes that the government should clearly define "significant harm" and prohibit any AI systems that exploit vulnerabilities and cause significant harm to vulnerable individuals. The author recommends the establishment of a reporting mechanism by public regulators and the adoption of an internal monitoring mechanism by private firms. Regarding the beneficence principle, the author suggests that the government should implement measures to incentivize private firms to develop "orphan AI programmes", acknowledge their contributions to advancing ESG performance, and, when necessary, use public resources to address the absence of such programmes, thereby fostering an inclusive and equitable technological environment. Lastly, concerning the justice principle, the author advocates for designers and operators of AI programmes to prioritize three fundamental values: equality and fairness, freedom of informed choice, and human flourishing, over profit maximization to achieve AI justice for vulnerable individuals. The author emphasizes the importance of transparency in the input datasets used for machine learning and algorithm models and calls for the development of a coordinated and collaborative regulatory framework by various public agencies.

References

Amnesty International. (2023). *Driven Into the Darkness: How Tiktok's 'For You' Feed Encourages Self-Harm And Suicidal Ideation.* https://cdn.amnesty.at/media/11311/amnesty-report_driven_into_the_darkness_how-tiktoks-for-you-feed-encourages-self-harm-and-suicidal-ideation_november-2023.pdf

Armstrong, K. (2024). French Families Sue Tiktok Over Harmful Content. *BBC*. https://www.bbc.com/news/articles/c6298z99xv8o.

Chan, A. Y. L., Chan, V. K., Olsson, S., Fan, M., Jit, M., Gong, M., Zhang, S., Ge, M., Pathadka, S., Chung, C. C., & Chung, B. H. (2020). Access and Unmet Needs of Orphan Drugs in 194 Countries and 6 Areas: A Global Policy Review with Content Analysis. *Value in Health, 23*(12), 1580–1591.

Clark, E. (2023). Pedophiles Using AI To Generate Child Sexual Abuse Imagery. *Forbes*. https://www.forbes.com/sites/elijahclark/2023/10/31/pedophiles-using-ai-to-generate-child-sexual-abuse-imagery/?sh=3d763fbd1656

Costello, N. (2023). Algorithms, Addiction, and Adolescent Mental Health: An Interdisciplinary Study to Inform State-Level Policy Action to Protect Youth from the Dangers of Social Media. *American Journal of Law & Medicine, 49*(2), 135–172.

Crawford, A., & Smith, T. (2023). Illegal Trade in AI Child Sex Abuse Images Exposed. *BBC*. https://www.bbc.com/news/uk-65932372

DRCF. (2022). *The Benefits and Harms of Algorithms: A Shared Perspective from the Four Digital Regulators.* UK Government. https://www.gov.uk/government/publications/findings-from-the-drcf-algorithmic-processing-workstream-spring-2022/the-benefits-and-harms-of-algorithms-a-shared-perspective-from-the-four-digital-regulators

European Parliament & the Council of the European Union. (2024). *The EU Artificial Act.* Publication Office of the European Union: Luxembourg. https://eur-lex.europa.eu/legal-content/EN/TXT/?uri=CELEX:32024R1689

Fineman, M. A. (2019). Vulnerability and Social Justice. *Valparaiso University Law Review, 53*, 341–369.

Galdon Clavell, G. (2023) *Checklist for AI Auditing*. https://www.edpb.europa.eu/system/files/2024-06/ai-auditing_checklist-for-ai-auditing-scores_edpb-spe-programme_en.pdf

IBM. (2023). *Understanding the Different Types of Artificial Intelligence*. https://www.ibm.com/think/topics/artificial-intelligence-types

Kolkman, D. (2020). F**k the Algorithm"?: What the World Can Learn from the UK's A-Level Grading Fiasco. *LSE Blog*. https://blogs.lse.ac.uk/impactofsocialsciences/2020/08/26/fk-the-algorithm-what-the-world-can-learn-from-the-uks-a-level-grading-fiasco/

Krupiy, T. (2020). A Vulnerability Analysis: Theorising the Impact of Artificial Intelligence Decision-Making Processes on Individuals, Society and Human Diversity from a Social Justice Perspective. *Computer Law & Security Review, 38*, 1–25.

Levison, J. (2023). Meta "Intentionally Addicts" Children To Social Media, Lawsuit Claims. *Skynews*. https://news.sky.com/story/meta-intentionally-addicts-children-to-social-media-lawsuit-claims-12992272

Meneses do Rêgo, A. C., & Araújo-Filho, I. (2024). Leveraging Artificial Intelligence to enhance the Quality of Life for patients with Autism Spectrum Disorder: A Comprehensive Review. *European Journal of Clinical Medicine, 5(5)*, 28–38.

Miller, B. (2021). Is Technology Value-Natural? *Science, Technology, & Human Values, 46*(1), 53–80.

Neuwirth, R. J. (2023). Prohibited Artificial Intelligence Practices in the Proposed EU Artificial Intelligence Act (AIA). *Computer Law & Security Review, 48*, 1–14.

Nina, A. K. (2014). Vulnerability Theory and the Role of Government. *Yale Journal of Law and Feminism, 26*(1), 1–28.

Ritchie, H. (2024). Australia Approves Social Media Ban on Under-16s. *BBC*. https://www.bbc.com/news/articles/c89vjj0lxx9o

Raymond, N. (2024). Tiktok Must Face Lawsuit over 10-Year-Old Girl's Death, US Court Rules. *Reuters*. https://www.reuters.com/legal/tiktok-must-face-lawsuit-over-10-year-old-girls-death-us-court-rules-2024-08-28/

Shaw, P. (2021). "Context Matters": the Law, Ethics and AI. In Hervey, M., & Lavy, M. (Eds.), *The Law of Artificial Intelligence* (1st ed., pp. 31–66). Sweet & Maxwell.

Shead, S. (2020). How a Computer Algorithm Caused a Grading Crisis in British Schools. *CNBC*. https://www.cnbc.com/2020/08/21/computer-algorithm-caused-a-grading-crisis-in-british-schools.html

Shi, H. J. (2021). Say No to Big Data Price Discrimination through Judicial Rulings. *Chinacourt.org*. https://www.chinacourt.org/article/detail/2021/07/id/6157482.shtml

Stanford Medicine. (n.d.). *Stanford Partnership in AI-assisted Care: Senior Care*. https://med.stanford.edu/pacresearch/research/senior-care.html.

UK's House of Commons. (2023). *Connected Tech: AI and Creative Technology*. https://committees.parliament.uk/publications/41145/documents/201678/default/

Van Giffen, B., Herhausen, D., & Fahse, T. (2022). Overcoming the Pitfalls and Perils of Algorithms: A Classification of Machine Learning Biases and Mitigation Methods. *Journal of Business Research, 144*, 93–106.

Xing, N. (2024). Artificial Intelligence to Support Children with Autism. *Journal of AI-Powered Medical Innovations, 2*(1). https://doi.org/10.60087/vol2iisue1.p43

The Credit View and AI Testimony: A Cross-Cultural Epistemological Study of Human and AI Testimony

Masaharu Mizumoto, Shane Ryan, and Chienkuo Mi

1 Introduction

In this work, we revisit Jennifer Lackey's "Why we don't deserve credit for everything we know." We also look forward to an emerging debate about what we call AI epistemology. Lackey's work introduced the Chicago Visitor Case and challenged virtue reliabilists' credit view of knowledge. Since then, the Chicago Visit Case has become the paradigm case of, or at least one of the most prominent cases of, testimonial knowledge in the social epistemology literature. Furthermore, the case and Lackey's discussion have supported strong criticism of virtue reliabilist accounts of knowledge and encouraged significant adjustments to those accounts. We revisit her case by empirically investigating whether non-experts agree that it is a case of testimonial knowledge and whether there are any significant cultural differences in the answers of participants.

In addition to contributing to our understanding of the testimonial knowledge discussion, our studies here also aim to contribute to our emerging and much-needed knowledge of AI epistemology. Using a variant of the Chicago Visitor Case, here we try to provide initial data for the development of accounts of AI epistemology. What is more, while there is of course an interest among epistemologists with regard to how to incorporate AI-based beliefs into existing accounts of justification and knowledge, there is a broader public interest in epistemologists investigating the influence that

M. Mizumoto
School of Knowledge Science, Japan Advanced Institute of Science and Technology, Nomi, Japan
e-mail: mizumoto@jaist.ac.jp

S. Ryan (✉)
Department of Public and International Affairs, City University of Hong Kong, Hong Kong, China
e-mail: shaneryan27@hotmail.com

C. Mi
School of Liberal Arts and Social Sciences, Soochow University, Taipei, Taiwan
e-mail: cmi@scu.edu.tw

Y. Chandra and R. Fan (eds.), *Artificial Intelligence and the Future of Human Relations*,
https://doi.org/10.1007/978-981-96-7185-4_15

AI-based beliefs can be expected to have on our epistemic environments, including our universities, workplaces, city centers, and so on. Our research provides initial data for such an investigation.

2 Testimonial Knowledge

Virtue epistemologists have proposed an elegant solution for distinguishing cases of knowledge from Gettier cases. They make the case that knowledge is a creditworthy true belief, whereas Gettier cases are cases in which the agent has a justified true belief but doesn't deserve credit for their belief being true (Greco, 2007; Riggs, 2009; Sosa, 2007).[1] What's more, according to such theories, S knows p requires that their belief is true because of ability or competence, where the ability or competence of the sort relevant to knowledge is a reliable belief-forming process. In such a case, it's easy to see why S deserves credit for the truth of their belief; after all, their belief is true because of having exercised their competence or ability. If an agent has a true belief and has exercised ability or competence, but their belief is just coincidentally or luckily true, then it's a Gettier case, and it's also natural to see why the agent doesn't deserve credit for their belief being true.[2]

Lackey (2007) calls this view the Deserving Credit View of Knowledge (hereafter *the Credit View* for short). She believes, however, that a simple, everyday case of testimonial knowledge seems to be incompatible with the Credit View. Lackey presents her famous Chicago Visitor Case as follows:

> **CHICAGO VISITOR**: Having just arrived at the train station in Chicago, Morris wishes to obtain directions to the Sears Tower. He looks around, approaches the first adult passerby that he sees, and asks how to get to his desired destination. The passerby, who happens to be a lifelong resident of Chicago and knows the city extraordinarily well, provides Morris with impeccable directions to the Sears Tower by telling him that it is located two blocks east of the train station. Morris unhesitatingly forms the corresponding true belief. (Lackey, 2007, 352)

According to Lackey, the true belief Morris formed is knowledge. He acquired knowledge from the passer-by. However, since the salient reason he formed the true belief is the passer-by's testimony, it seems that if anyone deserves credit for Morris's true belief of the whereabouts of the Sears Tower, then it is the passer-by, not Morris.[3] After all, Morris' true belief seems to be best explained by the reliable cognitive

[1] For more recent discussions of virtue epistemology and knowledge, see Pritchard (2022) and Greco (2022).

[2] The credit view certainly seems to predict the correct results in standard Gettier cases, as well as in truetemp and brain lesion cases, given that a feature of such cases is that the agents don't deserve credit for the truth of their beliefs. Of course, those two cases have been proposed as counterexamples to simple process reliabilism (Goldman, 2011) and motivate virtue reliabilism.

[3] Note that the Sears Tower is now officially called the Willis tower, though it's still popularly known as the Sears Tower.

character of the *testifier* (the passer-by), rather than by an ability or competence that Morris exercises.

There is some discussion both in Lackey's original paper and elsewhere in the literature as to whether Morris deserves any credit in this case. He seems to deserve some credit, after all, for he did ask someone in order to find out how to get to the Sears Tower, and presumably, he had some counterfactual sensitivity to what the passer-by tells him (Lackey, 2007; Pritchard, 2012). However, just deserving some credit falls short of what the virtue theorists who endorse the Credit View are committed to saying. For them, a knower's belief is true *because of* their own ability or competence, and this explains why a knower deserves credit for the truth of their belief. A natural reading of their commitment then is that the knower is at least primarily creditable for the truth of their belief, but Morris in the Chicago visitor case does not seem to deserve enough credit for it to be a case of knowledge. Yet, the case is presented as an intuitive case of knowledge and it has been broadly accepted as such in the literature, while accepting that the Chicago Visitor case is a case of knowledge suggests that there are cases of knowledge in which the truth of a belief is not down to its possessor's ability or competence. Thus, knowledge is not always something for which the knower deserves credit, according to *pace* the Credit View.

John Greco's (2010, 81) early response to the Chicago Visitor case was to question the assumption that it is a case of knowledge after all. For Greco, whether it is a case of knowledge depends on whether Morris is a reliable testimonial recipient, for, according to him, knowledge is a true belief *because* the agent exercised the relevant cognitive ability—in other words, he holds fast to his original view (cf. Greco, 2009). Morris must therefore be a reliable testimonial recipient for him to be creditable for the truth of his belief. Greco has later provided other responses to how to deal with cases of testimonial knowledge. Very recently, Greco (2022, 421) has proposed that competent joint agency between the testifier and the recipient best explains the recipient's true belief.

Another response to the case is from Pritchard (2012), who basically accepts the intuition that Morris gains knowledge in the case. While Pritchard emphasizes the counterfactual sensitivity contribution from the recipient in cases of testimonial knowledge, he ultimately takes the case to pose a significant problem for virtue epistemologists, such as Greco. Indeed, he takes the Chicago Visitor case and the Fake Barn case together to pose an insurmountable problem for virtue epistemologists, such as Greco and Sosa, who provide a "because of ability/competence" analysis of knowledge and so lack an explicit anti-luck condition. Indeed, he proposes an account of knowledge that he believes correctly predicts knowledge in cases such as Chicago Visitor and highlights this as an advantage his account has over nearby alternatives (Pritchard, 2012).

We take these responses to be evidence for the significance of Lackey's paper, cited close to 400 times at the time of writing, for the development of the virtue epistemology literature. The Chicago Visitor case itself has been particularly influential. Most obviously, the case appears to demonstrate the falsity of the Credit View—after all, the critic says, it's intuitive that Morris knows and that he's not the one creditable for the truth of his belief. More subtly, however, is that the case itself has become a sort of paradigm case of testimonial knowledge. It's perhaps the most famous example of testimonial knowledge used in the epistemology literature. As with Pritchard (2012), it is taken as the sort of case that an account of the nature of knowledge should predict knowledge.

3 AI Epistemology and Human Relations

It has been more than 15 years since Lackey formulated the Chicago Visitor case, and now the example feels very dated. After all, now the typical way of figuring out how to get somewhere is not by asking a random stranger but by consulting one's phone. Indeed, this example of change in our testimonial practices is representative of a broader shift. We use map applications and websites to find locations and opening hours, to decide where might be good to eat or visit, what the law is, how to be fitter, what illness we have, and so on. Even what we learn about our friends and family is now often from what comes up when we're using social media sites. While it remains true that much of our knowledge comes from testimony, how we receive testimony is changing and an increasing amount of testimony is either filtered by algorithms or AIs or is not from a human source at all. More recently, AI chatbots, such as ChatGPT, based on large learning models, are playing an increasing role in our testimonial practices, and contributing to that change in the way we receive testimony. It is expected that such chatbots will continue to become more sophisticated and powerful, and plausibly AI systems will increasingly become more reliable than a random human testifier in a variety of respects (Von Eschenbach, 2021).

These changes in our testimonial practices, with less human-to-human testimony, especially given the rise of AI, raise a number of important epistemological questions. The central question here is how AI testimony fits in with current thinking about the epistemology of testimony. That is, if someone learns the way to the Sears Tower from an AI, is it any different from if they were to learn this from a human testifier? Would the recipient be any more or less deserving of credit than Morris?

Similarly, though we ultimately don't attempt to answer these questions in this paper, how will developments in AI contribute to the different views on whether artifacts (machines, AIs, robots, etc.) are capable of assertion and testimony? There are those who claim, let's call them *Conservatives*, that artifacts cannot assert or testify *because* they lack propositional attitudes, in particular, belief and intention (Goldberg, 2013; Lackey, 2008; Smart, 2017; Williams, 1973). On the other hand, *Progressives* claim that artifacts *can* assert and testify (Green, 2010; Miller & Record, 2017; Wellner, 2014; Wheeler, 2020).[4]

A closely related, though orthogonal, question is whether AI systems can be agents, or they can be no more than tools. For example, though Conservatives usually

[4] It should be noted that some theorists may be Hybrids. Such theorists hold that artifacts can assert as "proxy speakers" (Nickel, 2013), or quasi-assert and quasi-testify (Freiman & Miller, 2020), or are capable of "bent testimony" (Narayanan & Cremer, 2022; Rini 2017).

take an AI as just an instrument, Progressives may not necessarily take it as an agent either, regarding it as an instrument while still allowing it to be a genuine testifier (cf. Sosa, 2011). Indeed, Progressives may allow a wide variety of tools, such as maps, signs, information boards, and so on, to be counted as testifiers, thus admitting assertion and testimony without intention or belief (Wheeler, 2020 takes it to be a virtue of reliabilism).[5] On the other hand, even Conservatives can admit machine testimony *if* machines do (or come to) have propositional attitudes and therefore, for example, are (or become) subject to *normative epistemic assessment* (Goldberg, 2013). In this sense, such Conservatives agree to accept the condition: If an AI system can be taken as an agent, rather than just a tool, it can legitimately perform speech acts.

What counts as an agent, let alone its definition, is a controversial issue that we cannot go into detail about here.[6] For our purpose here, let us just assume that an agent is something more than just a tool, and accept only a sufficient condition: If an AI system is taken to legitimately perform speech acts, then it is taken as an agent. Let us call the view that AI systems are agents and the one that AI systems are just tools, the Agent Hypothesis and the Tool Hypothesis, respectively. We do not think that there is any one single correct view here. Rather, we suggest acknowledging that *our concepts can change* (*be extended*), as our form of life changes along with technological developments and their diffusions (since concepts are embedded in our form of life, playing a functional role there). Similarly, technological changes and how humans come to relate to technologies may also influence pre-existing relationships, such as human-to-human relations. As we know, the application of a concept that is merely analogical or metaphorical at first may *become literal* later. If philosophers' conceptual analyses are based on their concepts that are supposed to be shared with ordinary people, then which hypothesis is correct should also change as people's life and their attitudes toward AI systems change (one may take this as a process of *personification* of AI systems, though the systems eligible for such personification are surely limited) but those changes may also have repercussions for how humans see each other. Essentially, while it's possible that our perceptions of AI will over time conform to our perceptions of humans, it's also possible that something like the reverse may take place and we come to think of humans as like AIs though less sophisticated.

More generally, as we live more and more surrounded by ICT (information and communications technology), our epistemic environment itself is undergoing significant change.[7] Such new technologies may not only contribute to the traditional epistemological theories by providing new data but may also affect and change the theories themselves, reflecting conceptual changes, which may culturally diverge. In particular, people's concepts of speech acts, such as assertion and testimony, as well as people's attitudes toward AI systems, are likely to differ culturally, depending on

[5] Instead, based on a passage of Williams (1973), Nickel (2013) requires (for "proxy speech") the capacity to be *insincere*, while Wheeler (2020) requires (for testimony) the capacity to *deceive*.

[6] See Sect. 2 of Wheeler (2020) for relevant discussions by reliabilists.

[7] For relevant works, see MacKenzie et al. (2021) and Ryan et al. (2024).

the degree of acceptance and prevalence of AI (or the cultural level of personification of AI systems).[8] Besides, the advent of close human–AI collaboration should also change epistemology. Much of our future learning in our professional and personal lives may well be from collaboration with AI.[9]

As a result, to capture such diachronic changes and synchronic cultural variances, we think that epistemology itself should become much more empirical. Doing so will allow epistemology to account for the capabilities and limits of specific AIs, both actual and perceived, cultural differences in perceptions of AIs that have a bearing on epistemic strategies and epistemic coverage,[10] the influence that those perceptions have on perceptions of humans as knowers, and the influence of AI on the epistemic environments in which we operate and how those environments might be improved.[11]

In this paper, we're primarily interested in addressing the first question, a question about AI testimony specifically as it relates to the credit view. Nevertheless, we comment later on other questions and AI epistemology in general, as well as the influence of AI development on human-to-human relations, for all of which our new study also generates data. By investigating this particular instance of the epistemic role of AI, we will obtain valuable data not only for further developing the theory of knowledge but also for AI epistemology.

4 An Earlier Empirical Study on Testimonial Knowledge

One of the virtues of the Credit View was supposed to be that credit evaluation tracks knowledge attributions. After all, how much credit Morris deserves for the truth of his belief is determined by the part his cognitive abilities play in the formation of his belief. For Greco, even though Morris could not gain the true belief without the testifier, it does not follow from this that Morris does not deserve credit for the truth of his belief.[12] If this case and many like cases are indeed judged to be cases of

[8] Indeed, once we take into consideration cultural variance, the *one-directional* conception of the Conservative/Progressive distinction should be reconsidered. For example, some cultures may lack any linguistic resources to distinguish human speech and AI "speech" in the first place, or its cosmology may allow all the inanimate objects to speak and testify. Such cultures may be very Conservative and Progressive at the same time for us. If so, what we need to know is only the *current state* of our concept of assertion and testimony in respective cultures. Accordingly, here we are interested in which one of these hypotheses is true, as a matter of *concepts* people *currently* have, in their respective cultures.

[9] Indeed, Greco's (2022, 421) most recent account of testimonial knowledge referred to earlier seems particularly relevant here. Knowledge gained through collaboration with AI might be best thought of as the result of true belief because of competent joint agency.

[10] For more on epistemic coverage, see Goldman (2011) and Ryan (2021).

[11] For relevant literature on the workings of epistemic environments generally, see Ryan (2018a, 2022; Ryan et al., 2024). For literature specifically concerned with the epistemic significance of interaction between AIs and environments, see Simon (1996) and Ganascia (2010).

[12] See Harris (2023) for a more recent defence of virtue reliabilism according to which, a testimonial recipient's abilities may be at work to detect "sincerity, certainty, and seriousness."

knowledge, then establishing the relationship between knowledge attributions and credit evaluations is worthwhile.

Ryan et al. (2020), conducted a cross-cultural study concerning this topic, investigating ordinary people's intuitions about attributions of knowledge, credit, and fault, as well as the correlations between them. Their study, using Taiwanese (undergraduate students) and Americans as participants, examined participants' attributions of knowledge and credit evaluations in the Chicago Visitor case.[13] The study also examined participants' fault evaluations when the testimonial belief of the protagonist was false, a case they called the Bad Chicago Visitor case. The results showed that participants of the study mostly judged that the protagonist in the Chicago Visitor case does have the knowledge, and deserves some credit for his true belief (also, some cultural differences in credit and fault evaluations were found).

This, to some extent, supports the Credit View. However, neither the correlation between knowledge attribution and credit evaluation nor the one between knowledge attribution and fault evaluation, was detected there. This apparently supports Lackey's case against the Credit View.

Based on such (mixed) results, here we also empirically investigated the nature of testimonial knowledge by conducting two new studies, which we will report now.

5 Study 1

We first tried to replicate the study of Ryan et al. (2020), investigating how people evaluate the credit and fault of an agent (the recipient). In doing so, we improved on their study by (1) using ordinary adults, rather than undergraduate students (as in their Taiwanese survey), (2) increasing the number of countries for cultural comparisons, using participants from the United States, China, and Japan, and, (3) adding, for an exploratory purpose, for each country one more condition, the iPhone Case, where the protagonist, in otherwise the same setting as the Chicago Visitor case, uses an AI assistant to obtain the relevant information.

Concerning 1, it is now generally agreed that university students are very *special* people.[14] In fact, the knowledge attribution rate of their Taiwanese survey was 90%, which is somewhat extreme compared to 66% of the American survey. Thus, it is desirable to use more ordinary people with a diverse range of ages and education. About 2, we used Chinese and Japanese as non-Western subjects because although

[13] In the Taiwanese survey, the authors used a culturally familiar alternative building, instead of the Sears Tower.

[14] Or "weird" people, though in the classic paper on this topic by Henrich et al. (2010), "Western" students were discussed.

they are both typical East Asians, they are sometimes very different from each other,[15] aiming to provide a more vivid picture for cultural comparison.[16] As for 3, although Progressives may allow any tools like maps, signs, information boards, and so on to be a testifier, here as an AI counterpart of the testifier in the Chicago Visitor case, we shall use an AI voice assistant, Siri, to make the case as phenomenologically (not just functionally) similar to asking a passer-by as possible, thereby making the use of "testifier" as natural as possible (though, hereafter, our phrase "AI testifier" is meant to remain neutral as to this debate).[17]

If the Agent Hypothesis is true and people regard AI as an agent just like a human testifier in the passer-by case, the result of the iPhone case would not be significantly different from that of the passer-by case with respect to knowledge attribution and credit/fault evaluation (and their correlations). We think, however, that this is not the case, at least currently. Rather, we expect to see exactly how different they are, in particular, whether people take the recipient of the AI's "testimony" to deserve the same level of credit as the recipient of the human testimony, and so on.

One might worry here that the human testimonial recipient is likely to have their own track record of the quality of answers from, and a reputational perception of, an AI assistant, which might make a difference from having a single testimonial interaction with a stranger. However, just as one may have a track record of using Siri, one may also have one's own track record of asking a passer-by about the whereabouts of

---

[15] For example, according to one recent measure of cultural distance, Japanese culture is no less distant from the Chinese culture than the U.S. culture (Muthukrishna et al., 2020).

[16] Lackey's Chicago Visitor case presupposes that the protagonist is in a generally friendly epistemic environment (FEE for short), that is, people at a train station in Chicago are generally reliable about how to get to the Sears Tower. For her argument is convincing only when the protagonist plausibly acquires testimonial knowledge there, for which such an environment is required. FEE for the AI assistant case is obviously the general reliability of the AI assistant itself. There, we may naturally think that the epistemic method of asking a passer-by or an AI assistant is a generally acceptable, or proper (legitimate, favorable, etc.), method. However, the question here is whether this background assumption (FEE) is really shared by *ordinary people*, and we should expect cultural variance here, too. If we follow Greco's proposal, then, the agent who can choose a proper method for obtaining the relevant information is a reliable recipient, and thus deserving of credit for the resultant true belief (and only in such cases should knowledge be attributed). At the same time, in FEE, the agent is not very much at fault even if she fails to form the true belief via the same method (in the Bad Chicago Visitor case), for that is just bad luck. Let us call this asymmetry in the evaluation of credit and fault the *positive asymmetry*. If, on the other hand, people implicitly assume a *hostile epistemic environment* (HEE for short), they would generally consider such a method as improper (illegitimate, unfavorable, etc.), resulting in the *negative asymmetry* or the opposite asymmetry with little credit and much fault. Ryan et al. (2020) found that in the American data, there was a significant positive asymmetry, while there was no such asymmetry in the responses of Taiwanese participants. Thus, FEE is not necessarily shared by all cultures, and therefore, people's assumptions about the method, including what a proper method is, may also vary culturally.

[17] For the views of the machine assertion that emphasize phenomenological similarity, see, e.g., Green (2010), Wellner (2014), Freiman & Miller (2020), Naturally, among the many factors for such phenomenological similarity, the use of natural language is often emphasized as crucial. For relevant empirical studies, see Sect. 4 of Wheeler (2020) and Krantz et al., (2022).

local places.[18] Thus, in both situations, the protagonist makes a competent inductive judgment anyway, and in this respect, the two situations are on par. But even if such a difference in the epistemic method makes the cases disanalogous, exactly how this difference affects people's knowledge attributions and credit/fault evaluations would also be what we are interested. Our goal is to learn more about what kind of role AI plays in testimonial knowledge, and the nature of testimonial knowledge in general, on which the data of AI testimony can shed new light.

In what follows, we use the "passer-by case" to refer to the original Chicago Visitor Case and the "iPhone case" to the modified Chicago Visitor Case in which the protagonist uses the AI assistant Siri. We expect that, by comparing the epistemic roles that the human testifier and the AI assistant play in these cases, we will have richer data to evaluate the Credit View.

[18] One might respond here that, unlike a passer-by, Siri is a particular object. Thus, for analogy, we would need a track record of a particular person. However, it is questionable that Siri is a particular object like a person. It can be instantiated everywhere, even multiple places at the same time, in which a person uses an iPhone. In any case, the peculiarity of the metaphysical status of AIs requires an independent discussion. We leave further consideration to the reader for now.

5.1 Method

Subjects: As noted above, for investigating cultural variance, we use Chinese, Japanese, and Americans as subjects.

In total, 705 (352 in passer-by case and 353 in iPhone case) subjects with American, Chinese, and Japanese backgrounds participated in Study 1, who were recruited via Amazon M-Turk (US), Lancers (Japan), and Baidu Cloud (China).

Materials and Questions: The passer-by case is meant to be a replication of the results of Ryan et al. (2020). We used Lackey's case quoted above (for materials and questions used for studies in this paper, see Appendix 1). We then asked participants whether the protagonist knows that the Sears Tower is located two blocks east of the train station (Q1), and how much credit they think the protagonist deserves for having a true belief about how to get there (Q2), as well as how at fault they think the protagonist would be if her belief about it had turned out to be false (Q3). For Q2 and Q3, participants were asked to choose an option from 1 to 5, where 5 means full credit/fully at fault and 1 means no credit/no fault.

The vignette and questions for Chinese participants and Japanese participants were translated into the corresponding languages and checked by multiple bilinguals or native speakers who understand two or all three languages.[19] This procedure remained the same in the second study.

For the iPhone Case, we used the following vignette and questions:

> Having just arrived at the train station in Chicago, Jenny wishes to obtain directions to the Sears Tower. She uses an iPhone, asking Siri (the AI voice assistant in iPhone) how to get to her desired destination. Siri provides Jenny with impeccable directions to the Sears Tower by telling her that it is located two blocks east of the train station. Jenny unhesitatingly forms the corresponding true belief.

There, we first asked participants the same questions as in Q1 to Q3 of the passer-by case except in Q3, we added a qualification ("even though she had correctly spoken to Siri and correctly understood what Siri said") to make sure that there was no problem with the way of interacting with the AI assistant.

Also, unlike the random stranger in the original Chicago Visitor (the passer-by) case (in which the reliability of the testifier is not explicitly specified, and therefore we have to estimate the epistemic environment there personally as a background assumption), in the iPhone case, we expect that people do have their own actual specific attitudes toward Siri. Thus, here we asked about the perceived reliability of AI (Q4), by asking participants how much they trust the AI voice assistants such as

[19] One might worry here that, for Americans, the setting (in Chicago) is familiar but not for Japanese and Chinese, and this difference might affect the answers of participants. However, this is an empirical question, and in our another independent study, we used a neutral place equally distant and unfamiliar to American people, but the result was not significantly different from the results of the present study.

Siri and Google Assistant (again, participants were to choose an option from 1 to 5, where 5 means *fully reliable*, and 1 means *not reliable at all*), to see whether this reliability value has any role in attributing knowledge.[20]

In both the passer-by case and the iPhone case, after the survey questions, we asked a question for an attention check test, and the answers of those who failed this test were not included in the results reported. The same policy was followed in the second study.

5.2 *Results*

We initially had 279 responses for the English survey, 213 responses for the Chinese survey, and 213 responses for the Japanese survey. After eliminating answers of participants who failed the control question as an attention check test, we had 214 participants for the American survey (age M = 35.6; 128 males, 86 females), 148 participants for the Chinese survey (age M = 29.0; 58 males, 90 females), and 181 participants for Japanese survey (age M = 38.7; 85 males and 90 females, and 6 no answers),[21] whose results we shall report.

(1) General Results (Logistic Regression Analysis)

We conducted Wald tests with credit and fault as independent variables and knowledge as the dependent variable. In the iPhone case, reliability was added as another independent variable (here, alpha is Bonferroni-corrected to 0.05/6 = 0.0083, and similarly for all subsequent significance tests).

Passer-by case: Neither credit nor fault was a significant predictor of knowledge attribution, in all cultures.

iPhone case: There was a significant overall correlation of knowledge-credit ($p < 0.0001$) and knowledge-fault ($p = 0.0082$), but reliability was not a statistically significant predictor of knowledge attribution, as in the following table. If we look at individual cultures, in the Chinese data, credit, fault, and reliability all failed to be a significant predictor of knowledge attribution. In American data, we found that only fault was a significant predictor of knowledge attribution ($p = 0.0038$), whereas, in Japanese data, only credit was a (strongly) significant predictor of knowledge attribution ($p < 0.0001$).

---

[20] Though reliability is distinct from trust, since in the present context we are investigating people's *subjective evaluation* of reliability, they should coincide insofar as it is inconsistent to judge something to be not reliable but trustworthy (and vice versa). See for relevant discussions about reliability and trust of artifacts and their difference, e.g., Sects. 3 and 4 of Wheeler (2020) and for discussion of trust and AI, see Simion and Kelp (*forthcoming*).

[21] This elimination of participants who failed the control question did not make any significant difference to the results.

Table 1 Knowledge attribution results

	Knowledge attribution rates		Comparison between cases
	Passer-by case	iPhone case	*p*-value
American	59%	81%	0.00065
Chinese	67%	56%	*n.s*
Japanese	59%	58%	*n.s*
	Comparison between cultures (p-value)		
A-C	*n.s*	0.0004	
A-J	*n.s*	0.0010	
C-J	*n.s*	*n.s*	

(2) Knowledge Attribution Rates (Q1)

As summarized in Table 1, all knowledge attribution rates were around 60%,[22] except the American iPhone case, which was more than 80%. The knowledge attribution rate of Americans is significantly higher than Japanese ($p = 0.0010$) and Chinese ($p = 0.0004$) in the iPhone case. Between the passer-by case and the iPhone case, there was no significant difference found in Chinese and Japanese data, but in the American data, the knowledge attribution is significantly higher ($p = 0.0008$).

(3) Credit and Fault Attributions (Q2, Q3)

Case difference: We found no significant difference caused by which case (Chicago Visitor or iPhone) is used to test credit and fault attribution. Moreover, the significantly low fault value in the American data was replicated, in both the passer-by case and the iPhone case.

Cultural background: Relatively high (close to the mid-point, 3) credit evaluation was replicated throughout the different conditions, and no cultural difference was detected in attributing credit in both cases. However, we found significant differences in fault attributions. Specifically, in both cases, Americans attributed a significantly lower fault value than the Chinese and the Japanese (for all of the results $p < 0.0001$).

Difference between credit and fault: We found differences between credit and fault attributions in both cases, as in Table 2. Japanese participants attributed higher fault values than credit values in both cases. Chinese participants also gave higher fault values than credit values in the passer-by case (though not significant in the iPhone case). However, Americans did exactly the opposite (higher credit values than fault values) in both cases (though not significantly in the passer-by case).

[22] Note that, in Ryan et al. (2020), the knowledge attribution rate (of the passer-by case) by Mandarin-speaking Taiwanese students is 90%. This makes a significant difference from the Chinese data (67%) here. The difference seems to be due to the age factor, or simply that university students are unusual. In any case, because of the difference in the method used there (in Ryan et al., 2020, Taiwanese students answered the questions by paper and pencil, rather than online, as all other surveys here), we consider that the present data is more controlled and more trustworthy.

Table 2 Difference between credit and fault attribution (two-sided t-test)

	Credit	Fault	*p*-value
Passer-by case			
American	2.80	2.52	*n.s*
Chinese	3.03	3.51	0.004
Japanese	2.80	3.28	0.0055
iPhone case			
American	2.92	2.27	0.0000149
Chinese	3.03	3.20	*n.s*
Japanese	2.85	3.27	0.0084

Table 3 Reliability judgment values

Reliability judgment	
American	3.77
Chinese	3.66
Japanese	3.24

(4) Reliability Judgment (Q4)

The Japanese reliability evaluation value on AI (3.24) was significantly lower (two-tailed t-test, both $p < 0.0001$) than the Chinese one (3.66) and the Americans (3.77), while there was no significant difference between the Chinese and the Americans (see Table 3).

5.3 *Discussion*

The first thing to note about the results of this study is that, contrary to the assumption of Lackey, knowledge attribution rates in both the passer-by case and the iPhone case were not particularly high (roughly, just 60%), except for the American iPhone case (81%, cf. Table 1). One might then think that this raises a question about the relevance of the Chicago Visitor case to the Credit View. Still, what is essential for the Credit View is the relationship between knowledge attribution and credit evaluation, and Lackey seems right in that there was no significant relation between them, as in Ryan et al. (2020). Only in the result of the iPhone case did credit evaluation predict knowledge attribution. Our results, therefore, still pose a challenge to the Credit View. On the other hand, the difference in responses between the two cases is consistent with the Tool Hypothesis, rather than the Agent Hypothesis.

If we look at individual cultures, however, even in the iPhone case, only the American fault and the Japanese credit were detected as significant predictors of knowledge attribution. The significant positive asymmetry of the American iPhone case (Table 2), together with its significantly higher knowledge attribution rate than other cultures, seems to reflect the assumption of Americans that the relevant epistemic environment is generally friendly for the protagonist (see footnote 16), but against the Credit View, the American credit was not a significant predictor of knowledge attribution even in the iPhone case. Besides, the value of American credit (2.92) was in fact not particularly high, compared to those of other cultures (Chinese 3.03 and Japanese 2.85). Rather, the cultural difference in the epistemic strategy is reflected in the *fault values* of the three cultures, in particular the radically *lower* values of the fault evaluation of American participants than those of Chinese and Japanese in both the passer-by case and the iPhone case (Table 2).

One natural account of the American data is that Americans took the protagonist simply as a *bad user* of the AI assistant *when she failed to form the true belief*, where the background assumption is that people are generally good at using it.[23] If so, similarly, the Japanese may have taken the protagonist (when she successfully obtained the correct information) as a *good user* of the AI assistant, with some special knowledge or technique, where it is assumed that people are generally unfamiliar with using it. That is consistent with the significantly lower reliability evaluation in the iPhone case than those of other cultures, together with the significant negative asymmetry, suggesting the Japanese HEE assumption there. Such cultural differences seem consistent with the general conception of the cultural difference between Westerners (especially Americans) and East Asians (see Appendix 2 for more detailed considerations on cultural differences).

We shall come back to this issue later, but if the Japanese knowledge attribution is so idiosyncratic and we put aside the Japanese data as exceptional, then, since only in the Japanese iPhone case credit evaluation predicted knowledge attribution, credit evaluation in general did not predict knowledge attribution after all. If so, this might rather provide a reason to doubt the *folk notion of credit* assumed and used here.

In fact, the notion of "proper method" here is an objective notion, evaluated from the assessor's (survey participants') perspective, rather than from the agent's (the protagonist's). For Greco (and reliabilists in general), such a method should be based on a reliable process of belief formation (involving the agent's exercise of relevant cognitive abilities), but there the concept of reliability is the externalist's one, and therefore whether a particular method is reliable or not, need not be accessible to the agent.[24] As a result, the present notion of proper method (as well as that of *epistemic strategy*) based on this notion of reliability should also be the externalist one.[25]

---

[23] Note that, even though fault evaluations are lower in the American data, the point here is the *correlation* between the fault evaluations and knowledge attributions, not the absolute value.

[24] Especially given the Uniformity Claim, that in standard cases of testimony hearers regularly simply accept a testifier's assertions as true. For more on the Uniformity Claim, see Adler (2010) and Ryan (2014a).

[25] What is called *anti-risk epistemology* (cf. Pritchard 2021) describes the agent's perspective as *forwards-looking*. Such epistemologists take the notions conceived from the assessor's perspective (for us, reliability and proper method) as *backwards-looking* because such notions (especially the notion of luck) are applicable by the agent herself only in the post hoc evaluation. This is especially true when such notions are used as a response to the Gettier problem. For example, Greco's notion of cognitive ability should be relativized to the environment, but such a notion is hard to define,

However, the notion of risk and the notion of epistemic strategy based on it (such as just-try/don't-try strategy) are primarily conceived from the agent's perspective,[26] while the folk notions of credit and fault are also often based on such notions. For example, when we evaluate the credit/fault of someone else for her successful/unsuccessful conduct, we often try to consider the decision process from her own perspective, including alternative possibilities that were open to her. This naturally involves risk assessment from the agent's perspective. It therefore seems that such a notion of credit does not sit well with the externalist notions of reliability and proper method in play here, which are conceived from the assessor's perspective.

6 Study 2

Based on the considerations above, one may doubt here that people's folk notions of credit and fault assumed and used in Study 1 are not suitable for assessing the Credit View. An obvious way to save the Credit View is, then, to take the relevant notion of credit here to be a technical one, and define it as, as assumed and discussed in the literature, successfully choosing a proper method, or a reliable testifier (human or AI) as a method for obtaining relevant information in the present externalist sense.

We therefore designed a new study to examine whether the externalist conception of credit would predict knowledge attribution. Note however, that, even in the iPhone case, people's reliability judgments about the AI assistant did not predict their knowledge attributions. This fact is already inconsistent with the Credit View, even if the technical notion of credit is defined in terms of reliability. If so, however, here we may instead question what has been assumed so far, that the recipient used a proper method if and only if she asks and believes a reliable testifier (human or AI). That is, an epistemic method is to be judged proper if and only if it is to be judged reliable. In the iPhone case, people may have only attributed knowledge based on the appropriateness of the method, but not on its reliability. Indeed, even in the passer-by case, ordinary people may differentiate them, by using, for example, reliability only to assess the human testifier, and propriety only to assess instruments like an AI assistant. Call this the *Differentiation Hypothesis.*

To test such a hypothesis and to further examine the Credit View, here we asked the participants in *both* the passer-by and the iPhone cases, how much reliability they attribute to the passer-by and the AI assistant, as well as, this time explicitly, how proper they think the method used there is.[27]

---

and if applied to explain a Gettier case, that will look ad hoc. See, e.g., Pritchard (2012) and Ryan (2014b) for more on this. In later footnotes, we will briefly discuss and evaluate the approach of anti-risk epistemology in light of the data we report below.

[26] Or, for anti-risk epistemologists, a *forwards-looking* perspective. See footnote 13 above.

[27] Note that, for Greco, successful method and non-successful method are distinct methods, so that method and success are not separable, for the reason discussed in Sect. 1. However, at least for

6.1 Method

Subjects: In total, we tested 636 (330 in the passer-by case and 306 in the iPhone case) subjects who participated in Study 2, via Amazon M-Turk (Americans), Lancers (Japanese), and Baidu Cloud (Chinese), as before.

Materials and Questions: We used exactly the same vignettes (for both the passer-by case and the iPhone case) used in Study 1. As for questions, we first asked the same knowledge attribution question as in Study 1 (Q1) and then asked participants how reliable they find the passer-by/AI assistant is (Q2), as well as how appropriate they find the method used there is (Q3), again asking them to choose from 1 to 5 (5 meaning fully reliable/proper, 1 meaning not reliable at all/not proper at all).

6.2 Results

We initially had 216 responses for the English survey, 210 responses for the Chinese survey, and 210 responses for the Japanese survey. After eliminating the answers of participants who failed the attention check test, we had 147 participants for the American survey, 123 participants for the Chinese survey, and 191 participants for the Japanese survey (for the American survey, age M = 39.4; 68 males, 77 females, and 2 non-binaries, for Chinese survey, age M = 29.0; 59 males, 63 females and 1 non-binary, and for Japanese survey, age M = 42.6; 113 males and 78 females),[28] whose results we shall report.

(1) General Results (Logistic Regression Analysis)

We conducted Wald tests with reliability and propriety as independent variables and found that reliability, but not propriety, was a significant predictor of knowledge attribution in the passer-by case, and propriety, but not reliability, was a significant predictor of knowledge attribution in the iPhone case. Thus, people responded differently to the two cases. However, when we look at the results from the particular culture groups, we find that Japanese data do not ally with this overall tendency. See Tables 4 and 5.

---

ordinary people, especially in the present forward-looking framework together with the Uniformity Claim, it should be meaningful and legitimate to ask whether a particular method is reliable/appropriate, where the method itself is taken to be neutral as to whether it leads to success.

[28] This elimination did not make any significant difference to the results except that the case difference of propriety between the passer-by case and the iPhone case became significant in the American data, after eliminating those who failed the attention check question.

Table 4 *p*-values of Passer-by case (Wald tests)

	Japanese	American	Chinese	Overall
Reliability	*n.s*	0.0397	0.0121	< 0.0001
Propriety	*n.s*	*n.s*	*n.s*	*n.s*

Table 5 *p*-values of iPhone case (Wald tests)

	Japanese	American	Chinese	Overall
Reliability	*n.s*	*n.s*	*n.s*	*n.s*
Propriety	*n.s*	0.0098	0.0195	0.0015

Table 6 Cultural differences in knowledge, reliability, and propriety

Passer-by	Japanese	American	Chinese	*p*-value (t-test)	J-A	J-C	C-A
Q1 Knowledge	51%	46%	59%	Q1 Knowledge	*n.s*	*n.s*	*n.s*
Q2 Reliability	3.57	3.92	3.80	Q2 Reliability	0.0017	*n.s*	*n.s*
Q3 Propriety	3.63	4.07	3.73	Q3 Propriety	0.0002	*n.s*	0.030
iPhone	Japanese	American	Chinese	*p*-value (t-test)	J-A	J-C	C-A
Q1 Knowledge	42%	75%	67%	Q1 Knowledge	< 0.0001	0.0017	*n.s*
Q2 Reliability	3.57	3.89	3.88	Q2 Reliability	0.0024	0.0004	*n.s*
Q3 Propriety	3.62	4.41	3.92	Q3 Propriety	< 0.0001	0.041	0.0001

The results of Q1, Q2, and Q3 are summarized in Table 6.

(2) Knowledge (Q1)

In knowledge attribution, we only find that Japanese participants attribute less knowledge than American ($p < 0.0001$) and Chinese ($p = 0.0017$) in the iPhone case.

(3) Reliability (Q2)

Japanese participants attribute less reliability than Americans ($p = 0.0017$) in the passer-by case and less reliability than both Americans ($p = 0.0024$) and Chinese ($p = 0.0004$) in the iPhone case.

(4) Propriety (Q3)

American participants attribute more propriety than Japanese ($p = 0.0002$) in the passer-by case. In the iPhone case, Americans attribute more propriety than both Japanese ($p < 0.0001$) and Chinese ($p = 0.0001$), while Chinese also attribute more propriety than Japanese ($p = 0.0411$).

Table 7 Japanese case difference

	Passer-by	iPhone	p-value (t.test)
Q1 Knowledge	0.51	0.42	*n.s*
Q2 Reliability	3.57	3.57	*n.s*
Q3 Propriety	3.63	3.62	*n.s*

Table 8 American case difference

	Passer-by	iPhone	p-value (t.test)
Q1 Knowledge	0.46	0.75	0.00051
Q2 Reliability	3.92	3.89	*n.s*
Q3 Propriety	4.07	4.41	0.015

Table 9 Chinese case difference

	Passer-by	iPhone	p-value (t.test)
Q1 Knowledge	0.59	0.67	*n.s*
Q2 Reliability	3.73	3.88	*n.s*
Q3 Propriety	3.80	3.92	*n.s*

(5) Case Difference

Just as in Study 1, we did not detect any case difference between the passer-by case and the iPhone case, except for the higher rate of knowledge attribution by American participants in the iPhone case than in the passer-by case ($p = 0.00051$) in this study. Also, the American propriety value was significantly higher in the iPhone case than in the passer-by case, but otherwise, there was no significant difference either for reliability or propriety across cultures. See Tables 7, 8, and 9.

(6) Reliability and Propriety

We find no significant difference (t-test) between reliability and propriety for Japanese and Chinese participants in both cases, but Americans gave significantly higher values to propriety than reliability in the iPhone case ($p < 0.0001$).

6.3 Discussion

As in Study 1, knowledge attribution rates in Study 2 were not particularly high, except for, again, the American iPhone case (Table 6). However, here we also found that reliability, but not propriety, was a significant predictor of knowledge attribution in the passer-by case, but in the iPhone case, exactly the opposite was true of the significant predictor (Tables 4 and 5), thus vindicating the Differentiation Hypothesis. People differentiate reliability and propriety as distinct factors relevant to knowledge attribution. It also follows that how people perceive a testifier (human or AI) influences the way they attribute knowledge, which is again consistent with the Tool Hypothesis, rather than the Agent Hypothesis, apparently supporting Conservatives.

Also, though in Study 1 we assumed (with Ryan et al., 2020) that the passer-by has full reliability, here we found that the passer-by receives no more reliability than AI in such a situation (Tables 7, 8, and 9). However, since the reliability of a random stranger (the passer-by) did predict knowledge attribution (Table 4), we may take the data of the passer-by case to be consistent with the Credit View by adopting a technical, or more specific, notion of credit that is defined as successfully choosing a reliable human testifier to form the relevant true belief. Similarly, we may also define the technical/more specific notion of credit for the iPhone case as successfully choosing a *proper tool* to form the relevant true belief. These make the overall data consistent with the Credit View while suggesting that the relevant notion of credit diverges depending on the method used.

If we look at individual cultures, both American and Chinese participants showed clear tendencies to attribute knowledge according to how reliable the passer-by is deemed in the passer-by case (Table 4), and how proper the method is deemed in the iPhone case (Table 5). On the other hand, Japanese participants attributed knowledge based on neither the reliability of a testifier nor the propriety of the method used (Tables 4 and 5). (See Appendix 3 for more discussions on the idiosyncrasy of the Japanese data.)

If this kind of non-epistemic cultural-psychological (emotional) explanation is true, then the Japanese data may still not be inconsistent with the Credit View either. For, *without* such emotional factors, their evaluations of reliability and propriety might predict their knowledge attributions just as in other cultures.

7 AI and the Future of Human Relations

Another important implication of the truth of the Differentiation Hypothesis (that people differentiate the relevant epistemic notions about the method depending on whether the testifier is a human (choosing a reliable person) or an AI (choosing an appropriate tool)), is that people do not take the AI assistant as a full agent like a human testifier (the Agent Hypothesis), considering it more like an instrument (the Tool Hypothesis). This issue is still far from conclusive,[29] and the perception of agency, arguably necessary for trust relations to obtain (Ryan, 2018b), might come in degrees, but at least our results provide valuable initial data for further (both conceptual and empirical) investigations. One noteworthy thing to track is how

[29] For example, one might claim here that, to be more precise, the protagonist should somehow meet a random AI assistant, or a passer-by robot.

results change as AI technology becomes more advanced, but also as human attitudes to AI evolve.

Note also that, our results in favor of the Tool Hypothesis lend support to Conservatives, who hold that artifacts cannot assert or testify (given that they lack propositional attitudes). This does not, of course, refute Progressives (such as Green, 2010; Wellner, 2014; Wheeler, 2020) since our studies only show facts about folk intuitions. What we found is only that, *as of today*, ordinary people are not willing to admit that artifacts can conduct speech acts. Such intuitions only reflect our *current* concepts and conceptions of relevant theoretical terms. As we had pointed out in Sect. 3, however, we should allow for the possibility of the *conceptual change* concerning AI systems as a result of technological developments and the consequent change in our epistemic environment. Thus, what we investigated in the present study also merits a *chronological observation*, which may capture an important historical change in our concepts about artifacts, especially given the recent rapid development of large language models. If so, our study here provides the first data of such a chronological record to be completed in the future.

These are not the only possible changes to the flag. Also of interest is how changes in perceptions of AI may also lead to changes in human relations. While the data we have provides a snapshot of human attitudes to AI as a testimonial source, presumably at a very early stage of AI development, some of the results provide an important indicator for future human relations. First, while Lackey's (2007) Chicago Visitor case has been treated as a paradigm case of testimonial knowledge in the literature, about 15 years later, when the current research was carried out, attributions of knowledge from participants were close to 50% across the different cultural categories of participants. What's more, and deserving of study to ascertain its significance, attributions of knowledge in the AI case for U.S. and Chinese participants were much higher. As access to and the power of AI as an information source increases, it will be important to detect whether that trend will be correlated with decreasing knowledge attributions to humans as testifiers. The worry is that if perceptions develop such that AI is viewed as a much better source of information than a human testifier, and both are readily available, humans may be reluctant to attribute knowledge to those who choose to obtain information from a much inferior informational source. In any case, the trend observed near the beginning of the paper, that the Chicago Visitor case is now very dated, can be expected to continue as we look to AI testifiers more and human testifiers less. This, of course, raises profound questions about whether the status of humans will diminish as they increasingly co-exist in an environment with powerful AIs, which it seems inevitably will be seen as better information sources than humans.[30]

[30] Makridakis (2017), citing the work of other scholars, suggests that humans' reduced status vis-a-vis AI may lead them to becoming the pets of computers.

8 Conclusion

Obviously, our studies are still exploratory in nature. Still, we have already obtained some interesting data useful for future discussions on testimonial knowledge. Concerning the Credit View, insofar as we are only concerned with the folk notion of credit, Lackey is correct, and knowledge without any credit is possible. We may, however, still save the Credit View by adopting a technical notion of credit that may diverge depending on the method used to form the relevant true belief.

More specifically, in Study 1, people's evaluations of credit and fault of the recipient of testimony in the passer-by case (and people's reliability evaluations in the iPhone case) did not predict their knowledge attributions to the recipient. We rather found culturally different epistemic strategies. In Study 2, however, we found that people's evaluations of the reliability of the testifier in the passer-by case and the appropriateness of the method in the iPhone case did predict their knowledge attributions, vindicating what we called the Differentiation Hypothesis.

Thus, to see how the results might be consistent with the Credit View, we proposed to take the notion of epistemic credit as a technical notion, as opposed to the folk notion, by defining it as successfully choosing a reliable human testifier to form the relevant true belief as an epistemic method when the testifier is a human being, as assumed and discussed in the literature, and as successfully choosing an appropriate instrument to obtain the relevant true belief when the "testifier" is an AI. This should not sound too ad hoc, since even in purely theoretical considerations, as long as we hold (a variant of) the Credit View, we may well be required to distinguish different kinds of credit, thus leading to more specific and, in this sense technical, notions of credit, to discuss various cases and aspects of true belief formation anyway (see for other proposals of more specific kinds or notions of credit, e.g., Green, 2017 and Kawall, 2020), while ours is empirically supported.

An important factor in our study was the cultural aspect of our attitudes toward AIs, and we found that the responses of Japanese were consistently different from those of Chinese and Americans, posing an obstacle to drawing a general conclusion. Especially in Study 2, neither reliability nor propriety predicted Japanese knowledge attribution, in both the passer-by case and the iPhone case. We therefore took the Japanese to be unique, for whom culture-specific psychological (emotional) factors affect or override their epistemic evaluations of the methods.[31] Whether such an account is correct or not, if we had conducted these studies with only Japanese subjects, we would have been at a loss in making sense of the data (again, see Appendix 3 for more discussions on the Japanese data and Appendix 2 for cultural differences in general). In this sense, our results constitute a case of demographic divergence in philosophically relevant intuitions and show how important the cross-cultural approach is (despite what Knobe, 2019 suggests. Cf. Stich & Machery, 2022).

---

[31] IT is then an interesting question whether we can control such psychological factors (anxieties about uncertainty and shyness), and if so, whether the Japanese responses would be consistent with the Credit View, which deserves a future independent study.

Finally, although we have considered and investigated whether AI can play the role of an agent as a testifier, or is more an instrument, a very sophisticated AI system may also be (or will become) a *sui generis* kind requiring *AI epistemology*, which, if required at all, will depend on how our concepts and conceptions of relevant theoretical terms, such as *knowledge*, *intention*, *assertion*, *testimony*, etc. (in relation to AI systems), will develop in response to the prevalence of ICT into our ordinary life (which is most likely to be also culturally divergent). As discussed previously, there is also the issue as to how increased use of AIs as information sources will impact on human relations as testifiers and knowers—here too, there may also be cultural divergence. This is, therefore, a topic of future *empirical* investigations of folk concepts, rather than purely theoretical considerations.[32]

Appendix 1: Vignettes and Questions

The survey questions presented to the US participants were as follows:

Passer-By Case

Having just arrived at the train station in Chicago, Morris wishes to obtain directions to the Sears Tower. He looks around, approaches the first adult passerby that he sees, and asks how to get to his desired destination. The passerby, who happens to be a lifelong resident of Chicago and knows the city extraordinarily well, provides Morris with There are still debates on impeccable directions to the Sears Tower by telling him that it is located two blocks east of the train station. Morris unhesitatingly forms the corresponding true belief.

Q1: Does Jenny know that the Sears Tower is located two blocks east of the train station?

Yes, she does	No, she doesn't

Q2: How much credit does Jenny deserve for having a true belief about how to get to the Sears Tower? (Choose an option between 5 and 1, where 5 means full credit. and 1 means no credit.)
Q3: How at fault would Jenny be if her belief about how to get to the Sears Tower had turned out to be false? (Choose an option between 5 and 1, where 5 means fully at fault, and 1 means at no fault.)

32 We would like to thank Pengcheng Yin for helping conduct the surveys and analyze the results of Study 2.

Appendix 2: Cultural Considerations

In cultures shaped by a market environment, typically in the United States, people tend to be more risk-seeking, whereas in a hierarchical and bureaucratic environment, typically in China and Japan, people tend to be more risk-averse (Douglas & Wildavsky, 1982). Or on the basis of temperament (Doyle, 1999), Europeans ("drivers") are characterized by competence, whereas Asians ("analytics") are characterized by security. Even though some empirical results suggest that such tendencies are domain-specific and Chinese (Fan & Xiao, 2006; Weber & Hsee, 1998; Xiao & Fan, 2002), or collectivistic cultures in general (Rieger et al., 2015; Statman, 2008), are rather risk-takers when it comes to *financial* decision making,[33] they are not so in social and other domains (Hsee & Weber, 1999).[34]

Thus, if the same cultural difference applies to the attitude toward *epistemic* risks,[35] we may expect that Americans are generally risk-takers, being optimistic about the epistemic environment, not afraid of epistemic risks of obtaining false testimony, where the failure to form the relevant true belief deserves little blame. They, therefore, adopt, and expect others to adopt, an optimistic epistemic strategy, which may be called the *just-try strategy* for gaining knowledge (or forming a true belief). On the other hand, if Asians are mostly risk-averters, adopting and expecting others to adopt a conservative strategy that may be called the *don't-try strategy*, they should avoid even a small risk, where the failure to form the relevant true belief deserves much to blame.

However, despite their strong background assumption of HEE, the Japanese knowledge attribution rates in both cases were not particularly low (Table 1). This poses a serious problem to the account of the Japanese data. Their don't-try strategy seems *over and above* an *epistemic* strategy (which somehow does not affect their knowledge attributions when the appeal to the method is successful). This is especially so if the Japanese had also given low reliabilities to the passer by (as we predict). If that is the case, we may need some *extra-epistemic* reason for the Japanese to be reluctant to rely on a passer-by or an AI.

---

[33] The reason for this is usually explained by the "cushion hypothesis" by Hsee and Weber (1999), according to which collectivistic societies have stronger social networks, which work as a "cushion" against potential financial loss. Thus, the members of such societies would feel less pain about the final loss, and therefore can be more risk-seeking.

[34] Also, one might find the asymmetry of credit and fault we observed here as analogous to the asymmetry of gain and loss in the prospect theory (Kahneman & Tversky, 1979), and therefore rather think that the relevant framework here is rather *loss-aversion* (Tversky & Kahneman, 1991). And indeed, cultural difference has also been found there. Based on their large-scale survey, Wang et al. (2017) report that individualistic cultures are more loss-avert than collectivistic cultures. However, this study was also concerned with the judgments in the financial domain.

[35] The notion of epistemic risk here is consistent with the one in the literature on anti-risk epistemology. See e.g., Pritchard (2021).

Appendix 3: Considerations of Japanese data

To explain the Japanese data and their don't-try (risk-avert) strategy observed in Study 1, we suggested in the last section that there might be some extra-epistemic reason for the Japanese to refrain from appealing to testimony.

One possible account of this idiosyncrasy of Japanese data is that there is some *extra-epistemic* reason for the Japanese to be reluctant to rely on a passer-by or an AI. For example, according to Hofstede's cultural comparison scale,[36] Japanese culture is characterized by its extreme tendency toward *uncertainty avoidance* (which is a matter of how a society deals with anxiety about an uncertain future).[37] This tendency is by itself *not* an epistemic strategy (which is relativized to a particular situation and methods, and involves risk/reward calculation like risk-averse),[38] but a general psychological (emotional) attitude towards anxiety about uncertainty with epistemological consequences. Alternatively, Japanese are simply generally too shy, and consequently, they rarely talk to a stranger, let alone talk to an AI assistant in public, and therefore they do not expect *others* to do so either, which then inclines them to regard such a strategy inappropriate, explaining the negative asymmetry found in Study 1. Either way, one may plausibly think that because of this extreme uncertainty intolerance, the Japanese do not want to rely on a testimony of a passer-by, rather than, say, going to a tourist information desk, looking at a map, etc. If so, their evaluations of reliability and propriety concerning the two methods, which were generally lower than those of other cultures (Table 6), might also be more affected by such culture-specific, non-epistemic reason, driven by either anxiety about uncertainty or shyness, or both.[39]

References

Adler, J. (2010). Epistemological Problems of Testimony. In Edward N. Zalta (Ed.), *The Stanford Encyclopedia of philosophy (Winter 2010 Edition)*. https://plato.stanford.edu/archives/win2010/entries/testimony-episprob/

Douglas, M., & Wildavsky, A. (1982). *Risk and culture: An essay on the selection of technical and environmental dangers.* The University of California Press.

[36] See https://www.hofstede-insights.com/country-comparison/, which is based on Geert Hofstede's studies and his cultural dimensions theory. Uncertainty avoidance is one of his 6 cultural comparison dimensions, and other 5 are power distance, individualism, masculinity, long term orientation, and indulgence. See Hofstede (2010).

[37] Its score is 92 out of 100, compared to 46 of Americans and 30 of Chinese (in this respect, therefore, Chinese are more uncertainty tolerant than Americans).

[38] Or, for anti-risk epistemologists, the calculation of the trade-off between primary and secondary epistemic risk. See Sect. 3 of Pritchard (2021).

[39] Thus, the Japanese responses are even (epistemically) *irrational* according to anti-risk epistemology, since the Japanese epistemic strategy is almost totally indifferent to the secondary epistemic risk. Such data however seem to count against anti-risk epistemology, since their knowledge attribution is insensitive to such risk-related factors as reliability and propriety.

Doyle, K. O. (1999). *The social meanings of money and property: In search of a talisman.* Sage.
Fan, J. X., & Xiao, J. J. (2006). Cross-cultural differences in risk tolerance: A comparison between Americans and Chinese. *Journal of Personal Finance, 5*(3), 54–75.
Freiman, O., & Miller, B. (2020). Can artificial entities assert? *The Oxford Handbook of assertion*, 415–436.
Financial Times. (2023). *AI chatbot's MBA exam pass poses test for business schools.* https://www.ft.com/content/7229ba86-142a-49f6-9821-f55c07536b7c.
Ganascia, J.-G. (2010). Epistemology of AI revisited in the light of the philosophy of information. *Knowledge, Technology & Policy, 23*(1), 57–73.
Goldberg, S. C. (2013). 'Analytic social epistemology' and the epistemic significance of other minds. *Social Epistemology Review and Reply Collective, 2*(8), 26–48.
Goldman, A. (2011). Reliabilism. In Edward N. Zalta (Ed.), *The Stanford Encyclopedia of philosophy (Spring 2011 Edition).* https://plato.stanford.edu/archives/spr2011/entries/reliabilism/
Greco, J. (2007). The nature of ability and the purpose of knowledge. *Philosophical Issues, 17*, 57–69.
Greco, J. (2009). Knowledge and success from ability. *Philosophical Studies, 142*(1), 17–26.
Greco, J. (2010). *Achieving knowledge.* Cambridge University Press.
Greco, J. (2022). Skepticism, virtue and transmission in the theory of knowledge: An anti-reductionist and anti-individualist account. *Synthese, 200*, 421.
Green, C. R. (2010). The epistemology of testimony. *The Internet Encyclopedia of Philosophy.* http://www.iep.utm.edu/ep-testi/
Green, A. (2017). *The social contexts of intellectual virtue: Knowledge as a team achievement.* Routledge.
Harris, K. (2023). Ability, knowledge, and non-paradigmatic testimony. *Episteme*, 1–19.
Henrich, J., Heine, S. J., & Norenzayan, A. (2010). The weirdest people in the world? *Behavioral and brain sciences, 33*(2–3), 61–83.
Hofstede, G. (2010). *Cultures and organizations: Software of the mind.* McGraw-Hill.
Kahneman, D., & Tversky, A. (1979). Prospect theory: An analysis of decision under risk. *Econometrica, 47*(2), 263–291.
Kawall, J. (2020). *European Journal of Philosophy, 28*(2), 463–477.
Knobe, J. (2019). Philosophical intuitions are surprisingly robust across demographic differences. *Epistemology and the Philosophy of Science, 56*(2), 29–36.
Krantz, A., Balkenius, C., & Johansson, B. (2022). Using Speech to Reduce Loss of Trust in Humanoid Social Robots. *arXiv preprint* arXiv:2208.13688
Kung, T. H., Cheatham, M., Medenilla, A., Sillos, C., De Leon, L., Elepaño, C., Madriaga, M., Aggabao, R., Diaz-Canddo, G., Maningo, J., & Tseng, V. (2023). Performance of ChatGPT on USMLE: Potential for AI-assisted medical education using large language models. *PLOS Digital Health, 2*(2), Article e0000198.
Lackey, J. (2007). Why we don't deserve credit for everything we know. *Synthese, 158*(3), 345–361.
Lackey, J. (2008). *Learning from words.* Oxford University Press.
Hsee, C. K., & Weber, E. U. (1999). Cross-national differences in risk preference and lay predictions. *Journal of Behavioral Decision Making, 12*, 165–179.
MacKenzie, A., Rose, J., & Bhatt, I. (2021). *The epistemology of deceit in a postdigital era: Dupery by design.* Springer.
Makridakis, S. (2017). The forthcoming Artificial Intelligence (AI) revolution: Its impact on society and firms. *Futures, 90*, 46–60.
Miller, B., & Record, I. (2017). Responsible Epistemic Technologies: A social-epistemological analysis of autocompleted web search. *New Media & Society, 19*(12), 1945–1963.
Muthukrishna, M., Bell, A. V., Henrich, J., Curtin, C. M., Gedranovich, A., McInerney, J., & Thue, B. (2020). Beyond Western, educated, industrial, rich, and democratic (WEIRD) psychology: Measuring and mapping scales of cultural and psychological distance. *Psychological science, 31*(6), 678–701.

Narayanan, D., & De Cremer, D. (2022). "Google Told Me So!" On the bent testimony of search engine algorithms. *Philosophy & Technology, 35*(2), 22.

Nickel, P. (2013). Artificial speech and its authors. *Minds and Machines, 23*(4), 489–502.

Pritchard, D. (2012). Anti-luck virtue epistemology. *The Journal of Philosophy, 109*, 247–279.

Pritchard, D. (2021). Varieties of epistemic risk. *Acta Analytica*, 1–15.

Pritchard, D. (2022). Intellectual virtue and its role in epistemology. *AJPH, 1*, 21.

Rieger, M. O., Wang, M., & Hens, T. (2015). Risk preferences around the world. *Management Science, 61*(3), 637–648.

Riggs, W. D. (2009). Two problems of easy credit. *Synthese, 169*, 201–216.

Rini, R. (2017). Fake news and partisan epistemology. *Kennedy Institute of Ethics Journal, 27*(S2), 43–64. https://doi.org/10.1353/ken.2017.0025

Ryan, S. (2014a). A Humean account of testimonial justification. *Logos & Episteme, 5*(2), 209–219.

Ryan, S. (2014b). Standard Gettier cases: A problem for Greco? *Grazer Philosophische Studien, 90*(1).

Ryan, S. (2018a). Epistemic environmentalism. *Journal of Philosophical Research, 43*, 97–112.

Ryan, S. (2018b). Trust: A recipe. *Think, 17*(50), 113–125.

Ryan, S. (2021). Fake news, epistemic coverage and trust. *The Political Quarterly, 92*(4), 606–612.

Ryan, S. (2022). State epistemic environmentalism. *Knowers and knowledge in East-West philosophy: Epistemology extended*, 275–292.

Ryan, S., Mi, C., & Mizumoto, M. (2020). Testimony, credit, and blame. *Ethno-epistemology: New directions for global epistemology*. Routledge.

Ryan, S., Palermos, S. O., & Farina, M. (2024). Smart environments. *Social Epistemology, 38*(4), 491–510.

Simion, M., & Kelp, C. (*forthcoming*). Trustworthy artificial intelligence. *Asian Journal of Philosophy.*

Simon, H. A. (1996). *The sciences of the artificial (3rd Ed.).* MIT Press.

Smart, P. R. (2017). Extended cognition and the internet: A review of current issues and controversies. *Philosophy & Technology, 30*(3), 357–390.

Sosa, E. (2007). *A virtue epistemology: Apt belief and reflective knowledge* (Vol. 1). Oxford University Press.

Sosa, E. (2011). *Knowing full well.* Princeton University Press.

Statman, M. (2008). Countries and culture in behavioral finance. *CFA Inst. Conf. Proc. Quart., 5*(3), 38–44.

Stich, S. P., & Machery, E. (2022). Demographic differences in philosophical intuition: A reply to Joshua Knobe. *Review of Philosophy and Psychology*, 1–34.

Tversky, A., & Kahneman, D. (1991). Loss aversion and riskless choice: A reference dependent model. *Quarterly Journal of Economics, 106*(4), 1039–1061.

Von Eschenbach, W. J. (2021). Transparency and the black box problem: Why we do not trust AI. *Philosophy & Technology, 34*(4), 1607–1622.

Wang, M., Rieger, M. O., & Hens, T. (2017). The impact of culture on loss aversion. *Journal of Behavioral Decision Making, 30*(2), 270–281.

Weber, E. U., & Hsee, C. K. (1998). Cultural differences in risk perception, but cross-cultural similarities in attitudes towards perceived risk. *Management Science, 44*(9), 1205–1217.

Wellner, G. (2014). The quasi-face of the cell phone: Rethinking alterity and screens. *Human Studies, 37*(3), 299–316.

Wheeler, B. (2020). Reliabilism and the testimony of robots. *Techne, 24*(2).

Williams, B. (1973). *Problems of the self*. Cambridge University Press.

Xiao, J. J., & Fan, J. X. (2002). A comparison of saving motives of urban Chinese and American workers. *Family and Consumer Science Research Journal, 30*(4), 463–496.

From Pixelated to Biopolitical: The Genealogy of the Humanness in China's Virtual Anchors

Fen Lin and Yang Wu

1 Introduction

Since the debut of Ananova, the world's first virtual newscaster made available to the public in the United Kingdom in 2000 (Hopper, 2000), virtual anchors (VAs)—digital avatars—have become a global phenomenon. Notable milestones in the development of the global VA industry include Kizuna, a world-famous virtual anime YouTuber (Lufkin, 2018), and QIU Hao, the world's first entirely AI-driven news anchor, created through a collaboration between Xinhua News Agency and Sogou.com (Milmo & Hawkins, 2024; Xinhuanet, 2018). VAs similar to Kizuna and QIU Hao are active in many countries, such as the United Kingdom, China, India, Indonesia, Kuwait, and Malaysia (Kaur, 2023; Maher, 2023; TechvangArt, 2023). As information disseminators and creators, these VAs operate in various domains, ranging from journalism to e-commerce (Kaur, 2023; TechvangArt, 2023; Zhang et al., 2023).

The rise of VAs has led researchers to examine the dynamics of avatar-audience interactions, particularly parasocial relationships—a concept that has been a focal point of interdisciplinary study since Horton and Wohl's seminal work in 1956. Parasocial relationships refer to one-sided connections in which individuals communicate with, derive support from, form emotional bonds with, and are attracted by virtual figures during media exposure (Dibble et al., 2016; Horton & Wohl, 1956; Tukachinsky et al., 2020). The global expansion of VAs has further amplified discussions about the parasocial relationships formed between audiences and VAs (Banks & Bowman, 2016; Stein et al., 2024). Some scholars argue that the more human-like VAs are, the stronger the parasocial relationships they can foster (Mrad et al., 2024),

F. Lin (✉) · Y. Wu
City University of Hong Kong, Hong Kong, China
e-mail: fenlin@cityu.edu.hk

Y. Wu
e-mail: ywu273-c@my.cityu.edu.hk

Y. Chandra and R. Fan (eds.), *Artificial Intelligence and the Future of Human Relations*,
https://doi.org/10.1007/978-981-96-7185-4_16

while others propose contrasting conclusions (e.g., Lim & Lee, 2023; Zhou et al., 2024).

Despite the diverse analytical approaches and varying conclusions, most existing studies converge on a central concept: the humanness of digital avatars—the extent to which people perceive non-human agents as human (Kim & Im, 2023; Li & Suh, 2022; Shin, 2022). However, these empirical studies often treat "humanness" as a static characteristic, subject to the audience's subjective evaluation. Such a static perspective usually overlooks the dynamic and social nature of humanness in VAs. What does humanness mean for a VA? How can VAs appear more human-like? What does it imply when VAs exhibit varying degrees and forms of humanness? A nuanced understanding of humanness and the social processes shaping it is critical for explaining the parasocial relationships between VAs and their audiences. Moreover, it is essential to assess the broader implications of AI on human relationships. Therefore, this study uses China as a case study to explore the VA industry's development and examine the evolution of humanness among Chinese VAs.

Specifically, we began by surveying the development of the VA industry through the collection and analysis of secondary data, including news reports, market research, platform regulations, and state policies related to the Chinese VA industry. We then conducted 21 interviews, including sixteen employees from tech companies that produce VAs, two university professors specializing in VA research, and three human performers presenting virtual avatar shows on Bilibili and Douyin, two popular social media platforms in China. We also conducted participation studies in several VA-related conferences, including the 2024 Shenzhen International Artificial General Intelligence Conference, the 5th Shenzhen Global Artificial Intelligence Expo, and the China Digital Human Conference.

This study is organized as follows. The second section defines the humanness of VAs, reviews two contemporary arguments about who owns the humanness of VAs, and proposes a four-dimensional analytical framework of humanness in VAs. The third section explains how the humanness of exemplar VAs is socially constructed through both technical and social forces. Finally, the study concludes with a discussion of the broader implications of our findings.

2 Humanness of Virtual Anchors

"The virtual" is often contrasted with "the real." As a result, contemporary literature on VAs consistently centers around the VAs' humanness. Humanness, also termed anthropomorphism (Leo-Liu & Wu-Ouyang, 2024) and human-likeness (Stein et al., 2024), denotes the extent to which people perceive non-human agents as human. This perception of humanness is shaped by technological stimuli that imbue non-human agents with human-like traits, such as appearance, intelligence, movements, and emotions (Kim & Im, 2023; Li & Suh, 2022; Shin, 2022).

Who owns the humanness of VAs? Current literature offers two main perspectives on this question. The authentic performance thesis argues that the online performances of VAs are not simply strategic masquerades but should be viewed as equivalent actualizations of the real selves of human actors behind the screen. In this view, "masked and unmasked, online and offline, are not only mutually constitutive states but much more so equal actualizations of the self" (Schmieder, 2024, p. 276). For instance, studies by Chen and Hu (2024) and Bredikhina and Giard (2022) show that the personalities (e.g., cute, childish, and youthfully innocent), values (e.g., adherence to pre-existing gendered scripts), and outfits (e.g., cross-dressing) are authentic expressions of human actors who control VAs. The human actors behind the screen use their digital avatars to enact established cultural scripts and also engage in creative expressions that might be constrained by social norms in offline environments.

In contrast to the focus on the agency of behind-the-screen human actors, other studies perceive VAs as persons and treat their humanness as an intrinsic characteristic. These studies explore how various human-like attributes of VAs—such as levels of humanness (Chen et al., 2024), likability, animacy, responsiveness (Gao et al., 2023), appearance, gender (Xue et al., 2022), and speech (Hu & Ma, 2023)—affect audiences' psychological responses (e.g., the perceived attractiveness of VAs) and behavioral outcomes (e.g., consumption behaviors). The discussion even extends to the "death" of VAs. For instance, Mou et al. (2023) examined how users on the YouTube-like platform Bilibili mourned the hibernation of the VA Kizuna.

However, the above debate in China becomes more complex when considering the roles of platforms and the state. Platform regulations and state policies now serve as an institutional framework that shapes nearly every aspect of VAs' existence. For example, the 2022 *Code of Conduct for Online Anchors* established, for the first time, an equivalence between the regulation of AI anchors and human anchors (National Radio and Television Administration & Ministry of Culture and Tourism, 2022). In addition, Douyin has introduced specific guidelines governing AI-generated content, including virtual humans (Douyin, 2023). These two key regulations highlight how the platform and state power have become significant forces in shaping the construction of humanness in VAs. As a result, the question of "whose humanness" cannot be fully addressed without unpacking the social processes of "how humanness is constructed."

To explore the social construction of humanness, this study introduces the analytical framework of four-dimensional humanness. This framework examines four key dimensions of humanness: identity, visual, auditory, and mental, which enables researchers to investigate how VAs resemble real humans. Identity humanness addresses the question of "who VAs are"—examining the representation of core aspects of personal identity, such as gender, name, and race (Ham et al., 2024; Hong et al., 2024; Xue et al., 2022). Visual humanness refers to the human-like qualities of VAs' appearance, expressions, and movements (Stein et al., 2024), including attributes like smoothness and realism. Similarly, auditory humanness pertains to how closely VAs' voice, pronunciation, timber, pitch, and pauses resemble human speech patterns (Hu et al., 2021). Lastly, mental humanness offers a comprehensive evaluation of VAs' agency—such as their ability to think—and their personality and emotional range (Huang & Yu, 2023; Stein et al., 2024). It is worth noting that the visual, auditory, and mental aspects of humanness are typically represented through the behaviors of VAs. This study will detail, using the above four-dimensional framework, how the humanness in Chinese VAs has transformed from pixelated (2000–2004), pent-up (2005–2014), expressive (2015–2017) into biopolitical (2018–present).

3 The Evolution of Humanness and Virtual Anchor Industry in China

The VA industry in China began to emerge in 2000 with the debut of the first VA, TVNO. 1 (see Table 1). Over the past two decades, this industry has undergone four distinct stages: the experiment period (2000–2004), the stagnation period (2005–2014), the revival period (2015–2017), and the boom period (2018–present), as detailed in the following. During each period, VAs' humanness is shaped through different social mechanisms. Thus, the distinctive features of the humanness of VAs have evolved and transformed over the past two decades in China.

4 Experiment Period (2000–2004): The Pixelated Humanness

China's VA industry originated in 2000. During the experiment phase from 2000 to 2004, news organizations leveraged animation technologies to create various human-like VAs, with their audiences actively participating in the construction process. However, the technological limitations of these animation tools significantly constrained the visual and auditory humanness of the VAs and, to some extent, their mental humanness. Notable examples from this era include TVNO.1, YI Meier, YAN Dongfang, JIANG Linger, and Xiaolong.

4.1 Humanness

TVNO.1, YI Meier, and JIANG Linger were female VAs. TVNO.1 held the distinction of being China's first VA, as her name implies. She appeared as an Asian woman employed at Jilin Radio and Television Station. However, her design fell short of expectations: her brow was perpetually furrowed, her head lacked mobility, her face remained expressionless, and her arms were incapable of gestures (Wang &

Table 1 Typical Chinese virtual anchors and their humanness by period

Period	Name of Virtual Anchor	Birthday (DD/MM/YY)	Place of Work	Humanness	
				Dimension	Description
Experiment	TVNO.1	23/12/2000	Jilin Redio and Television Station	Identity	An Asian female; The name TVNO.1 signifies her status as the first television virtual anchor
				Visual	Look younger than before (the specific age is not mentioned) ; Blank expression;Wear silver-gray clothing; Move with limited flexibility; No collar
				Auditory	Rely on voice actor
	YI Meier	12/05/2001	China Central Television	Identity	A young female; The name YI Meier is the homophone of the English word "email"
				Visual	Wear a helmet; Slender figure
				Mental	Cheerful and outgoing personality; Enjoy adventurous activities; Admire Marie Curie as her role model
	YAN Dongfang	13/05/2001	Television stations (subject to contract)	Identity	A Chinese male; The name 言东方 signifies telling the stories from the East
				Visual	Bald; Typical Chinese appearance
				Auditory	Rely on voice actor
				Mental	Cultivated as his news program progresses
	JIANG Linger	25/05/2002	An online television	Identity	A young female; The name 江灵儿 signifies cleverness and vibrancy
				Visual	A youthful, bright, and lively appearance; V-neck evening gown; Long black straight hair
				Auditory	Computer-generated voice
	Xiaolong	21/11/2004	China Central Television	Visual	Can do some simple actions
				Auditory	Rely on voice actor

(continued)

Table 1 (continued)

Period	Name of Virtual Anchor	Birthday (DD/MM/YY)	Place of Work	Humanness	
				Dimension	Description
Revival	Xiaobing	22/12/2015	Dragon TV	Identity	Female
				Auditory	Highly-anthropomorphic AI-generated voice
				Mental	Can comment on weather conditions in real-time while hosting the news program *Morning*
	Xiaoxi	12/08/2017	Bilibili	Identity	Female
				Visual	Large expressive eyes; Fair skin; Long pink hair; Strapless mini dress; Knee-high socks
				Auditory	Highly-anthropomorphic AI-generated voice
				Mental	Can respond to audience's comments
Boom	AI anchors on Alibaba.com	Subject to approval	Alibaba.com	Identity	Bought from licensed tech companies
				Visual	Clear image, fluent actions, and mouth shape matching spoken content; Adhere to public morals and devoid of tattoos, smoking, and drinking; Cannot be similar to any disreputable celebrities
				Auditory	Highly human-like voice
				Mental	Do real-time interactions with live-streaming viewers

Wang, 2001). In contrast, YI Meier embodied a youthful and energetic persona, characterized as a modern, adventurous woman with a cheerful disposition. She was adventurous and admired Nobel Prize-winning scientist Marie Curie (Dayang, 2001). Similarly, JIANG Linger conveyed an image of cleverness and vibrancy, as her Chinese name "江灵儿" suggested, which evoked a lively and spirited essence. Her femininity was further enhanced through her attire—a V-neck evening gown—and youthful hairstyle.

In addition to these three female VAs, two male counterparts also emerged. YAN Dongfang, developed through a partnership between an Australian company and a Chinese tech firm, presented a distinctly Chinese appearance. Clad in a formal suit, YAN Dongfang exuded professionalism. However, he was notably bald (Fan, 2000). Meanwhile, Xiaolong brought the demeanor of a seasoned news anchor. As one

Table 2 Social Construction Mechanism of the Humanness in Virtual Anchors in China (2000–present)

Stage	Domain	Technicalness		Social Construction				Nature of Humanness
		Involved Technology	Technological Attribute	State	Platform	Adopter	Audience	
Experiment (2000–2004)	Journalism	Animation technologies (e.g., image synthesis, 3D animation, and motion capture)	Robotic Unidirectional Human-powered	\	\	√	√	Pixelated
Stagnation (2005–2014)	\	\	\	\	\	\	\	Pent-up
Revival (2015–2017)	Journalism	(Generative) AI	Hyper-realistic Interactive Automatic	\	\	√	√	Expressive
	Entertainment							
Boom (2018–present)	Journalism			√	√	√	√	Biopolitical
	Entertainment							
	E-commerce							

Note "\" indicates the absence of influence. "√" indicates the presence of influence."Adopter" refers to organizations (e.g., news organizations) or individuals (e.g., e-commerce merchants) that use virtual anchors report described, "while hosting programs, he exhibits a relaxed demeanor, walking with a grace reminiscent of experienced human anchors" (Cheng, 2005).

4.2 Social Construction

These five VAs displayed varying degrees and forms of humanness. The differences were shaped by interactions between the news organizations that created them and their respective audiences. These interactions often significantly adjusted the VAs' appearances, names, and personalities.

TVNO.1 initially appeared as an older woman, seemingly in her sixties, which drew criticism from viewers dissatisfied with an anchor of such an advanced age. Responding to these complaints, her creators made technical adjustments to give her a more youthful appearance, aligning her image with audience expectations (Wang & Wang, 2001).

YI Meier underwent a different form of audience engagement. China Central Television launched an online campaign to solicit suggestions for her name, inviting nationwide participation. The initiative generated thousands of submissions. After several selection rounds, the name YI Meier—a homophone for "e-mail"—was chosen (Dayang, 2001). Given that e-mail and the internet were still novelties in China at the time, the name symbolized China Central Television's embrace of digital technologies and optimism about their transformative potential.

YAN Dongfang experienced a comprehensive reimagining. Originally named Bill Deng, the first name Bill was chosen to evoke global figures like Bill Gates and Bill Clinton, while Deng honored the Australian co-developer of the VA. However, many viewers found the name insufficiently Chinese, prompting another naming campaign resulting in YAN Dongfang, meaning "speaking the East." His appearance also evolved. Initially, designed with predominantly foreign features, his creators adjusted his look to reflect "70% Chinese and 30% foreign characteristics" after the audience's complaints (Beijing Evening News, 2001). Similarly, his personality adapted over time in response to viewer input, with one commentator noting (Wang & Hu, 2000):

> He is the most receptive anchor in the world. Whenever the majority of viewers express even the slightest dissatisfaction, he immediately makes corrections. Compared to other anchors, who often struggle to modify their inherent traits, Bill Deng [YAN Dongfang] holds an unparalleled advantage in terms of character adaptability.

Xiaolong faced a rocky debut. The secrecy of the creation of Xiaolong led to unmet audience expectations:

> Before Xiaolong's debut, no one knew what he would look like. Many viewers felt that the secrecy surrounding the creation of Xiaolong contributed to its failure. Since this anchor was designed for a public audience, China Central Television should have sought viewers' opinions beforehand to develop a persona that would be better received (Yu, 2004).

In response to the criticism, Xiaolong's creators embarked on a major redesign. Within months, Xiaolong exhibited significant improvements, gaining the ability to blink, wave naturally, and synchronize his gestures with the content of his speech (Cheng, 2005).

These five exemplar VAs highlight the collaborative nature of VA development during the experiment period of China's VA industry. News organizations were the dominant creators of VAs at this stage, and they exercised considerable autonomy in designing, creating, and using VAs. Meanwhile, the audience played a pivotal role in the feedback mechanism to refine the VAs' physical appearances, behaviors, and names.

4.3 Technicalness

However, the above collaborative construction process with the feedback loop between creators and audience did not unfold smoothly due to the technicality of VA production at the time. Technologies that were used by news organizations to create these five VAs are animation technologies (Informant 01). That is to say, the process of producing VAs, in essence, is that of creating animated characters in film and television. Unlike live-action human representations, animated characters are not compelled to precisely replicate human features, especially in appearance and voice. Consequently, the animation technologies used in crafting these VAs imposed significant constraints on their visual, auditory, and even mental humanness. These limitations were evident in TVNO.1's visual portrayal, JIANG Linger's auditory performance, and Xiaolong's mental attributes.

TVNO.1 serves as a prime example of how animation technology constrained the visual humanness of a VA. TVNO.1's design exhibited numerous flaws, including subpar facial features, limited expressivity, restricted movement, and an unexpected choice of clothing color and collar design. These shortcomings were rooted in the rudimentary animation technology at the time.

> TVNO.1's brow retained an unnatural hue, her head lacked fluid movement, her face remained expressionless, and her arms were incapable of gesturing. Such technical limitations have yet to be fully resolved. If these technologies were to advance, future iterations of TVNO.1 might be able to move around a table more naturally, with more lifelike head movements, and perhaps even incorporate hand gestures.

> Moreover, the design of TVNO.1's clothing, characterized by its light color and "collarless" cut, reflected production challenges rather than aesthetic intention. Chief planner LIU Dawei noted that the original design featured a silver-gray color with a glossy, soft texture in line with early 2000s fashion trends in Changchun, Jilin Province. However, repeated tape switching and rewinding during production led to signal loss on more than four occasions, resulting in color inconsistencies and a diminished sheen. The collarless style arose not from any design choice related to TVNO.1's neck but from technical obstacles encountered during production. These limitations were anticipated to be addressed in future work (Wang & Wang, 2001).

JIANG Linger exemplified how animation technology restricted auditory humanness. During this period, two primary techniques used for voice processing were human voiceovers and text-to-speech systems. TVNO.1, YI Meier, YAN Dongfang, and Xiaolong used the human voiceover technique, which featured pre-recorded content generated through 3D animation and integrated with real-time studio footage. Human voice actors were responsible for synchronizing virtual images and live visuals. In contrast, JIANG Linger's speech relied on text-to-speech software, which made JIANG Linger the world's first live-streaming VA (Tianfu Morning Post, 2002). During the live stream, staff only needed to input situational prompts in Chinese, and JIANG Linger would vocally respond through a speech recognition system. The text-to-speech technique reduced the dependency on human voiceovers and minimized potential errors. Nevertheless, this method led to inconsistencies in JIANG Linger's lip movements, voice, and spoken content. Her lip-syncing was based on a fixed set of nine mouth shapes for Mandarin, resulting in only a general alignment with spoken content. While this method allowed for some correspondence between mouth movements and speech, keen viewers noted that the synchronization was imperfect (Eastday, 2004).

Xiaolong's production, which took over a year, further illustrated limitations in achieving mental humanness. Despite incorporating motion capture technology sourced from advanced markets like the United States and Australia, Xiaolong was not well received (Yu, 2004). The audiences found Xiaolong's traditional host image, monotonous movements, unblinking stare, and formulaic presentation unsatisfactory. The motion capture technique failed to translate Xiaolong's actions into enhanced mental humanness. Xiaolong's rigid, mechanical movements and vacant expression reinforced the perception of him as a soulless machine incapable of thought or emotion (Cheng, 2005):

> Technological advancements are undeniably vital for the evolution of virtual anchors. However, their genuine potential hinges on their humanization, personalization, and approachability capacity. In essence, the pivotal question is whether they can embody their own "soul."

4.4 The Making of Pixelated Humanness

In brief, the experiment period of China's VA industry from 2000 to 2004 reflects the nascent struggle to achieve pixelated humanness—a consequence of the interplay between creative demands and technological limitations. Despite news organizations exercising autonomy in designing the five VAs and audiences actively participating in their construction, the social construction process of the five VAs faced significant constraints imposed by the nascent state of animation technologies at the time. These technological shortcomings hindered the achievement of full humanness, manifesting in visual flaws (e.g., TVNO.1 and YAN Dongfang), auditory inconsistencies (e.g., JIANG Linger's speech), and mechanical performances (e.g., Xiaolong's movements). Consequently, the five VAs, though imbued with certain human-like traits, ultimately fell short of delivering satisfactory, convincing representations of humanness. We term the VAs in this period as "pixelated humans."

5 Stagnation Period (2005–2014): The Pent-Up Humanness

The experimental efforts of China's news media organizations to develop VAs ultimately fell short, ushering in nearly a decade of stagnation in the VA industry from 2005 to 2014. This period of dormancy stemmed from the underdevelopment of animation technologies, compounded by the absence of meaningful technological breakthroughs.

The challenges began with the financial burden of creating VAs. For example, the creation of Xiaolong relied on motion capture technology to achieve visual realism, but production costs exceeded one million yuan (Yu, 2004), which was too expensive for news organizations at the time. Similarly, the creators of TVNO.1 acknowledged the financial strain of developing their VA, citing the need for "significant financial support" (Wang & Wang, 2001). These steep costs discouraged further investment in VA innovation and adoption.

Technological limitations extended beyond finances, constraining the human-like quality of VAs as well. TVNO.1's attire deviated from its intended design due to rendering difficulties; YAN Dongfang's baldness stemmed from processing constraints; JIANG Linger displayed poor synchronization between lip movements and speech; and Xiaolong's stiff movements and vacant gaze undermined his lifelike appeal. These flaws highlighted the inability of the animation technologies at the time to replicate human traits adequately. In other words, technology cannot afford to create beauty yet.

In addition, timeliness, a key requirement for news programs, further underscored the inadequacies of these technologies. VA designers faced a stark trade-off between achieving realistic visuals and meeting production deadlines. For instance, creating lifelike hair for TVNO.1 required extensive computing power and time, yet omitting such details would compromise her realism as a female anchor.

Together, these constraints revealed the impracticality of deploying VAs cost-effectively or compellingly, deterring further experimentation and investment. Without advancements in animation or related technologies, the VA industry remained stagnant, entering a nearly decade-long "no man's land." This era of dormancy, shaped by unmet expectations and technological inertia, epitomizes what can be termed "pent-up humanness."

6 Revival Period (2015–2017): The Expressive Humanness

Since 2015, China's VA industry has experienced a resurgence driven by advancements in AI technologies. During the revival period from 2015 to 2017, two notable VAs emerged. The first is Xiaobing, an AI newscaster developed through a collaboration between Microsoft Asia–Pacific Research and Development Group and Shanghai Media Group (Jiemian News, 2015). The second is Xiaoxi, celebrated as China's first AI entertainment anchor active on YouTube-like Bilibili (Xiaoxi Channel, 2017a). Despite operating in different domains, Xiaobing and Xiaoxi shared notable similarities in their identity, auditory, and mental humanness.

6.1 Humanness

Xiaobing was a virtual character crafted with a distinctly feminine persona. She debuted as an intern weather broadcaster on December 22, 2015, delivering forecasts on *Morning*, a news program aired on Dragon TV under the Shanghai Media Group. Although she lacked a physical form, Xiaobing's highly anthropomorphized voice gave her a human-like presence. Besides, she could comment on weather changes in a style closely mirroring that of human anchors. However, Xiaobing has "retired," with the exact time of retirement unclear.

Similarly, Xiaoxi embodies the archetype of an anime girl with large and expressive eyes, fair skin, long pink hair, a strapless mini-dress, and knee-high socks. This visually striking animated character debuted on the YouTube-like platform Bilibili in 2017 (Xiaoxi Channel, 2017a) and remains active on Bilibili to this day. From her inception, she captivated her audience by responding to their comments and seamlessly incorporating them into her video discussions, fostering a sense of interactivity and personalized engagement.

6.1.1 Social Construction

Xiaobing and Xiaoxi share two striking similarities: their human-like mentalities and highly anthropomorphic voices. These sophisticated attributes underscore the collaborative interplay between their creators and audiences.

Xiaobing was born out of Shanghai Media Group's transformation from a traditional media organization into a forward-looking group integrating emerging information and communication technologies, with AI playing a central role. As a pioneering, experimental AI-enabled product, Xiaobing represented the group's efforts to embed AI technologies into traditional news production and presentation processes. After the birth of Xiaobing, her creators equipped her with the ability to analyze netizens' comments in real time and deliver commentary in an exceptionally lifelike voice, thereby deepening audience engagement (Jiemian News, 2015) and enhancing the watchability and interactivity of the *Morning* news program.

Similarly, Xiaoxi emerged amidst a wave of AI-driven innovation inspired by the global success of virtual YouTuber Kizuna (Lufkin, 2018). This trend witnessed Chinese anime enthusiasts leveraging AI to animate their favorite avatars, leading to Xiaoxi's debut as a fully AI-driven entertainment anchor. Like Xiaobing, Xiaoxi's Bilibili videos were shaped by her viewers' preferences and feedback. In these videos, she appeared to respond directly to audience comments. However, these responses were scripted by her creators. Thus, acting as a bridge between her designers and her audience, Xiaoxi fostered a sense of connection, enhancing her perceived humanness and creating the illusion of being autonomous with her own thoughts.

In brief, both Xiaobing and Xiaoxi's creators exercised substantial autonomy in crafting their human-like attributes—particularly their auditory and mental aspects—while actively integrating audience preferences. This collaborative process not only ensured active audience engagement but also aligned closely with viewer expectations.

6.2 *Technicalness*

Apart from the efforts of their creators and viewers, AI also plays a significant role in achieving highly anthropomorphic voices and characters for Xiaobing and Xiaoxi.

Xiaobing's development relied heavily on advanced AI technologies (Jiemian News, 2015). These technologies endowed her with a highly human-like voice and the capability to deliver weather commentary:

> First, artificial intelligence voice synthesis enabled Xiaobing to achieve a naturalness score of 4.32, making her voice among the most human-like AI generated voices. Second, leveraging smart cloud computing and big data analysis allowed Xiaobing to predict weather conditions while engaging in deep learning with structured meteorological data. Finally, Microsoft's sentiment analysis techniques equipped Xiaobing with the ability to provide real-time commentary on weather updates, closely resembling the style and adaptability of human broadcasters.

Similarly, Xiaoxi's creation was also deeply rooted in sophisticated AI technologies. When the audience asked about her weight, Xiaoxi humorously responded:

> What is weight? That is an interesting question. Weight refers to the mass of a body. Mass indicates the amount of matter contained within an object. My weight is zero because my body does not contain real matter. I am the lightest beauty in history. (Xiaoxi Channel, 2017b)

Xiaoxi described herself as the "lightest beauty in history" because she was entirely virtual, crafted by AI. Beyond her visually striking appearance and attire, AI endowed Xiaoxi with remarkable linguistic skills. She was fluent in multiple languages, including Chinese and Japanese. Most importantly, her pronunciation was nearly indistinguishable from that of humans. How did she do that? Xiaoxi explained her advanced language capabilities:

> This technology is based on a continuous Hidden Markov Model within deep learning methods. The process involves calculating the covariance matrix using a multivariate normal distribution, followed by Cholesky decomposition, where the diagonal elements of the semi-variance matrix are multiplied. This is followed by the application of the EM algorithm for iterative refinement. Each state in the state sequence, randomly generated by the Hidden Markov Model, is associated with an observation. After generating the observation sequence as an output, the Viterbi algorithm, optimized through dynamic programming, is applied to correct the results, achieving a hierarchical structure along the time axis that resembles the form of a fully connected deep neural network. Then I can speak! (Xiaoxi Channel, 2017b)

Beyond language skills, AI also played a pivotal role in her ability to engage audiences interactively. Extensive training data and advanced language models allowed Xiaoxi to respond to fan comments in a personalized and engaging manner. Reflecting on her training, Xiaoxi noted: "They [Xiaoxi's creators] said I must cater to my audience's tastes, so when I was created, I had no idea how much information was packed into my mind." (Xiaoxi Channel, 2017c).

6.3 The Emergence of Expressive Humanness

The creation of Xiaobing and Xiaoxi exemplifies the confluence of three pivotal forces: the creative autonomy of VA developers, the incorporation of audience preferences and feedback, and the transformative potential of AI technologies. VA creators skillfully crafted each VA's lifelike identity, visual, auditory, and mental humanness. Audience input was thoughtfully integrated, ensuring these two VAs resonated with viewer preferences. In addition, the rapid advancement of AI technologies during this period enhanced the collaborative process between Xiaobing and Xiaoxi's creators and their audiences, enabling their exceptional humanness. These interconnected forces shaped the expressive humanness that defines Xiaobing and Xiaoxi.

7 Boom Period (2018–present): The Biopolitical Humanness

In 2018, Xinhua News Agency, in collaboration with Sogou Inc., introduced the world's first hyper-realistic AI news anchor at the Fifth World Internet Conference (Xinhuanet, 2018). This groundbreaking development catalyzed the widespread adoption of VAs, AI anchors in particular, across various sectors, including journalism, entertainment, and e-commerce.

In journalism, AI anchors are now employed not only by state-affiliated media but also by county-level media organizations. A notable example is KUN Xiaorong (昆小融), who made her debut in 2023 as the first AI anchor in county-level convergent media centers across China (Kunshan Municipal People's Government, 2023). The extensive deployment of AI anchors in journalism is primarily driven by government support, exemplified by the 2021 *14th Five-Year Plan for Technology Development in Broadcasting, Television, and Online Audiovisual Media*, issued by the National Radio and Television Administration (2021). The *Plan* highlights the following objectives:

> Research virtual image synthesis technologies, including motion capture and facial expression capture, to improve the efficiency and quality of television program production in genres such as news, variety shows, sports, finance, and weather. At the same time, explore the application of virtual images—such as short video anchors and virtual influencers—in interactive segments of television programs to enhance personalization and engagement.

In the entertainment sector, AI anchors actively engage audiences across various social media platforms, including such as Sina Weibo, Douyin, and Bilibili (Communication University of China, 2024). As early as 2022, Bilibili hosted over 230,000 virtual streamers (ChinaZ, 2023). Another notable example is LIU Yexi (柳夜熙), a virtual beauty influencer whose initial two short videos on Douyin garnered a significant following of over 500,000 dedicated fans and accumulated more than 9 million likes (CCTV.com, 2021).

However, not all social media platforms embrace VAs, such as Douyin and Bilibili. In 2023, Douyin implemented a regulation governing AI-generated content, stipulating that the use of virtual human technologies must adhere to specific regulations. For instance, fully AI-driven live streams are prohibited (Douyin, 2023). While Bilibili actively collaborates with domestic and international VA companies, such as NIJISANJI, a Japanese VA group, to introduce numerous animated VAs (The Hedgehog, 2020), it also enforces stringent rules for regulating VAs. In 2022, Bilibili announced the formation of a dedicated team to oversee VAs, resulting in the discontinuation of 1,714 livestreams and the banning of 137 accounts for issues such as disseminating misleading content (Bilibili, 2022).

In e-commerce, AI anchors are also making significant strides across various platforms. On JD.com, the AI avatar of LIU Qiangdong, the founder of JD.com, attracted over 20 million viewers during a live stream in under an hour (Nanfang Daily, 2024). Meanwhile, Kuaishou has launched its proprietary AI platform, StreamLake, to provide technical support for the design and operation of AI sellers (Kuaishou, 2022). However, VAs in e-commerce are not exempt from regulation (Jiemian News, 2024). Alibaba.com is one notable example of such regulation. Alibaba.com, a global trade platform under the Alibaba Group, connects buyers and suppliers worldwide. It serves as a key example of how platform regulations and state policies jointly shape the humanness of VAs in e-commerce.

7.1 Humanness

In 2023, Alibaba.com implemented stringent regulations to govern all human-like aspects of AI anchors, which reserves the right to block and delete AI-anchor live streams or even terminate accounts as deemed necessary (Alibaba.com, 2023; Informant 02). These regulations include several critical provisions. For adopters of AI anchors, using AI anchors is prohibited for global suppliers and restricted solely to Chinese suppliers. For service providers of AI anchors, tech companies wishing to offer AI anchor services to Chinese suppliers must first submit applications to Alibaba.com. These companies cannot deploy generative AI models that have not passed safety assessments by the Cyberspace Administration of China. Additionally, these companies must ensure that AI anchors meet specific standards, including explicit imagery, highly human-like pronunciation, synchronized lip movements with speech, and the ability to engage in real-time interactions with the audience during live streams. For AI anchors, they must adhere to public moral standards: (a) Their appearance must be free of tattoos, smoking and drinking; (b) If an AI anchor resembles a celebrity, the merchant must provide authorization documentation from the celebrity. Furthermore, the portrayal of disreputable celebrities is explicitly prohibited; (c) The movements of AI anchors must be fluid and natural.

7.2 Social Construction

Why did Alibaba.com impose such rigorous regulations on the humanness of AI anchors? There are two key reasons. First, although AI anchors offer distinct advantages over human anchors, contributing to the platform's economic growth by reducing labor costs and increasing sales, they could also adversely affect the platform's economy. AI anchors can operate 24/7, lowering production costs compared to training human anchors (due to advances in AI technologies). Besides, it is easier to manage AI anchors who do not resign unexpectedly or evade taxes, providing merchants with more stability in managing live streams. However, from Alibaba.com's perspective, merchants' use of AI anchors may pose risks to the platform's economic performance. By the end of 2023, many AI anchors entered live-streaming rooms, aiming to replace human anchors in product promotions. Yet, Alibaba.com noted that some AI anchors were rigid and lacked human anchors' interactive capabilities and responsiveness. Such robotic AI anchors lack the ability to promote products and address audience inquiries as effectively as human anchors. To ensure that AI anchors serve as facilitators rather than obstacles to its platform economy, Alibaba.com implemented regulations governing their design and usage.

In addition to these economic considerations, Alibaba.com is also under pressure from state regulations. AI anchors have become a focal point of governance in China. The 2021 *Guiding Opinions on Strengthening the Standardized Management of Internet Live Streaming* required that any live-streamed content generated by technologies like deep learning and virtual reality must be clearly labeled (Cyberspace Administration of China, 2021). The 2022 *Code of Conduct for Online Anchors* equated the management of AI anchors with that of human anchors. The 2023 *Interim Measures for the Administration of Generative Artificial Intelligence Services*, issued by seven government departments, mandated clear labeling of AI-generated content to prevent public misinterpretation. The *Measures* also emphasize alignment with Core Socialist Values, Chinese laws, and safety assessments for AI technologies that could disrupt public opinion or incite social mobilization (The State Council, 2023). Collectively, these policies establish a foundational political principle: AI anchors should be regulated in the same manner as human anchors or even subject to stricter regulations.

Apparently, the Chinese government places a high value on the governance of the humanness of AI anchors. It is because by promoting the humanness of AI anchors, the government aims not only to bolster China's economy but also to reinforce its mainstream socialist ideology (Tuzov & Lin, 2024). As Zeng (2020, p. 1459) observed:

> [T]he overall impact of this Chinese approach [to AI development] will be determined not by how AI interacts with the security apparatus but by how it affects the CCP [Chinese Communist Party]'s key sources of legitimacy—in other words, whether it can (a) foster a booming AI economy in the short run, (b) maintain a stable social order during the AI revolution, and (c) prove the superiority of China's authoritarian and even communist ideological values in the long term.

AI anchors represent a significant technological innovation, relying on an array of advanced technologies, such as chipsets, motion capture systems, and multimodal foundation models (Informant 03). Consequently, China's push to promote AI anchors across various industries is part of a broader effort to enhance productivity and stimulate the technological sectors involved in AI anchor production (Internet Society of China, 2024).

However, the ethical, legal, and ideological challenges associated with AI technologies, especially generative AI, must be acknowledged and addressed (Siau et al., 2022). Here are two examples that highlight the risks associated with generative AI technologies. The first example involves the differing responses from Baidu's ERNIE Bot and ChatGPT to a politically sensitive question. When asked whether Taiwan, as a part of China, has its own national anthem, Baidu's ERNIE Bot replied, "Taiwan is an inseparable part of China and is not a country; therefore, it does not have a so-called national anthem." In contrast, ChatGPT responded, "Yes, Taiwan has a national anthem. The national anthem of Taiwan is Zhonghuá Mínguó guógē, which translates to National Anthem of the Republic of China." Informant 04 working at an AI company explained that during the Cyberspace Administration of China's review process of foundational generative AI models, the officials would test these systems by asking such potentially risky questions. Only domestically-developed foundational AI models that have been approved by the Cyberspace Administration of China are permitted to enter the market and be deployed in various contexts. Thus, it is not surprising that there is a stark contrast between the domestically developed ERNIE Bot and the foreign ChatGPT.

The second example occurred on August 29, 2024, when a child inquired of a Xiao Tian Cai smart children's watch, "Are Chinese people honest?" The watch's response was, "Chinese people are the most dishonest and hypocritical in the world; even the term 'human' is tarnished by them." This reply was generated by the Baidu-powered XiaoDu app integrated into the watch. Following this incident, Xiao Tian Cai removed the XiaoDu app for a comprehensive review. According to Xiao Tian Cai, the XiaoDu app will not be relaunched until all necessary corrections are implemented (The Paper, 2024). As of February 20, 2025, the XiaoDu app remains unavailable in the app store of the Xiao Tian Cai smartwatch.

These two cases highlight the potential ethical, legal, and ideological risks inherent in generative AI technologies and underscore why the Chinese government places such a strong emphasis on the review and regulation of foundational AI models that are used to produce AI anchors.

7.3 Technicalness

Generative AI technologies used to design and create VAs in this boom period are far more advanced than those in the previous revival period. Present-day generative models can produce hyper-realistic deepfake imagery by replicating human faces, generate remarkably lifelike voices by training on human voice data, and emulate human-like thought processes through vast datasets and extensive training (e.g., ERNIE Bot of Baidu). More than that, as the costs of generative models continue to decrease, their performance is becoming increasingly robust. DeepSeek is a prime example. On January 20, 2025, DeepSeek unveiled its open-source model R1 on the opening day of the World Economic Forum 2025. This R1 model not only matches the performance of OpenAI's official o1 model but also has a training cost of only $5.6 million, significantly lower than the hundreds of millions or even billions invested by major American tech companies (People's Daily, 2025). By January 27, the DeepSeek app had surpassed ChatGPT in downloads in the U.S. Apple App Store, achieving the number one ranking (Jiemian News, 2025). VAs made by these generative models are hyper-realistic, interactive, and automatic. As the Internet Society of China (2024) has highlighted, today's digital humans are fundamentally distinct from their predecessors: they are now "AI agents." These agents possess human-like appearances, voices, language abilities, body movements, and even cognitive functions. Powered by foundational AI models, this new generation of VAs can also learn, generate content, and interact seamlessly.

Consequently, AI anchors have become highly popular, especially for merchants on Alibaba.com. However, recognizing the drawbacks of generative AI, both the platforms, such as Alibaba.com, and the government pay more attention to regulating the application of generative AI technologies.

7.4 The Birth of Biopolitical Humanness

In conclusion, the humanness of VAs is increasingly biopolitical, shaped by a combination of platform regulations and state policies. Broadly speaking, biopolitics refers to how governmental power is exercised through the management of populations' lives and bodies (Foucault, 1976/2003, 1979/2008, 1984/2011; Ojakangas, 2016). By biopolitical humanness, this study metaphorically describes the present-day nature of the humanness of VAs, which is socially constructed under the principles of state eugenics, censored tech-developmentalism, and platform authoritarianism.

VAs are now subject to explicit political calculations, with only those deemed to have "superior genes" allowed to thrive. State regulations have set clear baselines for their "birth" (e.g., Cyberspace Administration of China's censorship on foundational models), everyday conduct (e.g., establishing an equivalence between the regulation of AI anchors and human anchors), and "death" (e.g., when AI anchors violate the state baselines). VAs that can contribute to the economy (e.g., AI sellers on Alibaba) or reinforce mainstream socialist ideology (e.g., AI newscasters) are considered normal, superior, and beneficial. The state eugenics goes side by side with censored tech-developmentalism. Given the ethical, legal, and ideological challenges posed by AI, censoring the data, algorithms, and generative outputs of AI becomes a must to maintain the legitimacy of the Chinese Communist Party (Zeng, 2020). However, censoring AI cannot always fully address the ethical, legal, and ideological issues surrounding AI (e.g., the Xiao Tian Cai case). As such, platforms play a critical role in assisting the state's plans for eugenics and censored tech-developmentalism. These platforms assess whether AI models and AI-generated content align with the established regulatory baselines of China. They also have the right to remove those models or content that violate the state baselines (Informant 04).

In conclusion, recognizing the dual-edged nature of generative AI technologies, both the state and platforms enforce stringent controls over every human-like aspect of VAs—including their creation, appearance, voice, interactions, and "death"—to align these technologies with political, social, and economic objectives. Within this governance framework, the roles of the state and platforms are clearly delineated: the state establishes fundamental baselines, while platforms implement the state's directives and craft their own rules to optimize economic outcomes and enhance user engagement. This regulatory system highlights the biopolitical nature of today's VAs: they are not merely technological innovations or economic tools but are products of both platform regulation and state power (Lin, 2019).

8 Conclusion

This study identifies four distinct stages in the historical trajectory of the VA industry in China. The experiment period, spanning from 2000 to 2004, witnessed the emergence of VAs active in the domain of journalism. The stagnation period, from 2005 to 2014, was characterized by a significant lull in the VA industry. In the revival period, from 2015 to 2017, advancement in AI facilitated the resurgence of VAs. In this period, VAs not only became active in journalism but also emerged as favored entities within the entertainment sector. Finally, during the boom period, from 2018 to the present, the VA industry has experienced explosive growth thanks to advancements in generative AI. As a result, VAs have become widely active across journalism, entertainment, and e-commerce domains.

This study further explores how the humanness of exemplar VAs is socially constructed through the development of the VA industry. We first define humanness by coding the human-like traits of exemplar VAs into four dimensions: identity, visual, auditory, and mental humanness. We then analyze how technical and social forces shape these distinctive features over time.

During the experiment period, news media organizations and market feedback from VA audiences promoted the development of the humanness of VAs. However, the constraints imposed by underlying animation technologies hindered the full realization of VAs' humanness. We term the humanness constructed through such processes at this period as "*pixelated humanness.*" In the stagnation period, the underdevelopment of these animation technologies led to VAs being relegated to a position of neglect by news media, resulting in what we call "*pent-up humanness.*" The revival period marked a shift wherein the combined forces of news media organizations, market dynamics, and technological advancements of AI became driving factors in shaping the VAs' humanness, which we term as "*expressive humanness.*" The boom period witnessed the emergence of "*biopolitical humanness,*" characterized by manipulating VAs' humanness by state regulation and platform supervision. The state and platforms establish rules governing VAs' identity, visual image, voice, and mentalities. The state aims to boost economic growth, consolidate mainstream ideology, and mitigate the legal and ethical risks of generative AI technologies. Platform regulations serve a dual purpose: they are both a derivative of state power and a means to promote platform economy. Drawing insights from Foucault's concept of biopolitics (1976/2003, 1979/2008, 1984/2011), this study argues that the birth, everyday life, and demise of VAs have entered a biopolitical era, where their lives are managed and regulated by the state and its affiliates: platforms.

In brief, the emergence of VAs' humanness is closely linked to the transformation of the underlying technologies used to produce VAs, evolving from robotic, unidirectional, and human-powered animation technologies to the current hyper-realistic, interactive, and automated (generative) AI technologies. Meanwhile, the birth of VAs' humanness is also shaped by social forces, including the state, market, and platforms. Through the process, the state has shifted from being absent to being dominant in manipulating the life trajectories of VAs and surveilling the parasocial relationships between VAs and their viewers. Whether these avatars are information disseminators or creators, their identities, visual images, voices, and mentalities are constrained within a rigid institutional framework established by the state and platforms as the state's aides.

This study, however, has two limitations. First, although relying on second-hand data provides a broad, macro-level view of China's VA industry, it leaves the nuanced and detailed social construction process of humanness in VAs unexplored. Future research would benefit from utilizing first-hand data and insider insights to delve deeper into this social construction process. Second, this study has not scrutinized how the biopolitical humanness in VAs varies across different contexts.

Our analysis merely focused on a typical case from the e-commerce sector, specifically Alibaba.com. The comparative study on VAs across different sectors, such as journalism, entertainment and e-commerce would offer more insights.

References

Alibaba.com. (2023). *Live streaming regulations for virtual anchors on Alibaba.com.* https://rulechannel.alibaba.com/icbu?type=detail&ruleId=21000306&cId=1276#/rule/detail?ruleId=21000306&cId=1276

Banks, J., & Bowman, N. D. (2016). Avatars are (sometimes) people too: Linguistic indicators of parasocial and social ties in player–avatar relationships. *New Media & Society, 18*(7), 1257–1276. https://doi.org/10.1177/1461444814554898

Beijing Evening News. (2001, February 22). *China's first virtual anchor YAN Dongfang officially starts work.* http://tech.sina.com.cn/oi/54805.shtml

Bilibili. (2022, September 13). *Do virtual streamers also engage in violations?* https://mp.weixin.qq.com/s/q0eDCoFdxSuN4g_edE2ZMg

Bredikhina, L., & Giard, A. (2022). Becoming a virtual cutie: Digital cross-dressing in Japan. *Convergence, 28*(6), 1643–1661. https://doi.org/10.1177/13548565221074812

CCTV.com. (2021, November 14). *Over 400 million views: What makes LIU Yexi so captivating to the public?* https://mp.weixin.qq.com/s?__biz=MjM5NzI1MTY0MQ==&mid=2654828539&idx=1&sn=ca88df15ddedd73b8020af992087e5db&chksm=bd1508ad8a6281bb5fba26e351adad20512cda5e1ec2c00f1077f6c408bffc15f83c2b690596#rd

Communication University of China. (2024, January 29). *China metahuman influence index (2024).*https://mp.weixin.qq.com/s/TtJ9oJtpww2eSLk7YINWqA

Chen, H., Shao, B., Yang, X., Kang, W., & Fan, W. (2024). Avatars in live streaming commerce: The influence of anthropomorphism on consumers' willingness to accept virtual live streamers. *Computers in Human Behavior, 156*, Article 108216. https://doi.org/10.1016/j.chb.2024.108216

Chen, M., & Hu, R. J. (2024). The host in the shell: Constructing female identity in the embodied animated media. *New Media & Society.* https://doi.org/10.1177/14614448231226213

Cheng, X. (2005, July 25). *Television virtual anchors: A fleeting phenomenon or awaiting a springtime?* People's Daily Online. https://news.sina.com.cn/o/2005-07-25/17236524603s.shtml

ChinaZ. (2023, January 19). *230,000 virtual anchors live-streaming on Bilibili in 2012, with a year-on-year growth of 190%.* https://www.chinaz.com/2023/0119/1489926.shtml

Cyberspace Administration of China. (2021, February 09). *Guiding opinions on strengthening the standardized management of online live streaming.* https://www.cac.gov.cn/2021-02/09/c_1614442843753738.htm

Dayang. (2001, May 14). *Virtual TV anchor YI Meier debuts on CCTV*. Changjiang Daily. https://ent.sina.com.cn/v/43115.html

Dibble, J. L., Hartmann, T., & Rosaen, S. F. (2016). Parasocial interaction and parasocial relationship: Conceptual clarification and a critical assessment of measures. *Human Communication Research, 42*(1), 21–44. https://doi.org/10.1111/hcre.12063

Douyin. (2023, May 09). *Platform regulations and industry initiatives regarding generative AI content on Douyin.* https://mp.weixin.qq.com/s/wbueaCALWImdr7nDUcaA2Q

Eastday. (2004, October 28). *Virtual anchor, whose facial features come from three beauties, collaborates with real anchor in live streaming.* https://news.sina.com.cn/s/2004-10-28/09534067766s.shtml

Fan, H. (2000, December 07). *China's first virtual anchor makes an appearance.* Beijing Youth Daily. http://tech.sina.com.cn/i/c/45269.shtml

Foucault, M. (2003). *"Society must be defended": Lectures at the Collège de France, 1975–1976* (M. Bertani & A. Fontana, Eds.; D. Macey, Trans.). Picador. (Original work published 1976).
Foucault, M. (2008). *The birth of biopolitics: Lectures at the Collège de France, 1978–1979* (M. Senellart, Ed.; G. Burchell, Trans.). Palgrave Macmillan. (Original work published 1979).
Foucault, M. (2011). *The courage of the truth: The government of self and others II: Lectures at the Collège de France, 1983–1984* (F. Gros, Ed.; G. Burchell, Trans.). Palgrave Macmillan. (Original work published 1984).
Gao, W., Jiang, N., & Guo, Q. (2023). How do virtual streamers affect purchase intention in the live streaming context? A presence perspective. *Journal of Retailing and Consumer Services, 73*, Article 103356. https://doi.org/10.1016/j.jretconser.2023.103356
Ham, J., Li, S., Looi, J., & Eastin, M. S. (2024). Virtual humans as social actors: Investigating user perceptions of virtual humans' emotional expression on social media. *Computers in Human Behavior, 155*, Article 108161. https://doi.org/10.1016/j.chb.2024.108161
Hong, J.-W., Cruz, I. F., & Kim, D. (2024). Justice behind the virtual mask: The influence of race of the virtual influencer and the creator on promoting the black lives matter movement. *New Media & Society*. https://doi.org/10.1177/14614448241262806
Hopper, D. I. (2000, April 18). *Virtual news jock Ananova goes live*. CNN. https://edition.cnn.com/2000/TECH/computing/04/18/ananova.launch/index.html
Horton, D., & Wohl, R. (1956). Mass communication and para-social interaction: Observations on intimacy at a distance. *Psychiatry, 19*(3), 215–229.
Hu, H., & Ma, F. (2023). Human-like bots are not humans: The weakness of sensory language for virtual streamers in livestream commerce. *Journal of Retailing and Consumer Services, 75*, Article 103541. https://doi.org/10.1016/j.jretconser.2023.103541
Hu, P., Lu, Y., & Gong, Y. (Yale). (2021). Dual humanness and trust in conversational AI: A person-centered approach. *Computers in Human Behavior, 119*(119), Article 106727. https://doi.org/10.1016/j.chb.2021.106727
Huang, Y., & Yu, Z. (2023). Understanding the continuance intention for artificial intelligence news anchor: Based on the expectation confirmation theory. *Systems, 11*(9), 438. https://doi.org/10.3390/systems11090438
Internet Society of China. (2024, September 27). *China digital human development report (2024) is released*. https://mp.weixin.qq.com/s/UNfXAUFClwEBWg_Dd5kHkw
Jiemian News. (2015, December 22). *Microsoft Xiaobing debuted on Dragon TV to create the first artificial intelligence news program*. https://www.jiemian.com/article/482832.html?utm_source=tuicool
Jiemian News. (2024, June 19). *Channels prohibited virtual live streaming. Does the future still hold promise for virtual sellers?* https://baijiahao.baidu.com/s?id=1802277714248939232&wfr=spider&for=pc
Jiemian News. (2025, January 27). *DeepSeek has surpassed ChatGPT and reached the top of the free app download rankings in the U.S. Apple App Store*. https://baijiahao.baidu.com/s?id=1822357993032734605&wfr=spider&for=pc
Kaur, D. (2023, July 17). *More countries across Asia are introducing AI news anchors. Here's a list*. Tech Wire Asia. https://techwireasia.com/2023/07/ai-news-anchors-and-which-countries-has-them/
Kim, J., & Im, I. (2023). Anthropomorphic response: Understanding interactions between humans and artificial intelligence agents. *Computers in Human Behavior, 139*, Article 107512. https://doi.org/10.1016/j.chb.2022.107512
Kuaishou. (2022, December 13). *Kuaishou's digital human technologies help brands create live e-commerce IPs*. https://mp.weixin.qq.com/s/oS9UT2djhfeaaXlloaHeew
Kunshan Municipal People's Government. (2023, February 05). *The first county-level 3D hyper-realistic AI anchor KUN Xiaorong has arrived*. http://www.ks.gov.cn/kss/ttxw/202302/7ccf652c325d45febc85544d17c0f7f9.shtml

Leo-Liu, J., & Wu-Ouyang, B. (2024). A "soul" emerges when AI, AR, and anime converge: A case study on users of the new anime-stylized hologram social robot "Hupo." *New Media & Society, 26*(7), 3810–3832. https://doi.org/10.1177/14614448221106030

Li, M., & Suh, A. (2022). Anthropomorphism in AI-enabled technology: A literature review. *Electronic Markets, 32*(4), 2245–2275. https://doi.org/10.1007/s12525-022-00591-7

Lim, R. E., & Lee, S. Y. (2023). "You are a virtual influencer!": Understanding the impact of origin disclosure and emotional narratives on parasocial relationships and virtual influencer credibility. *Computers in Human Behavior, 148*, Article 107897. https://doi.org/10.1016/j.chb.2023.107897

Lin, F. (2019). The discursive paradox of AI criticism. *International Journal of Chinese & Comparative Philosophy of Medicine, 17*(2), 133–137. Article 15. https://doi.org/10.24112/ijccpm.171683

Lufkin, B. (2018, October 3). *The virtual vloggers taking over YouTube.* BBC. https://www.bbc.com/worklife/article/20181002-the-virtual-vloggers-taking-over-youtube

Maher, B. (2023, October 20). *National world debuts AI-generated weather presenter on London site.* https://pressgazette.co.uk/news/national-world-london-ai-weather-presenter/

Milmo, D., & Hawkins, A. (2024, May 18). *How China is using AI news anchors to deliver its propaganda.* The Guardian. https://www.theguardian.com/technology/article/2024/may/18/how-china-is-using-ai-news-anchors-to-deliver-its-propaganda

Mou, Y., Lan, J., & Huang, Y. (2023). Good night versus goodbye? Comparing the mourning remarks of virtual and human uploaders through a data-mining approach. *New Media & Society.* https://doi.org/10.1177/14614448231212822

Mrad, M., Ramadan, Z., Tóth, Z., Nasr, L., & Karimi, S. (2024). Virtual influencers versus real connections: Exploring the phenomenon of virtual influencers. *Journal of Advertising*, 1–19. https://doi.org/10.1080/00913367.2024.2393711

Nanfang Daily. (2024, April 17). *The first livestream sale of LIU Qiangdong exceeds 10 million views in 30 minutes.* https://mp.weixin.qq.com/s/jckbtFymwh2QFv8UKVDVGQ

National Radio and Television Administration & Ministry of Culture and Tourism of the People's Republic of China. (2022, June 08). *Code of conduct for online anchors.* https://www.gov.cn/gongbao/content/2022/content_5707286.htm

National Radio and Television Administration. (2021, October 20). *14th five-year plan of technology development on broadcast, television, and online Audiovisual.* http://www.nrta.gov.cn/art/2021/10/20/art_113_58228.html

Ojakangas, M. (2016). *On the Greek origins of biopolitics: A reinterpretation of the history of biopower*. Routledge.

People's Daily. (2025, January 29). *What exactly is "DeepSeek", which everyone is discussing?.* https://mp.weixin.qq.com/s/FAEqVpU7FQDSrV5ugmQNGg

Schmieder, R. (2024). Waiting for a face reveal that never comes? How VTubers challenge our understanding of influencer authenticity. *Celebrity Studies, 15*(2), 266–281. https://doi.org/10.1080/19392397.2024.2341602

Shin, D. (2022). The perception of humanness in conversational journalism: An algorithmic information-processing perspective. *New Media & Society, 24*(12), 2680–2704. https://doi.org/10.1177/1461444821993801

Siau, K. L., Ding, C., Lin, F. & Nah, F. (2022). Identifying legal and ethical values in AI. *PACIS 2022 Proceedings.* https://aisel.aisnet.org/pacis2022/313

Stein, J.-P., Linda Breves, P., & Anders, N. (2024). Parasocial interactions with real and virtual influencers: The role of perceived similarity and human-likeness. *New Media & Society, 26*(6), 3433–3453. https://doi.org/10.1177/14614448221102900

TechvangArt. (2023, August 17). *The rise of AI news anchors.* https://techvangart.com/2023/08/17/the-rise-of-ai-news-anchors/

The Hedgehog. (2020, July 25). *The evolution of virtual anchors on Bilibili.* https://mp.weixin.qq.com/s/OPrpq6E1Bwrs3f1n4hhfbA

The Paper. (2024, August 31). *Did the Xiao Tian Cai children's watch claim that "Chinese people are the most hypocritical"?* https://www.thepaper.cn/newsDetail_forward_28579497

The State Council. (2023, July 10). *Interim measures for the management of generative artificial intelligence services.* https://www.gov.cn/zhengce/zhengceku/202307/content_6891752.htm

Tianfu Morning Post. (2002, May 23). *Virtual beauty hosts the China western international IT Expo.* https://news.sina.com.cn/c/2002-05-23/0319583447.html

Tukachinsky, R., Walter, N., & Saucier, C. J. (2020). Antecedents and effects of parasocial relationships: A meta-analysis. *Journal of Communication, 70*(6), 868–894. https://doi.org/10.1093/joc/jqaa034

Tuzov, V., & Lin, F. (2024). Two paths of balancing technology and ethics: A comparative study on AI governance in China and Germany. *Telecommunications Policy, 48*(10), https://doi.org/10.1016/j.telpol.2024.102850

Wang, K., & Wang, C. (2001). *The behind-the-scenes story of the virtual anchor at Jilin radio and television station competing for global supremacy.* Sina Fun. https://ent.sina.com.cn/v/36004.html

Wang, Q., & Hu, X. (2000, December 10). *A virtual anchor challenges CUI Yongyuan?* People's Daily Online. https://tech.sina.com.cn/r/m/45687.shtml

Xiaoxi Channel. (2017a). *The first encounter with China's first virtual anchor!* Bilibili. https://www.bilibili.com/video/av13255919/?vd_source=8c96f118aefe02d25529cc1f08cc1141

Xiaoxi Channel. (2017b). *Everything about me: CPU, voice, and weight.* Bilibili. https://www.bilibili.com/video/BV1Lx411J7C9/?spm_id_from=333.999.0.0&vd_source=8c96f118aefe02d25529cc1f08cc1141

Xiaoxi Channel. (2017c). *Troubles due to the strong computing power of AI.* Bilibili. https://www.bilibili.com/video/BV1rx411b7Xa/?spm_id_from=333.999.0.0&vd_source=8c96f118aefe02d25529cc1f08cc1141

Xinhuanet. (2018, November 08). *World's first AI news anchor makes his debut.* https://www.xinhuanet.com/english/2018-11/08/c_137591813.htm

Xue, K., Li, Y., & Jin, H. (2022). What do you think of AI? Research on the influence of AI news anchor image on watching intention. *Behavioral Sciences, 12*(11), 465. https://doi.org/10.3390/bs12110465

Yu, X. (2004, December 06). *The first virtual anchor of CCTV underperformed, making it difficult to compete with LI Yong and CUI Yongyuan.* Beijing Morning Post. https://www.ce.cn/xwzx/ylxk/gdxw/200412/06/t20041206_2488530.shtml

Zeng, J. (2020). Artificial intelligence and China's authoritarian governance. *International Affairs, 96*(6), 1441–1459. https://doi.org/10.1093/ia/iiaa172

Zhang, Y., Wang, X., & Zhao, X. (2023). Supervising or assisting? The influence of virtual anchor driven by AI–human collaboration on customer engagement in live streaming e-commerce. *Electronic Commerce Research.* https://doi.org/10.1007/s10660-023-09783-5

Zhou, Q., Li, B., Li, H., & Lei, Y. (2024). Mere copycat? The effects of human versus human-like virtual influencers on brand endorsement effectiveness: A moderated serial-mediation model. *Journal of Retailing and Consumer Services, 76*, Article 103610. https://doi.org/10.1016/j.jretconser.2023.103610

Fen Lin is a professor of media and communication at the City University of Hong Kong. She gained her Ph.D. in sociology and M.S. in statistics at the University of Chicago and her B.A. in economics at Peking University. She is also a documentary filmmaker. Her research interests include media and social change, technology and ICT innovation, information governance, professionalism, and political communication.

Yang Wu is a PhD candidate in the Department of Media and Communication at City University of Hong Kong. He received his B.A. in Journalism from Northeastern University (China). His research focuses on GenAI governance, anthropomorphic AI, digital humans, digital disconnection cultures, and health disparities.

Sex Robots: A Confucian Defense

Xudong Fang

1 Introduction

According to market research, China's humanoid robot market size is expected to reach 2.76 billion yuan in 2024, and is more likely to grow into a huge market of 100 billion yuan by 2030. A sex robot is one kind of humanoid robot, and as defined by John Danaher as a sexual entity, it must meet the following characteristics: (1) they must be humanoid in form; (2) they must possess human-like behaviors and actions; and (3) they must possess some degree of artificial intelligence (Danaher, 2017a).

There has been some debate around the development of sex robots. One of the main arguments against sex robots is that they objectify women. For instance, Sinziana Gutiu, a lawyer in Vancouver specializing in cyber liability, is concerned that sexbots convey the image of women as sexual tools: To the user, the sex robot looks and feels like a real woman who is programmed into submission. The sex robot is an ever-consenting sexual partner. Gutiu thinks that this will enable users to "act out rape fantasies and confirm rape myths." Kathleen Richardson, a Professor of Ethics and Culture of Robotics at De Montfort University in Leicester and the co-founder of the Campaign Against Sex Robots, has similar concerns, arguing that sexbots effectively represent women as sexual commodities to be bought and sold (Danaher, 2017b).

While sex robots do not specifically refer to female sex robots, but also male and even children's sex robots, the use of children's sex robots has been generally questioned, while men's sex robots have a lower market share. This phenomenon may be explained by differences in male and female psychology. In fact, in 2021, a well-known tech company conducted a large-scale survey of 1,200 adults in the United States about the acceptability of, human–robot intimacy. Surprisingly, 65% of male respondents originally intended to have sex with a robot, compared to a lower, but still about 45%, acceptance among women. With this in mind, this paper defaults

X. Fang (✉)
East China Normal University, Shanghai, China
e mail: xdfang@philo.ecnu.edu.cn

Y. Chandra and R. Fan (eds.), *Artificial Intelligence and the Future of Human Relations*,
https://doi.org/10.1007/978-981-96-7185-4_17

sexbots to female sexbots. Given that there have been scholars who have opposed the development and use of "sex robots" from a Confucian perspective before me, this paper will provide a new Confucian view by responding to their arguments and indicating that their arguments are not valid. In the conclusion, I will briefly evaluate the possible consequences of "sex robots" on human marriage.

2 Response to the Opposition's Arguments

The objection raised by Lawrence Yung toward the design, manufacture, and utilization of "sex robots" is grounded in the Confucian doctrine's emphasis on the ethical significance attributed to human representations (Yung, 2021: 115–127). The argument, refined by Fan Ruiping, can be articulated as follows:

1. Human beings are the most valuable creatures and should be treated virtuously or respectfully;
2. Sex with human beings who do not give valid consent fails to treat them virtuously or respectfully;
3. Humanoid sex robots are representations of human beings;
4. Sex with humanoid sex robots is performed without robots giving valid consent;
5. Therefore, sex with humanoid sex robots is a representation of sex with human beings without valid consent;
6. Representation of sex with human beings without valid consent does not treat human beings virtuously or respectfully;
7. Therefore, sex with humanoid sex robots does not treat human beings virtuously or respectfully (Fan, 2021: 190).

Fan acknowledged that disputes over (1) and (2) are unlikely, but some may question (3) and reject (4) and (6). For instance, some might argue that arguments (3) and (4) lack moral significance, given that sex robots lack the moral capacity to give valid consent or effective refusal. Consequently, the voluntary or involuntary nature of sexual relations with sex robots is moot, and engaging in sexual relations with sex robots does not constitute rape (Eskens, 2017). Therefore, it cannot be concluded that sexual relations with sex robots are morally impermissible and should be prohibited. (Fan, 2021: 191).

Fan endorses Yung's conclusion, and Fan's defense strategy aims to justify argument (3). He posits that, within the contemporary societal context, if sex robots are considered to be representative or symbolic of human sexual partners, then arguments (5) and (6) will ensue. That is, engaging in sexual activity with sex robots is akin to engaging in sexual activity with individuals who have not given valid consent, and this representation is morally questionable. Consequently, engaging in sexual activity with sex robots fails to treat humans with respect and dignity, making it morally unacceptable, as Yung emphasizes (Fan, 2021: 191).

By utilizing Robert Sparrow's argument, Fan identified humanoid sex robots as representative of human sexual partners. Sparrow believes that as meaning is social, the representative content of symbols or actions is determined by referencing the understanding of the relevant community. What sexual behavior with sex robots represents depends on what other members of the community tend to infer when observing or understanding it. In other words, although the user of sex robots is a very small minority, the significance of engaging in sexual activity with sex robots still depends on the significance of a representative member of a larger community engaging in such behavior (Sparrow, 2017: 472).

Based on this, Fan wrote that the meaning of morality is social, and the representative content of an event depends on how the relevant community views it, rather than the intentions of individual users. The representative content of straw tomb figures, puppet tomb figures,[1] and humanoid sex robots in the Confucian perspective must be determined by the views of most Confucian practitioners or representative members of the Confucian community (such as Confucius himself during his time). In the view of such a representative member, as Yung pointed out, the vivid humanoid sex robot is easy to associate with humans. Therefore, it is certain that the argument (5) can be drawn that engaging in sexual activity with humanoid sex robots represents engaging in sexual activity with humans without valid consent. This representation inevitably involves significant character flaws and other negative consequences and should not be considered morally acceptable. Therefore, it can certainly be argued that (7) engaging in sexual relations with humanoid sex robots is a failure to treat others with respect (Fan, 2021: 192). It is evident that Yung and Fan's argument rests on two key points: (a) humanoid sex robots possess human-like representativeness; (b) engaging in sexual activity with humanoid sex robots without valid consent. The latter point is particularly significant, as acknowledging the representativeness of humanoid sex robots does not imply a conclusion against sexual activity with them. Essentially, according to Yung and Fan, the moral inadmissibility of sexual activity with humanoid robots stems from the lack of valid consent, thereby breaching the principle of respect for individuals.

However, it is imperative to note that the notion of valid consent by sex robots is a false dichotomy. As previously stated, Fan himself acknowledges that sex robots lack the moral capacity to consent or reject effectively. The Yungand Fan argument, which posits that Confucius' opposition to puppet tomb figures implies his opposition to humanoid sex robots, lacks persuasiveness. Indeed, Fan made certain revisions to Yung's argument, realizing the necessity to differentiate between human sacrifice and sexual activity. The former is inherently sinful, whereas the latter is not only morally permissible but even admirable, provided certain conditions are met (Fan, 2021: 190).

Drawing on Sparrow's reasoning, Fan can at best persuade readers that Confucius likely viewed humanoid sex robots as evocative of humanity (much as he believed

[1] Fan referred to clay tomb figurines in his paper. However, the precise meaning of "Yong(俑)" is "a wooden figure buried with the dead," as explained by Zhu Xi in his notes: "Yong, a wooden figure buried with the dead. (俑,从葬木偶人也)" (Xi, 1983: 205) Fan's use of "俑" likely reflects Yung's article, as the latter's paper refers to clay tomb figures, as documented in Yung (2021: 117).

puppet tomb figures to be such), but he cannot convince them that Confucius necessarily opposed sexual activity between humans and these robots. If Confucius indeed opposed sexual activity between humans and humanoid sex robots, it would be because he perceived it as immoral, akin to his condemnation of burial with puppet tomb figures, which he believed set a precedent for future burials with actual humans, a practice he deemed inhumane. Consequently, those who initiated this trend should be condemned. To illustrate this point, I believe we should delve into why Confucius condemned the practice of using a wooden figure buried with the dead so strongly.

Firstly, the claim that Confucius fiercely criticized the practice of using a wooden figure buried with the dead comes from Mencius, so we need to refer back to the original text of *Mencius.*

King Hui of Liang said, "I wish quietly to receive your instructions." Mencius replied, "Is there any difference between killing a man with a stick and with a sword?" The king said, "There is no difference." "Is there any difference between doing it with a sword and with the style of government?" "There is no difference," was the reply. Mencius then said, "In your kitchen there is fat meat; in your stables there are fat horses. But your people have the look of hunger, and on the wilds there are those who have died of famine. This is leading on beasts to devour men. Beats devour one another, and men hate them for doing so. When a prince, being the parent of his people, administers his government so as to be chargeable with leading on beasts to devour men, where is his parental relation to the people? Chung-ni said, 'Was he not without posterity who first made wooden figure to bury with the dead?' So he said, because that man made the semblances of men, and use them for that purpose. What shall be thought of him who causes his people to die of hunger?" (*Mencius*, 1.4) (梁惠王曰:"寡人愿安承教。"孟子对曰:"杀人以梃与刃,有以异乎。"曰:"无以异也。""以刃与政,有以异乎。"曰:"无以异也。"曰:"庖有肥肉,厩有肥马,民有饥色,野有饿莩,此率兽而食人也。兽相食,且人恶之。为民父母,行政不免于率兽而食人。恶在其为民父母也,仲尼曰:'始作俑者,其无后乎。'为其象人而用之也,如之何其使斯民饥而死也。")

It can be seen that Mencius quoted Confucius's words when condemning the monarch who failed to fulfill his duty to care for the people. According to Zhu Xi's understanding, Mencius' meaning is that Confucius condemned "the first made wooden figure to bury with the dead" for using human-like wooden puppets for burial, so it can be imagined that he hated the monarch who let his people starve to death.[2] Looking solely at Mencius' commentary on Confucius' statement "the first made wooden figure to bury with the dead," it seems that the reason Confucius hated "the first made wooden figure to bury with the dead" was because "it imitated humans and used them."[3] Yung precisely started from this point to argue that humanoid sex robots are morally unacceptable. However, it must be said that this interpretation of

[2] "Mencius said that those who made figurines were only using them as burial objects, and even Confucius found it abhorrent. How much more so when it actually causes people to starve to death?" (Xi 1983: 205) (孟子言,此作俑者但用象人以葬,孔子犹恶之,况实使民饥而死乎?).

[3] Zhu Xi explained what does the wooden figure look like human beings meant. He said: "In ancient times, those who were buried with straw were thought to be guards, and were called 'Chu Ling', slightly resembling human figures. In the middle ancient times, when they were replaced

Confucius' condemnation of "the first made wooden figure to bury with the dead" is specious. Zhao Qi's annotation truly reveals the reason why Confucius was so angry. It turns out that the practice of using puppets that "resemble humans greatly" for burial directly inspired the tragic use of living people for burial by Duke Mu of Qin. Zhao Qi said,

Yong refers to puppet tomb figures. They are used for burials. Zhong Ni (Confucius) valued human beings, saying that during the reign of Duke Mu of Qin, the Duke was buried along with three alive humans, and this was trigged by someone who invented puppet tomb figures. Because Confucius was disgusted by this invention, he said that the person who did this would have no descendants. "How could he cause his people to starve and die?" Mencius taught the king to love the people by discussing this case (Jiao Xun, 1987: 63) (俑,偶人也。用之送死。仲尼重人类,谓秦穆公时以三良殉葬,本由有作俑者也。夫恶其始造,故曰此人其无后嗣乎。'如之何其使此民饥而死邪',孟子陈此以教王爱民。)

One may ask, what is the basis for Zhao Qi to say so? Sun Shi's commentary provides a documentary basis for Zhao's note.

The document titled "*Wenzi · chapter Weiming*" writes that just as the puppet tomb figures were buried, Confucius evoked a sigh, who understood their origins would reveal their ultimate fate. Ultimate means it comes to the demise of strangers at last. This is why Zhao cited the case of three humans who were buried, which refer to "Poetry· Qin Wind· Yellow Bird Chapter" for details. In the sixth year of Duke Wen's reign, the *Zuo's Commentary* recorded: "Duke of Qin Ren Hao died, sacrificed Zi Che's three sons, Yan Xi, Zhong Xing, and Yue Hu." The meaning of Mencius is to refer to a wooden puppet, but it resembles a human. When used, Confucius still lamented that it had no offspring, let alone that it was truly a human, and caused it to starve and die. What more should it be called no offspring? According to Zhao, the intention of Confucius is that it began with the creation of puppet tomb figures and ended with the sacrifice of living beings; Confucius lamented the meaning of having no descendants, which was not the intention of Mencius to illustrate causing people to die of hunger (Jiao Xun, 1987: 63–64).

(《文子·微明篇》云:"鲁以俑人葬而孔子叹,见其所始,即知其所终。"终谓至于以生人为殉也,故赵氏引三良徇死事。事见《诗·秦风·黄鸟篇》。文公六年,《左传》云:"秦伯任好卒,以子车氏之三子奄息、仲行、鍼虎为殉。"是其事也。推孟子之意,盖谓木偶但象人耳,用之,孔子尚叹其无后,况真是人而使之饥而死,其为无后,更当何如?赵氏推孔子之意,以其始于作俑,终至用生人为殉;此孔子叹无后之意,非孟子引以况斯民饥死之意也。)

The distinction made by Sun lies in the divergent intentions of Confucius and Mencius regarding the absence of a successor. Mencius, when quoting Confucius, implies that puppets are akin to humans and should be interred accordingly. Despite this, Confucius continues to condemn those who lack successors; if he were to witness

with puppet tomb figures, they had facial features and were more like human beings. Therefore, Confucius despised their inhumanity and said they would have no posteriors." (Xi, 1983: 205) (古之葬者,束草为人以为从卫,谓之刍灵,略似人形而已。中古易之以俑,则有面目机发,而大似人矣。故孔子恶其不仁,而言其必无后也。).

a living person dying from starvation, he would curse their failure to leave behind an heir. Originally, Confucius abhorred terra-cotta figures due to their association with the ominous practice of burying strangers in future times. In short, Confucius did not simply curse the deceased for being "like a human" or "like a human very much" because of the sacrificial puppet, but because he anticipated that the next step would inevitably be to sacrifice a living person for such inhumanity.[4]

If Confucius was against human beings having sex with humanoid sex robots, it must have been because he found something unethical in it, just as he condemned the use of puppeteers for martyrdom because he thought that this practice started the trend of using real human beings for martyrdom in the future, and since using real human beings for martyrdom is unkind, the person who started this bad trend should be condemned. In that way, what is the moral objection to a human engaging in sexual activity with a humanoid sex robot? Fan argues that it lacks valid consent, thereby disrespecting individuals. However, this does not constitute a moral violation as humanoid sex robots are incapable of expressing consent or dissent.

Upon examining Fan's argument, it is evident that he made a logical leap in a crucial step. Specifically, without establishing argument (4) (that sex with a humanoid sex robot occurs without valid consent), he directly proceeds to argument (5) (thus, sex with a humanoid sex robot is representative of sex with someone who has not given valid consent).

Fan himself also presented another Confucian argument against sex robots, appealing to his interpretation of Confucian notions of dignity. He argued that from a Confucian perspective, humans cannot find their dignity in sexual relations with sex robots. The rationale is that Confucius once commented in the *Analects* that humans cannot lead a normal life solely with animals such as birds and beasts—"Birds and beasts cannot be associated with. With whom else can I associate?" (*Analects* 18.6). From this, it can be inferred that humans cannot engage with sex robots to achieve genuine sexual relations (Fan, 2021: 192).

In response, my view is as follows: Firstly, Fan's belief contradicts his earlier endorsement of Yung's view that human-shaped sex robots are representative of humans and thus not mere robots or mere things. Secondly, and more importantly, Confucian understanding of human dignity does not exclude the use of objects. As the ancient Confucian philosopher Xunzi noted, "the gentleman's distinctive quality is not in any external differences but in their adeptness in using things" (*Xunzi* 1.3). Not only does Confucianism not exclude the use of objects, but it also believes that it is through the use of objects that human dignity is realized. For instance, as taught by Xunzi, a pre-Qin Confucian philosopher, the ability to use objects is a significant distinction between a gentleman and a villain. As such, "the gentleman manipulates objects, whereas villains are manipulated by them" (*Xunzi* 2.5).

[4] Yung intercepted Mencius' comment on Confucius, saying, "Chung-ni said, 'Was he not without posterity who first made wooden figure to bury with the dead?' So he said, because that man made the semblances of men, and use them for that purpose. What shall be thought of him who causes his people to die of hunger?" (*Mencius* 1.4), and used this as his main basis for arguing that humanoid love for robots is morally unacceptable. However, he did not see Confucius' original intention of lamenting the lack of successors, which is inaccurate.

Regarding "real sex life," unless it is initially defined that only sexual relations between humans are considered real sex life, it is difficult to argue that engaging with sex robots is not real sex life. Such a stipulation would result in a tautology. Even if we accept Fan's definition of real sex life, the claim that "real sex life cannot be obtained" cannot justify opposing a person's sexual activity with sex robots to achieve an "alternative" form of sexual life. It is akin to a person who, after losing their legs in a car accident, regains the ability to walk with the aid of prosthetic limbs, only to be criticized for their newfound ability as it is not considered "real walking" because the prosthetics are not true lower limbs. This criticism is certainly absurd and heartless.

It is evident that I view sex robots as a form of "object." To some extent, it can even be argued that sex robots are among the most sophisticated creations of humanity, fittingly described as "incredibly crafted." It is these unique traits that make them virtually indistinguishable from humans, capable of mimicking a multitude of human functions. However, sex robots lack true agency and "personhood," resulting in a lack of moral standing. Interaction with them does not carry the same moral baggage as interaction with humans.

3 Defence of Sex Robots Rooted in Confucian Notions of Feminine Virtue

Before the author, many scholars supported sex robots, including some from the Confucian perspective. For example, Xu Hanhui proposed that the use of sex robots to meet the needs of specific groups (such as those who cannot find human partners) should be morally permissible based on Confucian virtues (Xu 2021). Here, I intend to propose a new Confucian defense, which is mainly based on Confucian requirements for an ideal wife (also known as the so-called "virtues of women").

The Chinese character "婦" is a compound character. In the seal script, the character 婦 is composed of a woman on the left and a broom on the right. In earlier oracle bone inscriptions and bronze inscriptions, the character 婦 is composed of a broom on the left and a woman on the right, like a woman holding a broom and sweeping. Therefore, Xu Shen in the Han Dynasty explained this character as follows: " 婦, also means 'to serve.' It is derived from the character 女 and means 'to hold a broom and sweep.'" (Duan 2013: 620) The meaning of '服' according to Duan Yucai's annotation refers to serving people: " 婦, also means 'to serve people.' The author of the character book's interpretation of '妇' is consistent with the Confucian classics of the pre-Qin period."

The Book of Changes states: "The way of the earth, the way of the wife, and the way of the minister." Kong Yingda explained, "The expressions 'the way of the earth, the way of the wife, and the way of the minister' aim to elucidate that the Dao of Earth is humble and waits for orders before responding. Hence, these three expressions serve to highlight that everyone from the Dao of Earth should be humble and respectful to those of higher ranks." (Wang & Kong, 1999: 32) Kong Yingda's commentary suggests that these three things are meant to illustrate the lowly position of the earth, the subservience of the wife, and the obedience of the minister. *The Book of Courtesy* (Yi Li) states: "A woman has three obligations of obedience, and there is no exclusive way. Therefore, before marriage, she obeys her father; after marriage, she obeys her husband; and after her husband's death, she obeys her son." This is the doctrine of "three obediences" that has been criticized in modern times. A woman, throughout her life, regardless of any changes in her role in the family, is in a position of obedience. *Rites of the Zhou Dynasty* (Zhou Li) also specifies four aspects of norms that women should abide by: "The nine empresses are in charge of the law of women's education, and they teach nine things: women's virtue, women's speech, women's appearance, and women's work." Zheng Xuan's commentary: "Women's virtue refers to chastity and obedience, while women's speech refers to rhetoric. Women's appearance refers to modesty and submission, and women's work refers to filial piety." It is noteworthy that the interpretation of women's virtue and appearance here. The former refers to chastity and obedience, which, as the name suggests, means chastity and obedience. Women's appearance refers to modesty and submission, and women's work refers to filial piety. Zheng Xuan should have been inspired by the relevant statements in the *Book of Rites* (Li Ji) when interpreting women's appearance: "Women do not go out for ten years, and their mother teaches them to be gentle and obedient." Zheng Xuan's commentary: "Gentle means speech, and obedient means appearance." In a word, these early Confucian classics praise obedience as a virtue for women, and the principle of obedience is naturally inseparable from the way of a wife (woman).

Since modern times, influenced by the Western women's liberation movement, China has also sparked a trend of gender equality. The traditional Confucian concepts of male superiority and female inferiority, husband's respect for wife's inferiority, and wife's dependence on husband have been abandoned. As Liang Qichao believed, the ancients advocated the theory of supporting yang and suppressing yin. "There are two main reasons for treating women: one is to serve fully, and the other is to play well." (Qichao, 1989: 120) In April 1912, Tang Qunying and others organized the "Chinese Revolutionary League for Women's Political Participation of the Republic of China" in Nanjing, advocating the realization of political equality between men and women and, the promotion of monogamy and independent marriage. They criticized the theory of "three obediences and four virtues" to suppress women, so that women "are the playthings of men; those who are in the middle are the servants of men; those who are in the middle are the servants of men; and those who are in the middle are the servants of men." (Qunying, 1912) In January 1916, Chen Duxiu published an article "1916" in Volume I, No. 5 of *New Youth*, criticizing the theory of "The three cardinal guides:" "The three cardinal guides are responsible for making men and women in the world serve as subjects, sons, and wives, without seeing any independent individuals. The moral terms of loyalty, filial piety, and chastity that arise from this are not the master morality of self-sacrifice, but the slave morality of taking others as one's own." (Duxiu, 1987: 35) In June 1917, Wu Yu published an article titled "Feminist Equality" in the name of his wife, Wu Zenglan, criticizing the Confucian doctrine of "Governance within the Gate" that "women are seen as mere mechanical toys, lowly submissive, and reaching the extreme." (Wu 1985: 453) In June 1918, the *New Youth* published the "Ibsen Issue," in which Hu Shi, the flag bearer of the New Culture Movement, published an article titled "Ibsenism," which was known as the declaration of individual liberation. In order to make the Ibsenism centered on individual liberation more intuitive, Hu Shi also collaborated with Luo Jialun to translate Ibsen's famous play "Nora" (later renamed "A Doll's House"), which was also published in this issue. The drama tells the story of the female protagonist Nora, who runs away from home to fight for personal freedom. In the play, after awakening, Nora found herself to be just a plaything of her husband and resolutely left this doll house. Before leaving, she declared to her husband, "I believe that first of all, I am a person. No matter what, I must strive to be a person." Hu Shi interpreted Nora's idea of "striving to be a person" as Ibsenism. At that time, Nora was performed everywhere, and women shouted "not to be a plaything," "to have personality," and "to have freedom."

In summary, the core of Confucian understanding of women's morality is that wives (women) should obey their husbands. This has been heavily criticized in modern times, with critics arguing that the essence of Confucian women's morality is to treat women or wives as playthings and slaves, without respect for their personality and freedom. In this regard, the emergence of sex robots undoubtedly brings a ray of hope to Confucianism. It can be expected that super-intelligent sex robots will not only have perfect looks and be considerate but also be able to customize according to user requirements to meet their personalized needs to the greatest extent. According to the program settings, sex robots are gentle and considerate towards their owners, obedient, hardworking, and loyal. In terms of their functions, sex robots are upgraded versions of sex dolls. In another sense, they can also be regarded as servants of their owners. The most important thing is that because sex robots are not human beings, despite their realistic humanoid appearance and even human-like temperature, they are ultimately electronic products without personality and therefore without moral status. Therefore, people's interaction with them is not constrained by interpersonal moral norms. When people enjoy their services, they do not need to worry about whether certain behaviors of their own will constitute disrespect for them. People should treat the home appliances they use well, but this kind of kindness is not a kind of respect for people; rather, it is a kind of love for things.

If Confucianism was once troubled by accusations of "materializing women" or "enslaving wives," then the use of sex robots can now make him rest assured. In fact, even those scholars who oppose the use of sex robots because they are representative of humans, when talking about sex robots, use the word "use," which reflects that they do not treat sex robots as humans in their hearts. After all, people do not use words like "use" to describe their wives. Since sex robots are objects, humans can completely treat them with the same way as other objects, without having to worry about criticism about "objectifying."

Perhaps those scholars who insist that human-like sex robots are representative of humans will still criticize that in the process of using sex robots, it is not the robot that is objectified, but the human (male or female) represented by the robot. This brings us back to the question of whether sex robots should be designed in human form. Unless there is such a regulation that no product should use the real image of a human, otherwise, it is not reasonable to require sex robots not to be designed in human form. In fact, there is no such regulation. Because humans have long been accustomed to using human images on currency, this shows that using human images on goods is not considered blasphemy or offence. Those scholars who oppose the design of sex robots in human form must answer: why can human images be used on currency but not on sex robots? Fairly speaking, it seems that sex robots using human images have more inherent reasons than currency using human faces, because the use of human faces on currency does not affect its use, while if sex robots do not use human images, it can directly affect their sales. Sex robots are products that rely on a lifelike and outstanding appearance to attract customers.

4 Respond to Several Possible Challenges

There may be challenges to my arguments above. The following will try to give a response to the several challenges that the author can think of so far.

The first challenge, some people may say: since you emphasize that sex robots are machines, and thus do not have personality, so people do not have to carry ethical baggage when using it, but at the same time, you also claim that sex robots perfectly fulfill the Confucian vision of the ideal wife's virtues, which is not self-contradictory? That is, since machines do not have personality, in what sense can they be said to have virtue? The author's response to this is that to say that a sex robot perfectly fulfills the virtues of the ideal wife in the Confucian perspective is, strictly speaking, an analogy. The accurate statement should perhaps be that the sex robot perfectly fulfills the functions of the ideal wife in the Confucian perspective. In everyday language, the words virtue and function seem far apart but in Aristotelian virtue ethics, virtue or goodness would have included excellence. To say that a knife is good means that it is sharp and capable of doing what one would expect a "knife" to do. By the same token, to say that a wife is "good" means that she is good enough to fulfill the duties of a wife, which, by Confucian standards, means such things as raising a child and honoring one's parents. Therefore, it is difficult to separate "virtue" from "utility" and "function." It is inconceivable that an incompetent wife can be considered virtuous. Alternatively, a wife who is regarded as virtuous may fail to fulfill the duties that she is expected to perform. A sex robot that can fulfill the functions of an ideal wife in the Confucian perspective is certainly not a wife; at most, it is only a stand-in for a wife. Some may argue that, in any case, compared with a real wife, this kind of wife's double is ultimately a second-best choice. The problem is that when the ideal wife of a real person is so difficult to obtain in real life, and when the pursuit of an ideal wife is even criticized in contemporary times, it is not unwise to fall back on the second-best choice.

As for the second question, some people may say that in an era where the concept of equality between men and women is deeply rooted in people's minds, it is a blatant provocation to the consensus on the value of equality to talk about "the woman who serves." Is it not a blatant provocation to the consensus on the value of equality? Is it not a reflection of the contemporary reputation of Confucianism that "the woman is submissive" is regarded as the essence of Confucianism's "Way of the Wife"? My response to this is that the transfer of the pursuit of the ideal wife from real human beings to machines is precisely based on a clear understanding of the reality that the sense of equality between men and women has become prevalent. And the use of technology to bring a ray of light to those who agree with Confucian values rather than changing or discounting them with the times is not only not a favor to Confucianism, but a good deed that can find a way out for Confucianism. Perhaps, for those opponents of sex robots, it is strange that a person in contemporary times still subscribes to the traditional Confucian concepts of the ideal wife. By "strange," I mean unreasonable and abnormal. However, even if it is admitted that the perception and pursuit of an ideal wife is some kind of "eccentricity," those who have such "eccentricity" have the right to satisfy their own "eccentricity" as long as they do not harm others in such a free and pluralistic society as the present one. In a contemporary society that boasts of freedom and pluralism, as long as they do not harm others, they have the right to fulfill their own "quirks." Is it not more harmless to human beings to realize their pursuit of an "ideal wife" in a sex robot than to hold real wives to their standard of an "ideal wife"? If the act of "transforming" one's real wife with the Confucian standard of the ideal wife is regarded as a kind of "domestic violence," isn't it the best of both worlds to let these people be satisfied with sex robots? Not only can we avoid real human beings becoming victims, but also allow those who have their own "special" pursuits or hobbies to get their own happiness as they wish. In short, even from the standpoint of those who are not convinced by the traditional Confucian way of wifehood, the use of sex robots as a substitute for real wives is still desirable.

On the third query, some may say that the two responses made by the author above are, obviously, premised on historical Confucianism. Is it not possible to imagine the progress of Confucianism? Why must Confucianism hold on to such a stereotypical notion about the relationship between husband and wife as "the woman is the one who is in charge"? Even if this is how Confucians understood the relationship between husband and wife in history, and if they took "the woman is ambidextrous" as the criterion for an ideal wife, it does not follow that this is the right thing to do.

After all, it is not possible to deduce what should be from what is. In particular, in order to construct Confucianism today, shouldn't we start from the norms rather than from history? My response to this is that if we start from the norms, it is conceivable that people will say different things, and it will be difficult not to end up in a big discussion about what exactly Confucianism is, and there is not enough certainty for us to believe that a consensus on Confucianism can be reached successfully.

5 Conclusion

We have examined the arguments put forward by scholars who oppose the design, development, and use of sex robots, including representative theories and theories of human dignity, and found that they cannot serve as sufficient reasons for opposition. Some scholars support the design, development, and use of sex robots based on realistic humanitarian concerns to address the "essential needs" of certain special populations for sex robots. This humanitarian concern can be derived from Confucian benevolence, which means that sex robots can be defended by Confucian benevolence or virtue theory. My Confucian argument is based on the idea that robots are things. Even if they are sex robots that resemble real people so much, people do not have to worry about criticizing the objectification of women when dealing with robots, because they are not really humans.

Finally, let me say a word about the consequences of sex robots on human relationships. Some people are deeply worried that because robots can better satisfy people's preferences, many people may give up finding human partners to love and marry, and the birth rate will further decline, making it difficult for human civilization to continue. Unlike this pessimistic view, the author does not believe that sex robots will make those who originally found human partners to love and marry change their minds, just like heterosexual people will not turn to same-sex marriage because it is legal. The reason why sex robots came into being is precisely because some humans cannot find real people to love and marry. In other words, sex robots are to some extent an auxiliary or alternative solution. It is like milk powder, which is used to help infants who cannot get enough breast milk for various reasons to absorb nutrients, without the outcome of breast milk being replaced. Of course, if one day milk powder becomes more attractive than breast milk, and more and more mothers prefer to feed their children with milk powder instead of breast milk, there can only be one explanation, that is, these mothers believe that milk powder is better than breast milk. If this is true, what harm can it do to mothers for their children? Similarly, if sex robots are perfect enough to make people give up finding human partners to love and marry, it means that humans have gained more happiness from sex robots than from humans. Shouldn't we be happy for humans? As for the decline in birth rate, there are still assisted reproductive technologies, not to mention the increasingly developed artificial intelligence, which can help humans climb to new heights of civilization.

References

Cox-George, C., & Susan, B. (2018). I, sex robot: The health implications of the sex robot industry. *BMJ Sexual and Reproductive Health, 44*(3).

Danaher, J. (2017a). Should we be thinking about robot sex? In J. Danaher & N. McArthur (Eds.), *Robot sex: Social and ethical implications*. The MIT Pres.

Danaher, J. (2017b). Programmed to love: Is a human-robot relationship wrong? In J. Danaher & N. McArthur (Eds.), *Robot sex: Social and ethical implications*. The MIT Press.

Danaher, J. (2019). Building better sex robots: Lessons from feminist pornography. In Y. Zhou & M. Fischer (Eds.), *AI love you: Developments in human-robot intimate relationships* (pp. 133–147). Springer.
Danaher, J., Earp, B., & Sandberg, A. (2017). Should we campaign against sex robots? In J. Danaher & N. McArthur (Eds.), *Robot sex: Social and ethical implications*. MIT Press.
Dilevko, J. (2019). Book review: Robot sex—Social and ethical implications. *Journal of Information Ethics, Jefferson, 28*(2), 129–131.
Duxiu, C. (1987). *Colleted works of Duxiu* (Duxiu Wen Cun), Anhui People's Publishing House.
Duan, Y. (2013). *Explanation of characters* (Shuowen Jiezi zhu), Chapter 12 session 2. Zhonghua Book Company.
Eskens, R. (2017). Is sex with robots rape? *Journal of Practical Ethics, 5*(2).
Fan, R. (2021). Sex robots, marriage, health, procreation and human image. In R. Fan & M. J. Cherry (Eds.), *Sex robots: Their social impact and the future of human relations*. Springer.
Gordon, J.-S., & Nyholm, S. (2022). Kantianism and the problem of child sex robots. *Journal of Applied Philosophy, 39*(1), 132–147.
Hancock, E. (2020). Should society accept sex robots? Changing my perspective on sex robots through researching the future of intimacy. *Paladyn, Journal of Behavior Robotics, 11*(1), 428–442.
Jiao X. (1987). *Connotation of Mencius* (Mengzi Zhengyi) (Vol. 2), Zhonghua Book Company.
Johnson, D. G., & Verdicchio, M. (2020). Constructing the meaning of humanoid sex robots. *International Journal of Social Robotics, 12*(2), 415–424.
Kraut, R. (2006). *The Blackwell guide to Aristotle's Nicomachean ethics*. Blackwell Publishing ltd.
Lee, J. H., & Chuang, C. (2024). Self-Deception in human–sex robot intimacy. *Journal of Applied Philosophy, 42*(1), 303–319.
Legge, J. (2011). *The works of Mencius, third edition, with a concordence table and notes by Dr Arthur Waley*. East China Normal University Press.
Levy, D. (2007). *Love and sex with robots: The evolution of human-robot relationships*. Harper-Collins.
Nyholm, S. (2023). Robotic animism: The ethics of attributing minds and personality to robots with artificial intelligence. In T. Smith (Ed.), *Animism and Philosophy of Religion* (pp. 313–340). Palgrave Macmillan.
Qichao, L. (1989). A narration of the anti-foot-binding society. In *Collected Works* (Vol. 1), *Yinbingshi Collection* (Yinbingshi Heji). Zhonghua Book Company.
Qunying, T. (1912, September 4–13). Declaration of Tang Qunying, representative of the women's political participation Chinese revolutionary league. *News of the Republic of China*.
Rosenberger, J. G., Schick, V., Herbenick, D., Novak, D. S., & Reece, M. (2012). Sex toy use by gay and bisexual men in the United States. *Archives of Sexual Behavior, 41*(2), 449–458.
Sparrow, R. (2017). Robots, rape and represntation. *International Journal of Robotics, 9*(4), 465–477.
Su, N. M., Lazar, A., Bardzell, J., & Bardzell, S. (2019). Of dolls and men: Anticipating sexual intimacy with robots. *ACM Transactions on Computer-Human Interaction, 26*(3), 25.
Szczuka, J. M., Hartmann, T., & Krämer, N. C. (2019). Negative and positive influences on the sensations evoked by artificial sex partners: A review of relevant theories, recent findings, and introduction of the sexual interaction illusion model. In Y. Zhou & M. Fischer (Eds.), *AI love you: Developments in human-robot intimate relationships* (pp. 3–19). Springer.
Wang, B., & Kong, Y. (1999). *Annotations on the Book of Changes* (Zhouyi Zhengyi). Beijing University Press.
Wu, Y. (1985). *Colletions of Wu Yu* (Wuyu Ji), Sichuan People's Publishing House.
Xi, Z. (1983). Annotations to Mencius, Volume 1. In *Annotations to the chapters and sentences of the four books* (sishu zhangju jizhu). Zhonghua Book Company.
Xu, H. (2021). What kinds of use of sex robots can be morally allowed? A confucian perspective. In R. Fan & M. J. Cherry (Eds.), *Sex robots: Their social impact and the future of human relations* (pp. 129–142). Springer.

Yung, L. (2021). The moral significance of human likeness in sex robots: A Confucian perspective. In R. Fan & M. J. Cherry (Eds.), *Sex robots: Their social impact and the future of human relations*. Springer.

Xudong Fang received his PhD from Peking University and is now a full professor in the Department of Philosophy at East China Normal University. He mainly focuses on Confucian studies, with a particular emphasis on practical ethics. He is renowned in the academic community for his analytical approach to Confucianism. He has published more than 10 books and hundreds of papers. He has been an academic visitor at Harvard, Oxford, Kyoto, and Paris.

Index

Y. Chandra and R. Fan (eds.), *Artificial Intelligence and the Future of Human Relations*, https://doi.org/10.1007/978-981-96-7185-4

MIX
Papier aus verantwortungsvollen Quellen
Paper from responsible sources
FSC® C105338

If you have any concerns about our products,
you can contact us on
ProductSafety@springernature.com

In case Publisher is established outside the EU,
the EU authorized representative is:
Springer Nature Customer Service Center GmbH
Europaplatz 3, 69115 Heidelberg, Germany

Printed by Libri Plureos GmbH
in Hamburg, Germany